GREEN
COMMUNICATIONS

GREEN COMMUNICATIONS
PRINCIPLES, CONCEPTS AND PRACTICE

Edited by

Konstantinos Samdanis, Peter Rost, Andreas Maeder,
Michela Meo and Christos Verikoukis

This edition first published 2015
© 2015 John Wiley & Sons, Ltd

Registered office
John Wiley & Sons Ltd, The Atrium, Southern Gate, Chichester, West Sussex, PO19 8SQ, United Kingdom

For details of our global editorial offices, for customer services and for information about how to apply for permission to reuse the copyright material in this book please see our website at www.wiley.com.

Library of Congress Cataloging-in-Publication Data applied for.

A catalogue record for this book is available from the British Library.
ISBN: **9781118759264**

Typeset in 10/12 TimesLTStd by SPi Global, Chennai, India.
Printed and bound in Singapore by Markono Print Media Pte Ltd

1   2015

Contents

List of Contributors

Muhammad Ali Imran, Institute for Communication Systems (ICS), University of Surrey, Guildford, Surrey, UK

Luis Alonso, Department of Signal Theory and Communications, Universitat Politècnica de Catalunya (UPC), Barcelona, Spain

Jesus Alonso-Zarate, Centre Tecnològic de Telecomunicacions de Catalunya (CTTC), Barcelona, Spain

Giuseppe Anastasi, Department of Information Engineering, University of Pisa, Pisa, Italy

Mayutan Arumaithurai, Institute of Computer Science, Computer Networks Group, University of Goettingen, Goettingen, Germany

Michael Bennett, Lawrence Berkeley National Laboratory, Vallejo, USA

Raffaele Bolla, Department of Electrical, Electronic and Telecommunications Engineering, and Naval Architecture (DITEN), University of Genoa, Genoa, Italy; National Inter-university Consortium for Telecommunications (CNIT), University of Genoa Research Unit, Genoa, Italy

Simone Brienza, Department of Information Engineering, University of Pisa, Pisa, Italy

Roberto Bruschi, National Inter-university Consortium for Telecommunications (CNIT), University of Genoa Research Unit, Genoa, Italy

Łukasz Budzisz, Telecommunication Networks Group, Technische Universität Berlin, Berlin, Germany

Alessandro Carrega, Department of Electrical, Electronic and Telecommunications Engineering, and Naval Architecture (DITEN), University of Genoa, Genoa, Italy; National Inter-university Consortium for Telecommunications (CNIT), University of Genoa Research Unit, Genoa, Italy

C. Cavdar, Communication Systems Department, KTH Royal Institute of Technology, Kista, Sweden

I. Cerutti, Institute of Communication, Information and Perception Technologies, Scuola Superiore Sant'Anna, Pisa, Italy

J. Chen, Communication Systems Department, KTH Royal Institute of Technology, Kista, Sweden

Min Chen, Huazhong University of Science and Technology, Wuhan, Hubei, China

Luca Chiaraviglio, DIET Department, University of Rome "Sapienza", Rome, Italy

Ken Christensen, University of South Florida, Florida, USA

Antonio Cianfrani, DIET Department, University of Rome "Sapienza", Rome, Italy

Angelo Coiro, DIET Department, University of Rome "Sapienza", Rome, Italy

Alberto Conte, Alcatel-Lucent Bell Labs, Centre de Villarceaux, Nozay, France

Franco Davoli, Department of Electrical, Electronic and Telecommunications Engineering, and Naval Architecture (DITEN), University of Genoa, Genoa, Italy; National Inter-university Consortium for Telecommunications (CNIT), University of Genoa Research Unit, Genoa, Italy

Mischa Dohler, Centre for Telecommunications Research, King's College London (KCL), London, UK

Dominique Dudkowski, NEC Europe Ltd., NEC Laboratories Europe, Heidelberg, Germany

Simon Fletcher, NEC Telecom MODUS Ltd, Surrey, UK

Vasilis Friderikos, Centre for Telecommunications Research, King's College London, London, UK

Xiaohu Ge, Huazhong University of Science and Technology, Wuhan, Hubei, China

Toru Hasegawa, Information Networking, Osaka University, Osaka, Japan

Peer Hasselmeyer, NEC Europe Ltd., NEC Laboratories Europe, Heidelberg, Germany

Nageen Himayat, Intel Corporation, Intel Labs, Santa Clara, USA

Tobias Hoßfeld, University of Würzburg, Communication Networks, Würzburg, Germany; University of Duisburg-Essen, Modeling of Adaptive Systems, Essen, Germany

Iztok Humar, Faculty of Electrical Engineering, University of Ljubljana, Ljubljana, Slovenia

Taewon Hwang, Yonsei University, Department of Electrical and Electronic Engineering, Seoul, Korea

Michael Jarschel, University of Würzburg, Communication Networks, Würzburg, Germany; Nokia Networks, Munich, Germany

Minho Jo, Korea University, Seoul, Korea

Thomas Kessler, Deutsche Telekom AG, Darmstadt, Germany

Younggap Kwon, Yonsei University, Department of Electrical and Electronic Engineering, Seoul, Korea

Andres Laya, KTH Royal Institute of Technology, Kista, Sweden

Marco Listanti, DIET Department, University of Rome "Sapienza", Rome, Italy

Giuseppe Lo Re, DICGIM, University of Palermo, Palermo, Italy

Andreas Maeder, NEC Europe Ltd, Heidelberg, Germany

Juan Antonio Maestro, Universidad Antonio de Nebrija, Madrid, Spain

Michela Meo, Politecnico di Torino, Torino, Italy

Guowang Miao, KTH Royal Institute of Technology, Communications Department, Stockholm, Sweden

A. Mohammad, Electrical Engineering Department, Linköping University, Linköping, Sweden

P. Monti, Communication Systems Department, KTH Royal Institute of Technology, Kista, Sweden

D. C. Mur, NEC Europe Ltd, Heidelberg, Germany

Zhisheng Niu, Department of Electronic Engineering, Tsinghua University, Beijing, China

Bruce Nordman, Lawrence Berkeley National Laboratory, Vallejo, USA

Timothy O'Farrell, University of Sheffield, Department of Electronic and Electrical Engineering, Sheffield, UK

Marco Ortolani, DICGIM, University of Palermo, Palermo, Italy

Hyunsung Park, Yonsei University, Department of Electrical and Electronic Engineering, Seoul, Korea

Manuel Paul, Deutsche Telekom AG, Berlin, Germany

Marco Polverini, DIET Department, University of Rome "Sapienza", Rome, Italy

G. Punz, NEC Europe Ltd, Heidelberg, Germany

Yinan Qi, Institute for Communication Systems (ICS), University of Surrey, Guildford, Surrey, UK

Kadangode K. Ramakrishnan, Riverside Computer Science and Engineering, University of California, Riverside, USA

Marco Di Renzo, Paris-Saclay University, Laboratory of Signals and Systems, CNRS – CentraleSupelec – University Paris-Sud XI, Gif-sur-Yvette, Paris, France

Pedro Reviriego, Universidad Antonio de Nebrija, Madrid, Spain

Peter Rost, NEC Europe Ltd, Heidelberg, Germany

Konstantinos Samdanis, NEC Europe Ltd, Heidelberg, Germany

Rahim Tafazolli, Institute for Communication Systems (ICS), University of Surrey, Guildford, Surrey, UK

L. Velasco, Department of Computers Architecture, Universitat Politècnica de Catalunya (UPC), Barcelona, Spain

Christos Verikoukis, Telecommunications Technological Centre of Catalonia, Barcelona, Spain

P. Wiatr, Communication Systems Department, KTH Royal Institute of Technology, Kista, Sweden

Rolf Winter, NEC Europe Ltd., Heidelberg, Germany

Adam Wolisz, Telecommunication Networks Group, Technische Universität Berlin, Berlin, Germany

L. Wosinska, Communication Systems Department, KTH Royal Institute of Technology, Kista, Sweden

Lin Xiang, Huazhong University of Science and Technology, Wuhan, Hubei, China

Jing Zhang, Huazhong University of Science and Technology, Wuhan, Hubei, China

Bi Zhao, Centre for Telecommunications Research, King's College London, London, UK

Sheng Zhou, Department of Electronic Engineering, Tsinghua University, Beijing, China

Preface

Energy efficiency and green communications aim at addressing the quest for sustainability regarding power resources and environmental conditions. For telecommunication service providers, energy efficiency merely means cost reduction, in terms of capital and operational expenditures. For government bodies and regulators, energy efficiency and green communications is a duty to strengthen corporate responsibility towards the environment and motivate an ecological generation of network equipment and systems.

Energy efficiency evolved into a significant parameter of equipment design, architecture and management of telecommunication systems but has not been taken into account until the Kyoto Protocol early in 1997, which raised concerns regarding global warming. Initial efforts concentrated on network edge equipment and peripherals, communication protocols and then progressively on wireless radio and cellular systems as well as on fixed networks. Nowadays, after a steep increase of studies, innovation and practice, energy efficiency and green communications are entering a mature phase, with established solutions addressing particular aspects of a telecommunication system.

Despite such momentum, the potential for energy conservation is still huge especially since advanced services and applications are increasing the complexity of network usage and the demand for enhanced capacity, speed and network resources, driving the growth of network infrastructure deployment. In addition, such earlier approaches prepared the ground for further advance contributions that consider different equipment design features, a combination of communication protocols and holistic network mechanisms that are more sophisticated and comprehensive since they take into account several diverse aspects and cross-layer issues of a telecommunication system.

In structuring the material contained in this book, our goal is to elaborate the fundamentals of energy efficiency and green communications exploring the main challenges, mechanisms and practice considering both wireless and wireline systems. Wireless and wireline communications are organized into two different corresponding sections that address equipment, management, architecture, communication protocols, applications and different deployment aspects. Each chapter is organized in a tutorial nature that contains well-established solutions and the main associated findings in a way that is easy for the reader to follow, providing also a list of references for the interested reader to explore further. In closing each section, a dedicated chapter summarizes the current advances related to the main standardization bodies and list all related specifications and studies in an effort to enhance the view of the reader regarding the adoption and exploitation of such research technologies into industry products and solutions and to provide the basics for standardization engineers who wish to enter the field.

Often a choice had to be made about including certain concepts, evolving research areas and mechanisms, but given the limited space, the focus remained on material that enable the reader to understand the basics in order to innovate the development of more advance solutions. We hope that this book can serve the reader as a first orientation and as a tool for experts to dive deeper into this new vigorous and fascinating area.

Konstantinos Samdanis
NEC Europe Ltd

List of Abbreviations

1G	1st Generation of wireless telephone technology
2G	2nd Generation of wireless telephone technology
3G	3rd Generation of mobile telecommunications
3GPP	3rd Generation Partnership Project (3GPP)
3GPP2	3rd Generation Partnership Project 2
4G	4th Generation of mobile telecommunications
5G	5th Generation of mobile telecommunications
AAA	Authentication Authorization Accounting
AC-DC	Alternating Current –Direct Current
ACI	Adjacent Channel Interference
ACPI	Advanced Configuration and Power Interface
ADC	Analogue to- digital converter
ADSL	Asymmetric Digital Subscriber Line
A-ESR	Ant colony based-Energy Saving Routing
AF	Amplify-and-Forward
ALR	Adaptive Link Rate
ALTO	Application Layer Transport Optimization
AMC	Adaptive Modulation and Coding
AMR	Automatic Meter Reader
AON	Active Optical (access) Network
AP	Access Point
API	Application Program Interface
APP	Application
ARP	Address Resolution Protocol
ARPU	Average Revenue Per User
ATIS	Alliance for Telecommunications Industry Solutions
ATM	Asynchronous Transfer Mode
AUC	Authentication Centre
AWGN	Additive White Gaussian Noise
BAN	Body Area Network
BB	Baseband
BBF	Broadband Forum,
BCCH	Broadcast Control Channel
BBU	Baseband Unit

BLE	Bluetooth Low Energy
BS	Base Stations
BSC	Base Station Controller
BER	Bit Error Rate
BSS	Basic Service Set
BTS	Base Station Sites
CAGR	Compound Annual Growth Rate
CAPM	Context-Aware Power Management
CAPEX	Capital Expenditure
CCN	Content-Centric Networking
CDF	Cumulative Distribution Function
CDMA	Code Division Multiple Access
CDN	Content Distribution Network
CLI	Convergence Layer Interface
CN	Core Network
CO	Central Offices
CoC	Code-of-Conduct
CoMP	Coordinated Multi-Point
COPSS	Content-Oriented Publish/Subscribe System
CPE	Customer Premises Equipment
CPRI	Common Public Radio Interface
CPU	Central Processing Unit
CR	Cognitive Radio
CR mode	Continuous Reception mode
CS	Cell zooming Server
CSI	Channel State Information
CTMC	Continuous-Time Markov Chain
CUBS	Coordinated Upload Bandwidth Sharing
D2D	Device-to-Device
DA	Dynamic Adaptation
DC	Data Center
DAC	Digital-to-Analogue Converter
DAG	Data Acquisition and Generation
DAISIES	Distributed and Adaptive Interface Switch off for Internet Energy Saving
DDoS	Distributed Denial-of-Service attack
DE	Deployment Efficiency
DER	Distributed Energy Resource
DF	Decode-and-Forward
DHCP	Dynamic Host Configuration Protocol
DiR	Differentiated Reliability
DLC	Digital Loop Carrier
DNS	Domain Name Systems
DP	Dynamic Programming
DPA	Doherty Power Amplifiers
DPD	Digital Pre-Distortion
DR	Demand Response

DRAM	Dynamic Random-Access Memory
DRX	Discontinuous Reception
DSP	Digital Signal Processing
DSL	Digital Subscriber Line
DSLAM	DSL Access Multiplexer
DTIM	Delivery Traffic Indication Message
DTN	Delay Tolerant Network
DTX	Discontinuous Transmission
DWDM	Dense Wavelength Division Multiplexing
EASes	Energy-Aware States
EAT	Energy-Aware Traffic engineering
EA-RWA	Energy-Aware RWA
EC	European Commission
ECI	Energy Consumption Index
ECR	Energy Consumption Ratio
EDGE	Enhanced Data Rates for GSM Evolution
EE	Energy Efficiency
EEE	Energy-Efficient Ethernet
EEF	Energy Efficiency Factor
EEER	Equipment Energy Efficiency Ration
EIA-RWA	Energy and Impairment aware RWA
EIR	Equipment Identity Register
EIRP	Effective Isotropic Radiated Power
EMAN	Energy Management
eNB	evolved Node B
EPAR	Energy Profile Aware Routing
EPC	Evolved Packet Core
EPON	Ethernet passive optical network
ERG	Energy Reduction Gain
ES	Energy Saving
ESACON	Energy Saving based on Algebraic Connectivity
ESIR	Energy-Saving IP Routing Strategy
ES-TE	Energy-Saving Traffic Engineering
ESM	Energy Saving Management
ETG	Energy Throughput Gain
ETSI	European Telecommunication Standardization Institute
eUTRAN	Evolved Universal Terrestrial Radio Access Network
EV-DO	Evolution-Data Optimized
FE	Forwarding Element
FIB	Forwarding Information Base
FTP	File Transfer Protocol
FFT/iFFT	Fast Fourier Transform/inverse Fast Fourier Transform
FTTH	Fiber To The Home
FSU	Femto SU
FWM	Four Wave Mixing
GABS	Gradient Assisted Binary Search

GAL	Green Abstraction Layer
GBNB	Green Backbone Networks with Bundled links
GEPON	Gigabit Ethernet PON
GERAN	GSM EDGE Radio Access Network
GGSN	Gateway GPRS Support Node
GHG	Global Greenhouse Gas
GMPLS	Generalized MPLS
GMTE	Green MPLS Traffic Engineering
GPON	Gigabit Passive Optical Network
GPS	Global Positioning System
GPRS	General packet radio service
GR	Green Radio
GRiDA	Green Distributed Algorithm
GreenTE	Green Traffic Engineering
GSI	Green Standard Interface
GSM	Global System for Mobile
GSMA	GSM Association
GtCO2e	Gigatons of CO2 equivalent
GTP	GPRS Tunneling Protocol
HARQ	Hybrid-Automatic Repeat Request
H2H	Human-to-Human
HDD	Hard Disk
HDLC	High-level Data Link Control
HetNet	Heterogeneous Networks
H-GW	Home Gateway
HLR	Home Location Register
HSDPA	High Speed Downlink Packet Access
HSPA	High Speed Packet Access
HSS	Home Subscriber Server
HW	Hardware
IA-RWA	Impairment Aware-RWA
ICMP	Internet Control Message Protocol
ICN	Information Centric Networking
ICT	Information Communication Technology
IEEE	Institute of Electrical and Electronics Engineers
IETF	Internet Engineering Task Force
IL	Idle Logic
ILP	Integer Linear Programming
IMPEX	Implementation expenses
IoT	Internet of Things
IP	Internet Protocol
IRTF	Internet Research Task Force
ISD	Inter-Site Distance
IT	Information Technology
ITU	International Telecommunication Union
ISIS	Intermediate System to Intermediate System

ISP	Internet Service Providers
KPI	Key Performance Indicators
LAN	Local Area Network
LC	Line Card
LCA	Life Cycle Assessment
LCPs	Local Control Policies
LFA	Least Flow Algorithm
LLDP	Link Layer Discovery Protocol
LNA	Low-Noise Amplifier
LoM	LAN on Motherboard
LPI	Low Power Idle
LSDB	Link State Database
LSI	Large-Scale Integration
LSP	Label Switched Paths
LTE	Long-Term Evolution
LTE-A	LTE-Advanced
M2M	Machine-to-Machine
MAC	Medium Access Control
MAN	Metropolitan Area Network
MCFP	Maximum Conditional Failure Probability
MCS	Modulation and Coding Scheme
MGW	Media Gateway
MIB	Management Information Base
MILP	Mixed-Integer Linear Programming
MIMO	Multiple-Input-Multiple-Output
MME	Mobility management Entity
mm-wave	millimeter-wave
MNO	Mobile Network Operator
MPA	Most Power Algorithm
MPLS	Multi-Protocol Label Switching
MSAN	Multi-Service Access Node
MSC	Mobile Switching Centre
MSU	Macro SU
MU-MIMO	Multiuser-MIMO
MSAN	Multi-Service Access Node
MTC	Machine Type Communications
MU	Mobile User
NAPS	Non-intrusive location-Aware Power management Scheme
NAS	Non-Access Stratum
NAT	Network Address Translation
NCM	Network Control Module
NCP	Network Connectivity Proxy
NCPs	Network-wide Control Policies
ND	Network Device
NDN	Named Data Networking
NFV	Network Function Virtualization

NGMN	Next Generation Mobile Network
NIC	Network Interface Card
NM	Network management
NoA	Notice of Absence
NW	Network
OAM	Operation, Administration and Maintenance
OCP	Optical Control Plane
OFDM	Orthogonal Frequency-Division Multiplexing
OFDMA	Orthogonal Frequency-Division Multiple Access
OLT	Optical Line Terminal
ONU	Optical Network Unit
OPEX	Operating Expenditure
OS	Operating System
OSP	Optimization Service Protocol
OSPF	Open Shortest Path First
OSPF-TE	Open Shortest Path First-Traffic engineering
OSI	Open Systems Interconnection
OTN	Optical Transport Network
P2P	Peer-to-Peer
PA	Power Amplifier
PAPR	Peak-to-Average Power Ratio
PC	Personal Computer
PCE	Path Computation Element
PCRF	Policy and Charging Rules Function
PD	Powered Device
PDA	Personal Digital Assistant
PDCP	Packet Data Convergence Protocol
PDH	Plesiochronous Digital Hierarchy
PDN GW	Packet Data Network -Gateway
PDP	Packet Data Protocol
PDU	Power Distribution Unit
PIT	Pending Interest Table
PLC	Power Line Communication
PLI	Physical Layer Impairments
PLMN	Public Land Mobile Networks
PMP	Power Management Primitives
PS	Packet Switching
PSE	Power Source Equipment
PSIRP	Publish Subscribe Internet Routing Paradigm
PS-Poll	Power Save Poll
PSM	Power Saving Mode
PtP	Point-to-Point
RE	Resource Element
PHY	Physical
RN/RS	Relay Nodes/Relay Stations
PoE	Power over Ethernet

PON	Passive Optical Network
PPP	Point-to-Point Protocol
ProSe	Proximity Services
PSU	Power Supply Unit
PU	Primary User
PUA	Power per Unit Area
PUE	Power Usage Effectiveness
QoE	Quality of Experience
QoS	Quality of Service
QoT	Quality of Transmission
RAN	Radio Access Network
RAR	Random Access Response
RAT	Radio Access Technology
RAU	Routing Area Update
RF	Radio Frequency
RFID	Radio Frequency Identification
RH	Radio Head
RLC	Radio Link Control
RNC	Radio Network Controller
RoD	Resources-on-Demand
RP	Router Processor
RPT	Reverse Path Tree
RRC	Radio Resource Control
RRH	Remote Radio Head
RSSI	Received Signal Strength Indicator
RTPGE	Reduced Twisted Pair Gigabit Ethernet
RSVP-TE	Resource Reservation Protocol - Traffic Engineering
RWA	Routing and Wavelength Assignment
Rx	Receiver
SCaaS	Small Cells as a Service
SC-FDMA	Single Carrier - Frequency Division Multiple Access
SDH	Synchronous Digital Hierarchy
SDN	Software-Defined Network
SDO	Standards Development Organizations
SE	Spectral Efficiency
SGW	Serving Gateway
SIEPON	Service Interoperability in Ethernet Passive Optical Networks
SINR	Signal-to-Interference-and-Noise Ratio
SIP	Session Initiation Protocol
SIR	Signal-to-Interference Ratio
SISO	Single-Input-Single-Output
SLA	Service Level Agreement
SNMP	Simple Network Management Protocol
SON	Self-Organizing Network
SONET	Synchronous Optical Networking
SSR	Sleep Server

ST	Subscription Table
SU	Secondary User
SU-MIMO	Single-User MIMO
TA	Timing Alignment
TAU	Tracking Area Update
TDM	Time Division Multiplexing
TE	Traffic Engineering
TIA	Transimpedance Amplifier
TCAM	Ternary Content Addressable Memory
TCP/IP	Transmission Control Protocol/Internet Protocol
TEER	Telecommunications Energy Efficiency Ratio
TLB	Table Lookup Bypass
TLS	Transport Layer Security
TPG	Throughput Gain
TTU	Transmission Time Unit
TVWS	TV White Spectrum
TWT	Target Wake Time
Tx	Transmitter
U-APSD	Unscheduled Automatic Power Save Delivery
UE	User Equipment
UDP	User Datagram Protocol
UMTS	Universal Mobile Telecommunications System
UPC	Uplink Power Control
UPnP	Universal Plug and Play
URI	Uniform Resource Identifier
UTP	Unshielded Twisted Pair
UTRAN	Universal Terrestrial Radio Access Network
UWB	Ultra-Wideband
VLAN	Virtual LAN
VM	Virtual Machine
VoIP	Voice over IP
VPN	Virtual Private Network
WCDMA	Wideband Code Division Multiple Access
WDM	Wavelength Division Multiplexing
WFA	Wi-Fi Alliance
Wi-Fi	Wireless Fidelity
WiMAX	Worldwide Interoperability for Microwave Access
WLAN	Wireless LAN
WoL	Wake-on-LAN
WPA-OR	Weighted Power-Aware Optical Routing
WSS	Wavelength selective switch
WXC	Wavelength Cross Connect
XPM	Cross Phase Modulation

1

Introduction

Konstantinos Samdanis[1], Peter Rost[1], Michela Meo[2], Christos Verikoukis[3] and Andreas Maeder[1]

[1]NEC Europe Ltd, Heidelberg, Germany
[2]Politecnico di Torino, Torino, Italy
[3]Telecommunications Technological Centre of Catalonia, Barcelona, Spain

1.1 Origins of Green Communications

Climate change and energy crisis are not just a vague future problems, as their effects are becoming apparent at a rapid pace. Greenhouse emissions, such as CO_2 and methane, cause global warming with catastrophic consequences on planet and on the human society as documented in Ref. [1]. The Kyoto protocol to the United Nations Framework Convention on Climate Change (UNFCCC) established in 1997 indicates that mainly the developed countries are responsible for the current levels of greenhouse gases, followed by a strong objective to take action against global warming. In the Conference of the Parties (COP) 17 climate change conference in Durban, 2011 scientists raised concerns that the measures taken so far are not sufficient to avoid global warming beyond 2°C (a limit established in the G8 meeting in L'Aquila in June 2009 to avoid unpredictable environmental damage) and more urgent action is needed. Besides environmental concerns, the energy crisis becomes apparent with claims stating that approximately 50% of the world petroleum resources are already exploited [2], creating major obstacles for power supply with negative consequences on the economy. In general, energy pricing influenced by fuel prices is showing an increasing trend according to the forecast study performed by the Energy Information Administration (EIA) of the US Department of Energy [3].

As the use of Information Communication Technology (ICT) becomes an essential part of human daily lifestyle allowing social and business interaction on a local and global scale, the pace of development and integration of new technologies is performed on an accelerating rate. According to ITU estimations, Internet users reached around 2.7 billion by the end of 2013, almost 40% of the world's population, whereas mobile cellular subscribers approach 7

Table 1.1   Traffic projections between 2010 and 2020 for the mobile access, wireline access and core networks modeled by GreenTouch for the Mature Market segment [7]

Year	Mature market traffic projections (PB/month)		
	Mobile access	Wireline access	Core network
2010	161	7,727	10,707
2015	3,858	33,879	45,402
2020	14,266	74,462	103,085
2020/2010	89x	9.6x	9.6x

billion, with the mobile broadband being the most dynamic market with 2.1 billion subscribers globally [4]. Effectively, such enormous adoption of ICT is accompanied by a massive growth of the number of user devices, wireless, indoor, transport, core and data center networks and services, raising significant cost and sustainability concerns. In particular, ICT across a wide range of applications currently accounts for 5.7% of the world's electricity consumption and 1.8% of CO_2 emissions [5], stressing the need for enhancing energy efficiency in ICT products. Data from different telecommunication operators globally confirms their huge consumption of electricity as summarized in Ref. [6].

To comprehend the energy consumption of modern telecommunication systems considering the expected scale of increase, a summary of historic and near-future global traffic volumes for the mature markets over the decade 2010–2020 regarding mobile access, wireline and core networks is illustrated in Table 1.1 [7]. From the traffic growth projections, it becomes evident that there is a significant increase on mobile access, but with the wireline access and core networks still carrying the majority of user traffic. Multimedia applications and particularly video is one of the main contributors of such significantly higher traffic volumes, with Cisco Internet traffic projection forecasting that in 2015 the video consumption will reach one million video minutes per second, which is equivalent of 674 days [8]. Such tremendous data growth also as a consequence of the increasing population of users that can afford ICT services, raises new economic and sustainability challenges for network operators, which face the need of expanding their deployed network infrastructure in order to cope with the continuously increasing traffic volumes. Furthermore, the introduction of novel architectures and technologies, for example, machine-to-machine communications, smart grid, automotive, social media, smart cities, adds an extra degree of complexity to the network design, generating a compulsory need for holistic end-to-end approaches for the network operation.

The telecommunication networks' greenhouse emissions growth rate is expected to continuously increase as illustrated in Figure 1.1. Wireless and wireline networks accounted for an equal share on greenhouse emissions between 2002 and 2011 with a footprint of 0.13 and 0.20 gigatons of CO_2 equivalent (GtCO_2e), respectively. However, the remarkably high data volumes in the mobile communications sector and the launch of fourth-generation (4G) long-term evolution (LTE) networks have the consequence of increasing the wireless greenhouse footprint to 0.16 GtCO_2e compared to a wireline equivalent of 0.14 GtCO_2e anticipating the adoption of fiber optics that significantly lower the energy consumption.

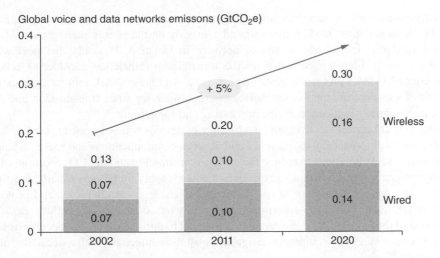

Global voice and data networks emissons (GtCO$_2$e)

Figure 1.1 Telecommunication system emissions between 2002 and 2020 showing wireless and wire-line contributions [9]

Besides economic and environmental matters, energy efficiency in telecommunication systems is particularly crucial for developing countries providing a means for bridging the digital divide. In such cases, cost is the main barrier and hence energy efficiency may reduce the operational expenses provisioning affordable ICT services. In addition, green communications relying on alternative energy sources may provide the means of equipping remote areas without or with limited power supply creating an environmental friendly ICT adoption in developing countries. Furthermore, energy efficient mechanisms and practices could prove critical for telecommunication systems in disaster situations, where the power generation or the distribution infrastructure is damaged. Specifically, network equipment that supports a low-power mode may sustain a basic operation available only for critical services.

1.2 Energy Efficiency in Telecommunication Systems: Then and Now

The problem of energy efficiency in telecommunication systems is not entirely new. Indeed, energy efficiency has always been a significant issue for maximizing the battery lifetime of handheld devices and wireless sensor nodes that operate autonomously without a power grid supply. Early solutions considering the field of cellular and portable communications focus on wireless and mobile terminals, notebooks, and personal digital assistants (PDAs). The main objective was to establish power saving modes or states, in where devices either operate with minimal power consumption or simply sleep, that is, freeze their prior operating state. Common examples are the idle mode of global system for mobile (GSM) communication devices or the hibernate/standby mode of laptops and PDAs.

Similarly, studies on ad hoc and wireless sensor networks have devoted major efforts on energy optimization with the goal of extending the entire system lifetime. A plethora of different energy saving techniques including device sleeping patterns, clustering and head selection for minimizing the energy expenditure of the ad hoc or sensor system as well as energy-aware routing and data aggregation protocols have been extensively analyzed [10, 11]. However, these studies concentrated primarily on the energy management of each device considering the ad hoc or sensor network in isolation. Probably the most widely studied energy problem concentrates on the transmission efficiency considering medium access control (MAC) protocols, error correction, and radio channel gain techniques [12]. Again the focus concentrates on the device energy efficiency since transmission and radio optimizations simply reduce the device processing and forwarding.

These early efforts attempt to extend the lifetime of devices without a power supply. Fundamentally, this is different from the current trend of green communications and energy efficiency in ICT, which seeks solutions to reduce the energy consumption and the CO_2 footprint of network equipment and ICT systems irrespective of power supply limitations usually at off-peak time or when equipment are not in use. Effectively, strategies, mechanisms, and protocols, which stretch along home and enterprise networks, wireless access and cellular networks, transport, and the Internet, are considered with the goal to introduce adaptive operations based on load variations and new equipment design and system architectures with reduced the energy consumption needs.

Such a green communication vision has also got momentum within ICT government and regulatory bodies, especially since the launch of the Kyoto protocol, which aimed to create policies and recommendations for developing and using ICT equipment and systems. Early efforts from the US environmental Protection Agency concentrated on energy efficiency of personal computers (PCs), peripherals, and monitors, establishing a voluntary labeling program referred to as Energy Star in 1992. The European Commission (EC) recognized since 1999 the need for further actions toward energy efficiency and green communications, introducing the code-of-conduct [13] to drive policies and recommendations for reducing CO_2 emissions considering power supplies, digital TV, broadband equipment, and data centers. A number of different government acts and industry initiatives on green ICTs are documented in the survey by the Organisation for Economic Co-operation and Development (OECD) in Ref. [14]. Out of over ninety proposals only 20% suggest measurable targets, with government programs driving the majority of them. A representative example illustrating different targets and initiatives regarding green communications and energy efficiency from various telecommunication operators around the globe is shown in Figure 1.2.

Besides encouraging the adoption of green ICT, government and regulation policies drive energy efficiency standardization efforts by inquiring key standards organizations to provide indications for upper bound power limits of network equipment. The goal of such efforts is to develop power saving features, mechanisms, and protocols; examples are the mandate M-462 introduced by the EC that was communicated to European Telecommunication Standardization Institute (ETSI), European Committee for Standardization (CEN), and European Committee for Electrotechnical Standardization (CENELEC) [15]. Apart from standards, the EC has initiated the ICT for Energy Efficiency (ICT4EE) forum [16] since 2010 to address green communications and energy saving, bringing together high-tech industry from Europe, Japan, and America including the DIGITALEUROPE, Global e-Sustainability Initiate (GeSI), the Japanese Business Council Europe (JBCE), and Tech America Europe.

Furthermore, green communications and energy efficiency in ICT has motivated important initiatives, such as GreenTouch [7], a consortium of industry, academic, and nongovernmental research bodies and experts sharing a vision of increasing the total energy efficiency of ICT by the scale of 1000. The goal of the participants is to transform data and wireless communications as well as the Internet to reduce significantly the CO_2 footprint of network equipment, system

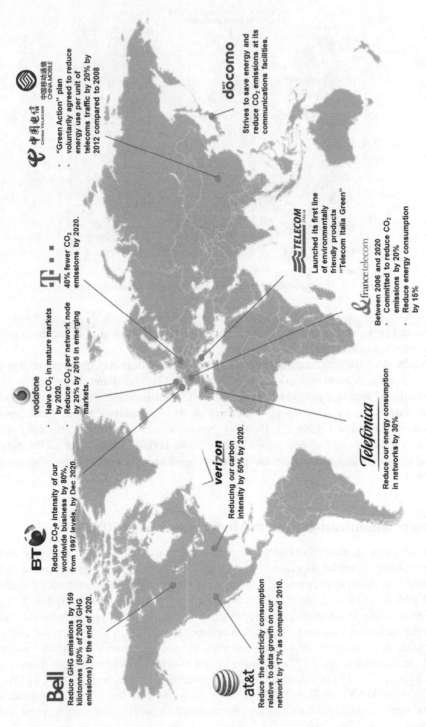

Figure 1.2 Green and energy efficiency ICT targets introduced by the major telecommunication operators around the globe

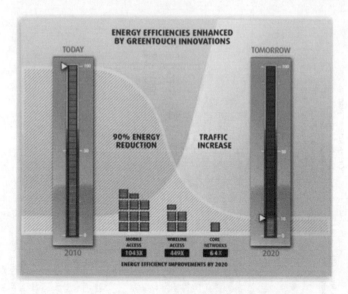

Figure 1.3   Energy efficiency forecast by GreenTouch innovation [7]

platforms, and networks, employing novel, smart, and sophisticated architectures, mechanisms, protocols, and algorithms.

In particular, GreenTouch recently announced the potential of reducing the net energy consumption in communication networks up to 90% by 2020 [7]. This dramatic net energy reduction in wired and wireless networks, despite the foreseen increase in traffic, is fueled by significant improvements in the energy efficiency of network equipment and the component networks as illustrated in Figure 1.3. Hence, it becomes clear that energy efficiency should be studied in a holistic end-to-end network perspective, taking into account all the network layers, and corresponding mechanism and protocols considering a variety of possible network architectures.

1.3   Telecommunication System Model and Energy Efficiency

As mentioned, telecommunication networks need to carry exponentially increasing data traffic while avoiding a similar increase of energy consumption. A telecommunication system is structured into multiple network parts or regions which employ a diverse set of technologies, each posing different challenges for reducing energy consumption. Figure 1.4 shows the architecture of a telecommunication system including the different type of networks, which comprise the subject of this book. The access network provides connectivity to the end users through either fixed network or wireless network technologies including digital subscriber line (DSL) technologies, optical fiber, cellular systems such as Universal Mobile Telecommunications System (UMTS) or wireless broadband technologies such as IEEE 802.11 (Wi-Fi), IEEE 802.16 (WiMAX), or 3GPP LTE. Wireless broadband technologies allow access for mobile end user terminals such as laptops, mobile handheld device, smartphones and tablets, or sensors.

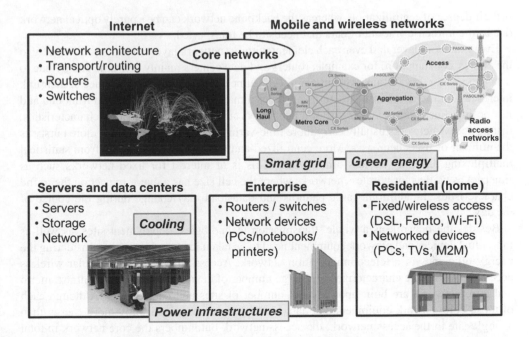

Figure 1.4 Telecommunication system model

Fixed access networks provide connectivity to residential and enterprise networks, which support a number of devices including PCs, peripherals, sensors and Wi-Fi access points, user-deployed femtocells, or operator-deployed base stations. Data offloading solutions or indoor networks employing Wi-Fi or picocells, for example, in shopping malls or train station, as well as coverage extensions via small cells or capacity booster cells can enhance the access providing service diversity for mobile and fixed operators.

Traffic from the access network is aggregated at different stages usually in a tree-structured network which is referred to as backhaul network. It is difficult to draw a clear line between access and aggregation network, for example, if an end user deploys a Wi-Fi access point, the DSL connection may also be part of access network or already part of the aggregation network. The aggregation network is connected to the metro network, which provides connectivity between the individual network equipment within a metropolitan area. Usually, such a network has a ring structure and provides eventual access to the core network. The size of a metro network may vary significantly from a few ten kilometers up to a few hundred kilometers. In metro networks, either optical technologies, such as synchronous optical networking (SONET)/synchronous digital hierarchy (SDH) or wavelength division multiplexing (WDM), are deployed or high-throughput Ethernet technologies such as Carrier-Ethernet.

Finally, the core and backbone network connects different metro networks nationwide, and it connects the telecommunication network with the Internet as well as other telecommunication networks. It is composed of switching and routing equipment, gateways as well as subscriber information, authentication, policing, mobility, charging, and Operations Administrations and Management (OAM) functions. Data centers are typically a part of the core and backbone networks, hosting network services and user applications. Core networks can vary significantly in their degree of centralization, whereas the backbone network can be a purely optical network due to the immense amount of data that needs to be transported.

Every network layer and even each element within a network layer contributes differently to the energy consumption, for example, routers and data centers mainly consume energy due to data processing, large-scale switches mainly consume energy due the maintenance of their link interfaces, and wireless infrastructure consumes a major portion for their radio front end and its cooling. Furthermore, each part of a network is subject to different traffic characteristics, that is, access networks usually carry more time-variant traffic which becomes more bursty as the number of users decreases. More centralized and aggregated traffic gains from statistical multiplexing and may be less time variant. This is of interest for fixed networks such as Ethernet as well as for wireless networks where the cell size has a major impact on the traffic characteristics. In addition, traffic patterns and data load may change during the course of the day.

Besides the network as a whole, transmission technologies, deployment sites, as well as individual devices and network equipment need to be taken into account when considering the energy efficiency of a telecommunication network. Access networks, in particular wireless access networks, are characterized by a large number of devices and sites. By contrast, metro and core networks are built upon a small number of sites and data centers. Although each of these core network equipment and data centers consumes significantly more energy than a single site in the access network, the access network outnumbers the core network in total energy consumption.

In Ref. [6], the energy expenditure of the access is estimated to account 70% of the entire network consumption, whereas the transport and core is responsible for the remaining 30%, despite the fact that the access network equipment consume around 1000 W/hour, whereas the transport and core consume six and ten times more, respectively. Therefore, the energy efficiency of devices and sites within the individual network layers needs to be considered differently. Another major challenge for energy efficiency is to scale the energy consumption to the actual demand for resources. Many networks consume significant energy resources for being ready to provide peak throughput although only a fraction of it is used.

End-user terminals, such as PCs and handheld devices as well as network peripherals and wireless sensor devices, compose the first layer. As stated already, an improvement of energy efficiency for terminals by introducing a sleep or idle state rather improves the lifetime of wireless devices and operational cost of fixed ones than having a major impact on the overall energy consumption of the telecommunication network. Similarly, proxy devices are used in fixed-line networks, which take care of the network communication and keep connections alive while the host computer or other peripherals are turned off or hibernated. In the access network, the main challenge is the interaction of wireless network infrastructure with the mobile backhaul network. Both use different technologies and require continuous protocol signaling exchange. This complicates turning off a wireless access point as it needs to be re-initialized both on the air interface and toward the backhaul and core network. Similarly, in reducing the power consumption of router and switches, there is a need to perform the corresponding protocol updates for coordinating the network topology.

In general, wireless access networks are less energy efficient in terms of throughput per Joule than wired networks, but wireless networks allow for more deployment and usage flexibility. Furthermore, deploying one wireless access point allows for connecting a large number of end users, whereas trenching new cables for a wired may be more expensive and energy intensive, at least in the short term. This relation is considered by a concept called embodied energy [17], which not only accounts for the energy consumed during the operation but also to build and deploy a communication technology. Apparently, this complicates the evaluation of energy efficiency and makes it very specific to individual scenarios.

Metro and core networks are increasingly adopting optical technologies, alongside asynchronous transfer mode (ATM), Internet Protocol (IP), and Carrier-Ethernet. Optical transport consumes the least energy during operation because of reduced data processing and forwarding effort. Such efficiency alongside the very high data rates makes optical technology a good candidate for highly reliable and high-throughput networks. The largest part of energy consumption in optical networks is caused by the conversion from optical to electrical domain and back, that is, from optical to IP routing or vice versa. Therefore, purely optical switches or cross-layer optimization solutions are of particular interest in the research community and industry as they significantly reduce the overall energy consumption. Data centers constitute an elementary part of telecommunication networks but are rather in scope of computing energy efficiency, for example, higher silicon integration. A promising option for data centers is to use renewable energy in order to reduce the CO_2 footprint. In this book, data centers' internal architecture is not in focus, but only the networking related issues such as routing and network topology, which provide access and connectivity to data centers.

Within each network type, energy efficiency can also be considered on the individual Open Systems Interconnection (OSI) protocol layer, being subject to different constraints and causes of energy consumption. For instance, on PHY layer, energy efficiency has been of particular interest in the area of wireless networks. In wireless networks, the radio front end and the baseband processing consume a major portion of the energy. In fact, more than 50% of the energy at a base station is consumed by the power amplifier, most of it as dissipated heat. Hence, engineers and researchers are investigating ways to improve the amplifier efficiency, to introduce micro-sleeps during which the radio front end is turned off, alongside cooperative algorithms, advanced multi-antenna schemes, and encoding/decoding algorithms, which have already received significant attention. Similarly for fixed networks, managing the power of particular interfaces by applying sleeping either on a temporarily, regular, or event basis is significant for the energy consumption of switches, routers, and other core network equipment.

On data link and network layer, scheduling, packet segmentation as well as feedback algorithms are subject of investigation for improving energy efficiency. For instance, in the area of wireless networks, intercell interference coordination, load balancing, and efficient segmentation processes may have a significant impact on the energy efficiency. For wired networks, the primary focus is on developing routing and switching protocols and algorithms to handle energy efficiency on network equipment while taking care of synchronization and service performance issues.

Considering the transport and application layer the main energy saving efforts concentrate on transport protocols, for example, Transmission Control Protocol/Internet Protocol (TCP/IP), content distribution, codecs and digital signal processing (DSP) as well as on context adaptation and location information. Finally, besides individual improvements on each protocol layer, cross-layer considerations have an even stronger impact on energy efficiency, for example, switching off access points is only possible if efficient protocols on higher layers are available which allow disrupting connections.

1.4 Energy Saving Concepts

When analyzing energy saving and green communications it is important to firstly understand what characterizes a communication system as energy efficient or green. Energy efficiency is associated with the provision and operation of resources that enable a service corresponding to a user demand, with reduced energy consumption. It is typically defined as the ratio of a functional unit to the energy required to deliver such a functional unit, with higher values indicating greater energy efficiency [18]. Network equipment and communication systems are typically provisioned to handle peak-time traffic demands, which differ significantly from the average load during other times. In fact, even when idle, that is, not in use, ordinary nonenergy aware network equipment and communication systems consume a significant amount of power, unless energy efficient mechanisms are employed.

In principle, reducing the energy consumption of telecommunication networks can be achieved in the following five fundamental ways:

- *Network planning*: optimize the physical placement of resources and enable the potential to power off network equipment.
- *Equipment re-engineering*: introduce low-energy network equipment and devices; redesign internal equipment architectures to accommodate energy saving requirements.
- *Network management*: optimize the operation of network equipment as well as network-wide protocols and mechanisms by adjusting network resources based on the users' demand.
- *Renewable energy*: reduce the energy consumption from power grid by supplying energy from alternative sources, for example, solar and/or wind, which are also environmental friendly.
- *Social awareness*: educate users to avoid wasting energy.

Network planning is an offline activity, a first step that aims to accommodate energy saving by considering the dimension of the network topology, equipment, and deployment aspects. For instance, the consideration of small cells and heterogeneous networks within the network planning phase can increase the spectral efficiency of a wireless system, while reducing power amplifier needs and cooling for network elements, therefore improving the network-wide energy efficiency. Similarly for wireline networks, energy efficiency can be realized by provisioning nodes, that is, switches or routers, and links forming a network topology by considering also the deployment strategy in terms of the transport technology, for example, fiber, microwave, etc., and the corresponding routing policy. Additional opportunities for energy saving in network planning are obtained via network virtualization and node consolidation, which can achieve significant savings through the use of overlay networks by integrating functions and services into less network equipment.

Energy efficiency in telecommunication systems relies mainly on the power consumption of individual components that comprise them and on the way that different components and mechanisms influence one another. For instance, new power operations on network equipment may affect and introduce new features on communication protocols. Perhaps the most fundamental issue is to consider energy-efficient hardware improvements on the devices and equipment, which may allow an operation with reduced power. The use of advanced materials may reduce cooling, whereas new technologies, for example, fiber optics, can cut down significantly data processing and forwarding power. In addition, enhancing the internal equipment architecture design to accommodate energy saving needs is another important hardware attribute, which can contribute to scale down the need of the power amplifier, a component that consumes a high portion of energy. It can also enhance energy efficiency in data processing, by improving the energy usage profile of equipment hence achieving energy proportionality, that is, scaling the energy consumption with profitable use.

Hardware advancements may also create opportunities for a more efficient energy saving management of devices by controlling the energy state of equipment reflecting traffic dynamics according to the following two fundamental mechanisms:

- Power off where equipment is not operational:
  - (i) A cyclic-operation of being powered on and powered off during particular time periods.
  - (ii) A soft-sleep or dossing where equipment maintain a minimal components or devices operational to allow rapid wake-up.
  - (iii) An extended or deep-sleep where equipment need a significant amount of time to become fully operational again.
- Slow down the operation of equipment:
  - (i) Modular where selected functions are reduced, while the equipment is partially operating.
  - (ii) Rate adaptation reducing the operational rate of processing and/or forwarding traffic.

In managing the energy operations, it is important to ensure that network equipment are profitably under a power saving state or mode for a sufficient time period avoiding oscillations and the associated costs (i.e., signaling, power, computation) for changing the power state or mode of network equipment. Hence, monitoring of network traffic load is essential as suggested in Ref. [19] to identify long-term off-peak time periods and avoid local traffic load minima as illustrated in Figure 1.5.

Such traffic load monitoring assists the network management operation which can control the energy state or mode of individual network equipment considering a network-wide view. The corresponding energy control actuators may reside locally on network equipment or can be centralized.

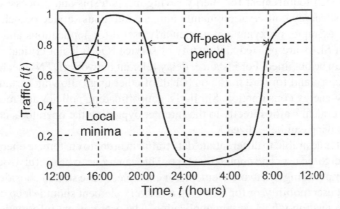

Figure 1.5 Example of normalized traffic variation over a daily period

Network-wide mechanisms that rely on network management decisions can mainly influence traffic shaping and steering realizing the following energy saving opportunities:

- Regulating traffic forwarding between selected network equipment allowing a soft-sleep or dossing within particular time periods, for example, discontinuous transmission (DTX) in LTE, Energy-Efficient Ethernet (EEE), and fast sleep in passive optical networks (PONs). In addition, scheduling or filtering may shape traffic volumes reducing the network resource utilization providing energy efficiency.
- Traffic aggregation in selected routes or parts of the network allowing underutilized network equipment, for example, base stations, routers, switches, interfaces, to enter or maintain a reduced power consumption operation. Cell zooming and energy partitions are common examples for wireless communications, whereas routing, traffic steering, and transport mechanism may achieve the equivalent for wireline systems.

Additionally, network-wide protocols and content applications can further improve the energy efficiency associated with network resources and operations via:

- Reducing the volume of traffic across the backhaul and core networks, enabling users to retrieve content at closer locations by introducing content distribution network (CDN) and cache mechanisms at selected edge positions. Similarly, the use of information centric networking (ICN) may reduce the data traffic processing and forwarding on network equipment within the backhaul and Internet, while peer-to-peer and location aware applications may provide more efficient information exchange for both wireless and wireline communications.
- Introducing lightweight protocols, which require less control plane signaling and protocol advancements that result in reduced energy consumption, for example, MAC protocols, or adapt considering power saving states or modes of network equipment.

A critical attribute when providing energy efficiency in networking is the effect on service level agreements (SLAs) bound to particular users and applications. Typically, it is desired to maintain SLAs, irrespective of network power conservation circumstances and hence this is a significant design requirement for energy saving mechanisms and processes. Mechanisms which can assist energy conservation maintaining a continuous service on behalf of sleeping equipment, for example, proxying, are particularly useful for applications and other network services. When SLAs are relaxed and quality of service (QoS) is not critical, further energy conservation can be obtained. For instance, delay tolerant network (DTN) mechanisms enable users to store, carry, and forward traffic on behalf of other users allowing certain base stations to maintain low-energy consumption. Similarly, data offloading policies and mechanisms may steer non-QoS critical traffic directly to the Internet bypassing the operator's core network in where energy saving can be realized.

In practice, it is desirable to differentiate data traffic in order to customize energy saving policies with respect to the user and application type. This can prove to be useful to determine when it is worth re-empowering a network element, for example, a base station considering the application type and user mobility, or for how long a network element should sleep considering the delay caused in relation with a certain application. The user context information may further enhance energy saving considering the user activity/behavior and movement, for example, a home base station may only operate when the user initiates an application or a user's movement away or toward his premises may power off or re-empower his home base station. Similarly, enterprise networks can manage efficiently the energy of printers, PCs, wireless access points, and so on, based on information about the incoming users' profiles and permissions. Evolving applications, such as proximity services and social applications, may optimize device communication considering location information allowing energy saving in radio and core network.

Nevertheless, as energy efficiency may reduce the operational costs of telecommunication networks making networking commodities cheaper, the user may consume higher amounts of services and applications creating a rebound effect as suggested by the Khazzoom–Brookes postulate [20]. Hence, it is also of utmost importance to raise social awareness, introducing policies that reward users who conserve energy on behalf of the network operator. Charging models may also reflect energy consumption to a certain extend encouraging users to saving while receiving better bills.

1.5   Quantifying Energy Efficiency in ICT

The definition of quantitative metrics for the characterization of the energy efficiency of network equipment is fundamental for comparing solutions and technologies, estimating their effectiveness, taking proper decisions. Moreover, metrics and indices to quantify energy efficiency can be used as variables in the internals of algorithms and strategies such as those devoted to resource allocation or resource sharing. However, despite the intense research and development activity performed in the field of energy-efficient communications, a definite established agreement on the most appropriate metrics is still open. The manifold aspects related to energy efficiency make the definition of proper metrics so complex that the efficiency of a given system cannot be described only by a single energy-efficient indicator but has to be expressed as a set of metrics or a hybrid metric that relate in different ways the work done by the system, in terms of the amount and quality of a provided service, with its cost in terms of the amount of energy that is needed to produce that work.

Thus, several different metrics have been proposed that are specific of a system or scenario, depending on the kind of work or application that is provided. For example, indices that aim at measuring the efficiency of a processor unit relate the amount of executed operations with the amount of used energy; indices that target the efficiency of a transceiver system relate the emitted power with the power drained by the system; and for devices that carry telephone calls efficiency is measured in terms of the number of Erlang of carried traffic per unit of energy. Moreover, for the same device or system, depending on the specific interest and objective of an evaluation, the metrics might be different. For instance, in a wireless systems, the amount of power that must be drained to obtain a given transmitted power from the antenna might be relevant, as well as the energy needed to provide coverage of a given unit area.

For core network nodes, ETSI defines in Ref. [21] a general metric, referred to as energy efficiency ratio as the fraction of useful output, intended as the capacity of service of a core network node, to the requested power. The useful output is then, depending on the functions of the node, expressed as the number of Erlang, for a device targeting voice calls, or as packets per second, or number of simultaneously attached users. The measure can then be derived for different load levels and for various traffic mixes. More complex is the definition of proper energy efficiency metrics for wireless networks.

In these cases, the efficiency of the network devices themselves is not enough for many investigations. In fact, in addition to the quantity of capacity that is carried by a device, the concept of coverage and, hence, service provisioning, add a new dimension to the space of the possible metrics. When two technologies are compared, for example, the planning strategy, in terms of number and kind of devices deployed in a given area, can make an important difference; similarly, the need to cover areas even if the carried traffic is negligible can really change the evaluation of the efficiency of a solution. In Refs. [22, 23], the following main metrics are proposed for wireless access networks:

- The *energy per information bit* [J/b] or [W/bps] represents the energy needed to transfer a bit of information.
- The *power per unit area* [W/m²] relates the average power used by a device that provides connectivity to the size of the covered area.

For cellular systems, the first metric is particularly relevant for urban areas, where traffic is high, and cellular networks are capacity limited. In these cases, one of the most important characteristics of a system is the possibility to provide high capacity; the amount of work corresponds to number of bit transfers that can be done with a unit of energy. For this reason, this metric is widely used to evaluate and compare transmission techniques over an individual link. This metric is less relevant for systems under low traffic conditions, when the work that is actually needed by the system is low. In these cases, the second metric, the power per unit area, is more suited, since it targets the energy needed to provide service provisioning, regardless the actual usage of the service itself. This metric allows also to consider different system configurations. For example, given an area, the coverage might be provided by different kinds of devices, with different coverage and, thus, the metric is also well suited to give a view of heterogeneous network scenarios.

In Ref. [24], ETSI proposes two metrics that introduce, besides service provisioning, the concept of quality with which the service is provided:

- The ratio between the throughput of users obtaining a minimum specified (service-dependent) QoS within the served area and the total power consumed by the base stations offering service in the same area [b/J].
- The ratio between the number of users obtaining a minimum specified QoS within the served area and the total energy consumed by the base stations offering service in the same area during the observation time [J⁻¹].

QoS is then defined based on the kind of considered service. For circuit-switched services, like voice or constant bit rate data, blocking and dropping are the only QoS measures. For data services, an additional QoS metric is the throughput that is further distinguished in guaranteed throughput for streaming-like services and best effort.

The efficiency of devices alone does not include the efficiency of the site infrastructure that is needed to operate the device. If the perspective of a study is the overall consumption of a network node, the evaluation should therefore include also some aspects related to the site itself. Derived from the case of data centers, where the physical layout of the site, with the need of a power-hungry cooling infrastructure, backup power supply and other support systems, has a large impact, a measure of the efficient use of the energy in the whole site is the power usage effectiveness (PUE) defined by the ratio of the amount of power entering a site (or a data center) over the power needed to run the site.

The source of energy that empowers ICT is also significant since it has a different effect toward the environment. Usually a variety of energy sources is available, for example, coal, hydro, geothermal, nuclear, solar, and are typically mixed when empowering a system. Hence, a metric that can capture the CO_2 emissions per transmitted data volume may provide an insight of the environmental impact related to the corresponding energy source.

Finally, measuring the energy consumption of ICT systems is significant to gain understanding of their efficiency. Obtaining energy measurements is challenging in terms of accuracy, timeliness, overhead, and so on and can be performed instantaneous or with a degree of aggregation, either directly on network equipment or through a model-based approach [25]. Direct measurements can be carried out on hardware using power meters, by software means via Application Program Interfaces (APIs) or using management protocols such as Simple Network Management Protocol (SNMP), for example, in the case of power over Ethernet. Model-based approaches monitor certain traffic and equipment characteristics in order to derive the energy consumption and similarly to direct measurements can be carried out on hardware, for example, on a switch that may monitor the packet rate or traffic volume of an attached network equipment, by software, for example, PC's CPU load obtained via Windows Performance API or through SNMP queries, for example, querying the CPU load of a switch. A framework for monitoring and managing the energy consumption of network equipment including internal devices is detailed in Ref. [26] presenting an information model, which specifies energy objects and the means to monitor and control them.

1.6   Conclusions

Energy efficiency and green communication is getting momentum due to environmental and cost reasons, with government bodies creating energy regulations and policies for network equipment and systems. However, energy efficiency is not a new problem, but currently it is considered in relation with telecommunication systems and network equipment, not merely for user devices or sensor networks. Providing energy efficiency is a complex process that should consider all operational aspects and parts of a system in coordination providing a holistic approach. Network equipment should support one or a combination of energy saving features being able to power off, slow down, and regulate traversing traffic. Communication systems are expected to adopt energy efficiency in the planning and deployment phase as well as in network management, which ensures that network equipment are profitably saving energy. Besides equipment and systems, energy-aware protocols can handle communications, steer traffic, and retrieve content with minimum energy consumption, while considering the expected user QoS. However, the contribution and effect of these aforementioned methods and mechanisms is not quantified unless the appropriate metrics and measurements process are not in place. Hence, it is important to create such metrics and processes for measuring energy efficiency with respect to particular network equipment and specific system deployment scenarios.

This book aims to provide a detail insight into energy efficiency methods and mechanisms considering different operational layers and part of a telecommunications network. It initially explores wireless communications starting from a review of the fundamental aspects in Chapters 2–4, which contain a categorization and modeling of energy efficiency, an analysis of energy metrics, performance trade-offs, and the embodied energy consumption considering the system life cycle. The following four chapters are then considering the different aspects of a cellular system starting from the base station, the fundamental elements of a radio access network in Chapter 5, the network planning and design in Chapter 6, the radio considerations in Chapter 7, and the network management perspective in Chapter 8. The next three chapters explore energy efficiency studying specific wireless networks, with Chapter 9 concentrating on home and enterprise networks, Chapter 10 on delay tolerant and vehicular networks, and Chapter 11 on machine type communications (MTC) and Internet of Things (IoT). Finally, the last chapter for the wireless communication part, Chapter 12, overviews the current state of standards analyzing the efforts of 3rd Generation Partnership Project (3GPP), European Telecommunications Standards Institute (ETSI), and IEEE 802.11/Wi-Fi Alliance.

The remaining of the book analyze energy efficiency in wireline communications, considering in the first four chapters the typical technologies for fixed networks, with Chapter 13 concentrating on routing, switching, and transport, Chapter 14 on Energy Efficient Ethernet, Chapter 15 on optical communications, and Chapter 16 on data center routing and networking. The next three chapters consider different system attributes with Chapter 17 concentrating on network management, emphasizing the use of the emerging software-defined network (SDN) and network function virtualization (NFV) technologies, Chapter 18 analyzing energy aware communication protocols including TCP/IP, peer-to-peer, proxying, and context aware power management, and finally Chapter 19 considering content-based networking and ICN. The last chapter of the wireline communication part, Chapter 20, overviews the standardization efforts considering the Internet Engineering Task Force (IETF), Institute of Electrical and Electronics Engineers (IEEE), Broadband Forum (BBF), ETSI, and Alliance for Telecommunications Industry Solutions (ATIS). Finally, Chapter 21 provides the conclusions of the book and also presents the vision for energy efficiency in future deployments analyzing the evolving service requirements and analyzing the most promising areas for saving energy.

References

[1] D. Guggenheim, "An inconvenient truth: a global warning", an Academy Award-winning 2006 documentary starring Al Gore.

[2] J. Leggett, Half Gone: Oil, Gas, Hot Air and the Global Energy Crisis, Portobello Books Ltd, 2006.

[3] EIA Report: DOE/EIA-0383(2014), Annual Energy Outlook, US Energy Information Administration, 2014.

[4] ITU, ITU Releases Latest Global Technology Development Figures, Press Release, 2013.

[5] SBI Bulletin: energy efficiency technologies in information and communication industry 2005–2015.

[6] R. Bolla, R. Bruschi, F. Davoli, F. Cucchietti, "Energy efficiency in the future Internet: a survey of existing approaches and trends in energy-aware fixed network infrastructures", IEEE Commun. Surv. Tutorials, vol. 13, no. 2, 2011.

[7] GreenTouch, GreenTouch Green Meter Research Study: Reducing the Net Energy Consumption in Communication Networks by Up to 90% by 2020, 2013.

[8] Cisco, Global Internet Traffic Projected to Quadruple by 2015, Visual Networking Index (VNI) Forecast (2010–2015), 2011.

[9] GeSI SMARTer 2020: The Role of ICT in Driving a Sustainable Future, Global e-Sustainability Initiative and The Boston Consulting Group, Inc., 2012.

[10] G. Anastasi, M. Conti, M. Di Francesco, A. Passarella, "Energy conservation in wireless sensor networks; a survey", Ad-Hoc Netw. J., vol. 7, no. 3, 537–568, 2009.

[11] L.M. Feeney, M. Nilsson, "Investigating the energy consumption of a wireless interface in an ad-hoc networking environment", IEEE INFOCOM Anchorage, 2001.

[12] K. Langendoen, G. Halkes, "Energy-efficient medium access control", Embedded Systems Handbook, CRC Press, 2005.

[13] European Commission, Code of conduct on energy consumption of broadband equipment, Version 5.0, 2013.

[14] OECT, Towards green ICT strategies: assessing policies and programmes on ICT and the environment, Full Report: OECD Working Party on the Information Economy, 2009.

[15] EC, Standardization Mandate Addressed to CEN, CENELEC and ETSI in the field of ICT to enable Efficient Energy use in Fixed and Mobile Information and Communication Networks, M462, 2010.

[16] ICT and High Tech Join forces to Tackle Energy Efficiency and low Carbon Economy, 2010.

[17] H.A. Udo de Haes, R. Heijungs, "Life-cycle assessment for energy analysis and management", Applied Energy, vol. 84, no. 7–8, 817–827, 2007.

[18] ITU-T L.1310, "Energy efficient metrics and measurement methods for telecommunication equipment", 2011.

[19] M.A. Marsan, L. Chiaraviglio, D. Ciullo, M. Meo, Optimal Energy Savings in Cellular Access Networks, IEEE ICC Workshops, Dresden, 2009.

[20] H. Herring, "Does energy efficiency save energy? The debate and its consequences", Appl. Energy, vol. 63, no. 3, 209–226, 1999.

[21] ETSI ES 201 554, Environmental engineering (EE); measurement method for energy efficiency of core network equipment, v1.1.1, 2012.

[22] The EARTH project, "Most suitable efficiency metrics and utility functions", 2011.

[23] ETSI TS 102 706, Environmental engineering (EE) measurement method for energy efficiency of wireless access network equipment, V1.2.1, 2011.

[24] ETSI TR 103 117, Environmental engineering (EE); principles for mobile network level energy efficiency, v1.1.1, 2012.

[25] D. Dudkowski, K. Samdanis, "Energy consumption monitoring techniques in communication networks", IEEE CCNC, Las Vegas, 2012.

[26] J. Parello, C. Claise, B. Schoening, J. Quittek, "Energy management framework", IETF Internet-Draft, 2014.

2

Green Communication Concepts, Energy Metrics and Throughput Efficiency for Wireless Systems

Timothy O'Farrell[1] and Simon Fletcher[2]

[1]*University of Sheffield, Department of Electronic and Electrical Engineering, Sheffield, UK*
[2]*NEC Telecom MODUS Ltd, Surrey, UK*

2.1 Introduction

In the last decade, the communications industry has overcome some extraordinary challenges transforming the way people communicate and access information. Interestingly, this revolution in communication, which touches every aspect of society from the economy, through health care to leisure, has been achieved by an evolutionary approach to enhancing the communications network infrastructure. Over the years, greater access to information has been motivated by the introduction of faster broadband connections (both wired and wireless), versatile and highly portable data centric devices (e.g., smartphones and tablets), and new cloud-based services, which are globally appealing (e.g., social media and YouTube). Importantly, devices such as smartphones have introduced a platform concept that allows third parties to develop and distribute applications (Apps), which has sparked the rapid take-off of mobile broadband. As well as supporting core telecommunication services, network operators, including mobile network operators (MNOs), must provide reliable and inexpensive access to the Internet while the demand for data continues to grow rapidly. Against this backdrop, the industry is already turning its attention to evolving the next generation of broadband infrastructure. For example, the 5G mobile radio access network (RAN) will be characterized by even higher data speeds, prolific amounts of machine-type communication and extremely low latency. Industry is viewing cell densification, heterogeneous networks (HetNets), and massive MIMO technologies as potential approaches to meet this demanding specification. However, this road map comes

Green Communications: Principles, Concepts and Practice, First Edition.
Edited by Konstantinos Samdanis, Peter Rost, Andreas Maeder, Michela Meo and Christos Verikoukis.
© 2015 John Wiley & Sons, Ltd. Published 2015 by John Wiley & Sons, Ltd.

with a considerable prices tag in terms of energy and environmental cost as well as capital expenditure costs.

Accommodating the growth of data services by enhancing transmission speeds and increasing the amounts of communication infrastructure presents several key energy consumption challenges which can no longer be discarded because of their impact on network running costs and the environment. Operators are keen to reduce their long-term energy costs [1] given the recent rises in energy prices [2]. From 2009 to 2011, Vodafone reported that the impact of entering new markets together with upgrading existing network infrastructure resulted in a 35% increase in total energy consumption primarily because the number of base station sites owned was doubled during this period [3]. Equally, the challenges imposed by climate change have taken centre stage because scientific studies suggest that global warming directly correlates with Green House Gas (GHG) emissions, in particular CO_2 emissions. As energy consumption and GHG emissions are closely related, a reduction in the former influences strongly the latter. Accordingly, the European Commission has predicated three significant targets for its member states to be fulfilled by 2020; these are 20% reduction in GHG emissions, 20% reduction in total energy consumption, and 20% of energy generation from renewables [4]. This has resulted in companies publically committing to reducing their carbon footprint. For example, by 2020, Vodafone has committed to reduce carbon emissions by 50% from levels in 2008 [5]. Other companies have made similar commitments. While ICT as a whole is responsible for approximately 2% to 3% of global CO_2 emissions [6], it is increasingly evident that ICT is being heavily adopted as a tool to reduce CO_2 emissions in other industries, for example, the electricity supply (green grid) [7], transport, and building management systems, to name a few.

Recently, significant resources have been committed to understanding and remedying the high energy consumption of large-scale communication systems. Since 2008, a number of dedicated international research projects have addressed this issue predominantly focused on cellular mobile systems. For example, the Green Radio (GR) Core 5 Research Program in Mobile VCE [8, 9] has looked at a broad range of energy reduction techniques across the protocol stack and their integration. The GR program focused on intermediate to long-term solutions to achieve energy reductions in mobile networks. Other significant programs that targeted more near-term energy reduction solutions were the Celtic Initiative project OPERANet-I (Optimizing Power Efficiency in Mobile Radio Networks) [10, 11] and the European Framework Program 7 project EARTH (Energy Aware Radio and Network Technologies) [12]. The material presented in this chapter is mainly focused on an analytical framework developed in the GR project that addresses the evolution of an MNO's RAN based on realistic scenarios and traffic data.

The chapter provides an accessible analytical framework that can be used to compare the energy consumption and throughput characteristics of different RAN configurations based on the Long Term Evolution (LTE) family of standards (i.e. LTE and LTE-Advanced). The technical approach is useful for exploring how an LTE RAN may be upgraded in an energy efficient manner to meet the expected growth in mobile data traffic. The approach also allows particular energy saving techniques to be evaluated, such as deployment techniques (e.g., cell size and RAN topology) and power state techniques (e.g., sleepmodes). Drawing on the open literature, the analysis uses accurate energy consumption models of base station and access point equipment in order to account for the overhead, or load independent, energy consumption as well as the load dependent consumption. Therefore, the results reflect the overall operational energy consumption characteristic of a RAN and not just the radio frequency (RF) characteristic. Also, the Energy Consumption Gain (ERG) and Throughput Gain (TPG) figures of merit, developed by the authors in the GR project, are used to compare RAN configurations. In particular, a new figure of merit called the Energy Throughput Gain (ETG) is identified. The ETG reliably accounts for the difference in energy consumption and throughput between two distinct RAN configurations when transporting application data bits.

The chapter is organized as follows. Section 2.2 provides an overview of the evolution of broadband access networks focusing on radio access. Traffic growth, network upgrading strategies and the role of heterogeneous networks (HetNets) are discussed. Section 2.3 describes the prevailing energy consumption models for base stations and access points and the future trends in the energy consumption of these key subsystems whereas Section 2.4 details the metrics adopted for comparing energy consumption and throughput in distinct RAN configurations. Section 2.5 evaluates the energy consumption and throughput performance of a greenfield deployment scenario based on a regular, homogeneous RAN that uses different cell sizes and base station types. The study is augmented by exploring energy savings obtained from sleep-modes and HetNet deployments. Section 2.6 discusses the key findings drawn from the chapter.

2.2   Broadband Access Evolution

A key factor that has ignited the concern of the energy consumption in broadband access networks, in particular mobile access, is the consistent and rapid growth in broadband traffic. The highly cited Cisco Visual Networking Index (VNI) has reported that in 2012, global average internet traffic increased by 34% while global mobile data traffic grew by 70% [13]. Noticeably, the volume of data traffic consumed per smartphone is increasing rapidly. The Cisco VNI predicts that while portable devices such as smartphones and tablets accounted for 3% of global IP traffic in 2012, this will rise to 22% of IP traffic in 2017. As indicated in Table 2.1, the underlying cause of the exponential growth in mobile data traffic is the combined increase in the number of mobile broadband subscribers, the devices per subscriber, and the traffic consumed per device.

Traffic loads within an RAN can exhibit significant temporal and spatial variations. Though the traffic at any given base station may vary hugely over short time intervals (e.g., seconds or minutes), when averaged over longer time periods (e.g., hours, days or weeks), useful insights emerge that correlate strongly with user behavior. For example, users commuting between the residential and commercial districts of a city results in different busy hour traffic for each district. In commercial areas, the busy hour occurs in the morning while in residential areas, a busy hour is observed between 9 and 10 pm when most traffic is generated indoors by stationary users [14]. This phenomenon is illustrated in Figure 2.1, where normalized traffic intensity versus hour of the day is depicted for a European residential district [15, 16]. Typically, the density of base stations within a city is higher in commercial districts than residential areas. As such, the traffic is unevenly distributed with a significant imbalance between busy and non-busy hours, along with very large variations across the different parts of a city. Despite this temporal and geographic variation of traffic, base stations typically consume a fixed amount of overhead energy commensurate with operating the RAN during the busy hour period. It is widely recognized that this approach to base station deployment and operation is not sustainable from an energy consumption point of view in the face of predicted traffic growth rates.

MNOs have used a number of pathways to evolve their mobile access networks to meet growing traffic demands. Historically, network evolution and equipment upgrades have been

Table 2.1 CAGR for devices and traffic per device

Predicted device and traffic growth	2011–2016 Mbit/month CAGR (%)	2011–2016 Device No. CAGR (%)
M2M module	30	42
Smartphone	77	24
Tablet	52	50
Laptop	27	17
Portable gaming device	27	56

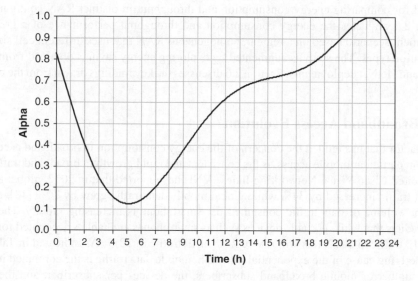

Figure 2.1 Load activity factor α for residential area

driven by capacity and coverage needs without considering network energy consumption constraints. The most significant network evolution pathways have been the introduction of new radio access technologies (RATs) through four successive technology generations. Since the transition in the early 1990s from the 1G analogue mobile systems to 2G digital systems, such as GSM, there has been a constant drive to connect mobile users to the Internet with faster data rates. For this purpose, existing GSM cell sites were upgraded with GPRS and EDGE, and in the early 2000s, a new 3G standard was introduced. Though deployed mostly in densely populated regions, the 3G network in Europe, called UMTS, was deployed with additional cell sites as well as new equipment using newly acquired licensed bands. However, the emergence of handheld devices with enhanced features to access digital content accelerated the upgrade of UMTS to HSPA and HSPA+ with significantly increased downlink peak data rates up to 21 Mbit/s. In 2009, the LTE standard was ratified in 3GPP Release-8 (Rel-8). Commonly referred to as 4G, LTE uses OFDM technology to achieve peak data rates up to 300 Mbit/s if 4×4 MIMO technology is used. Even higher peak data rates will be achieved with LTE-Advance, which is currently being standardized. Recent efforts to reduce the energy consumption of mobile cellular systems, for example, by enabling sleep modes, are planned for future revisions of the LTE specification.

The availability of multiple RATs each with a variety of configurations provides an MNO with the opportunity to deploy a RAN, which is energy-efficient as well as meets capacity and coverage requirements. While the existing 2G and 3G cell sites provide a blueprint on which to build a 4G RAN rapidly, the higher operating frequency of LTE at 2600 MHz together with its higher data rates will necessitate more cell sites in order to achieve seamless coverage, especially in urban areas. Though the lower frequency bands of LTE at 1800 and 800 MHz offer improved coverage, the preferred option to fill the coverage holes between macrocell sites in urban areas is to deploy additional small cells (e.g., micro-, pico-, and metro femto-cells), which boost the capacity within a small area. As an example of a HetNet, this layered approach of using macrocells over small cells can also enhance capacity by deploying small cells in traffic hotspots. The traffic served by the small cells releases resources for the macro-cell sites. Further, as small cells operate with less transmit RF power, they typically generate less interference while individual units consume less operational energy. For example, small cells using picocell base stations typically operate with RF transmit powers up to 5 W and a 100 m diameter cell. In a HetNet, small-cell deployment outdoors may be realized using an alternative RAT such as WiFi. By operating on a diferent frequency band, WiFi allows macro-cells to offload traffic to WiFi access points without increasing the interference level within the macrocell network.

Attempting to serve users located indoors using the outdoor network is problematic owing to the additional high path loss experienced by radio waves passing through walls. The penetra-tion loss can attenuate the radio signal by 10 to 30 dB, causing significant outage. One option to mitigate the effect is to increase the transmit power of outdoor base stations, but this approach is prohibitive in terms of power consumption for the transmission gains achieved. An alternative approach is to deploy femtocells in the form of very low RF transmit power (e.g., 1 mW) base stations operating over very short ranges (e.g., 10 m). By avoiding outdoor-to-indoor path loss, femtocells achieve good indoor coverage to a small group of users. In an enterprise setting, an MNO can plan the deployment of femtocells to meet capacity and coverage needs while in a domestic setting, the operator has no control over the placement of a base station, which can lead to problematic interference conditions. A similar concern exists in domestic WiFi networks where too many collocated WiFi access points induce capacity and latency issues.

Like WiFi access points, each femtocell requires a backhaul connection such as an xDSL connection in homes or an Ethernet connection in offices. In general, the energy consumption of a backhaul connection does not scale linearly with cell size, which results in a high residual backhaul overhead per cell in small-cell deployments. The issue affects outdoor small cells too and is a potential bottleneck as the dense deployment of indoor and outdoor small cells will significantly increase the number of backhaul connections and, therefore, the associated energy consumption.

This highly complex deployment scenario creates a significant challenge for an MNO to plan a RAN, which is efficient in terms of energy, capacity, and coverage. The evolution and upgrade options available to an MNO are subject to tight CAPEX, IMPEX, and OPEX cost constraints. When coupled with the rising cost of energy, there is a pressing need to leverage network planning approaches that minimize energy consumption and maximize capacity and coverage in a trade-off between these three critical design features. It is not enough to rely on Moore's law to reduce energy consumption with equipment upgrades. A proactive network planning methodology that designs for energy efficiency as well as capacity and coverage is needed when faced with exponentially growing traffic volumes. The methodology starts with a consideration of the energy consumption in network equipment, in particular the base station which is the entity that consumes the most energy in a RAN.

2.3   Cell Site Power Consumption Modeling

The mobile network can be divided into three logical categories: the core, the RAN, and the user devices. Though the number of network elements increases exponentially when enumerating from the core to the edge of the network, the highest proportion of network energy is consumed by the base stations and typically lies between 60% and 80% of the RAN. This fact is significant in RAN evolution because capacity is generally enhanced by adding more base stations. Between 2011 and 2012, Vodafone UK recorded 14,321 operating base stations [17], and this number is set to increase with the introduction of 4G.

Base station sites can vary considerably depending on the shape, size, and composition of the serving cells. However, each base station site has typically seven common key subsystem elements defined as: antennas and feed cables; power amplifier (PA); small-signal RF transceiver, processing unit, backhaul; cooling system; and the mains AC-DC power supply unit that connects to the power grid. This structure can be extended to different categories of base stations including macro, micro, and pico base stations. Auxiliary equipment located at the base station site can include battery backup and lighting.

Feeder cables provide a transmission link between the radio units, typically located at ground level with the base station equipment, and the antennas located at the top of the transmission mask. The cables introduce additional losses, which increase with cable length and RF carrier frequency. Typically, a loss of 3 dB is attributed to feeder cables, which means only half the power from a radio unit reaches the antenna. This relatively high loss can be alleviated by using remote radio head (RRH) technology, which involves locating a sector radio unit in close proximity with the antenna on the mask head. The RRH connects to the system unit, still located at ground level, by fiber but a power supply cable is required to drive the RRH. While a number of RRH configurations are possible, the feeder losses are typically reduced to about 1 dB.

The PA and RF transceiver together constitute the RF module, which can account for approximately 60% of the base station site power consumption. The efficiency of a PA is highly dependent on the peak-to-average power ratio (PAPR) of the signal format being transmitted. In OFDM-based systems like LTE, this can translate into low PA efficiencies of 15% to 35% over a 20 MHz bandwidth. Typically, a dedicated RF module is deployed for each sector and each antenna of a base station site making the RF power overhead a significant portion of the overall site power consumption. By operating a PA close to its saturation region, higher PA efficiencies can be achieved but at the expense of increasing distortion through nonlinear effects. Linearization methods such as Cartesian feedback, feedforward, and digital predistortion [18, 19] alleviate the distortion, allowing PAs to realize maximum efficiencies of 40% to 50%.

The processing unit consists of baseband (BB) signal processing, radio resource management, and site control functionalities. The module also includes digital conversion to and from the RF transceiver. A dedicated system module is usually deployed for each base station sector only. The power consumption of the BB signal processing has continuously benefitted from Moore's law and enhanced processing architecture over time leading to reduced power consumptions for this unit when equipment is upgraded or a new RAT emerges.

From a network planning perspective, the backhaul is an important subsystem element that is frequently omitted from overall RAN energy assessment studies. The backhaul provides a high-capacity connection between the base station site and the core mobile network. This connection may be provided using either fiber optic, microwave, or copper links, depending on the site location. While most sites use a microwave backhaul (48% in the United Kingdom), the proportion of fiber-based backhaul connections is growing. The advent of LTE and small-cell technology to enhance capacity is leading to significantly more base stations being deployed and the impact of this on overall RAN energy consumption needs to be quantified.

Site cooling is an energy-intense facility required at a base station site in order to maintain an appropriate operating temperature inside the base station cabinet. The power consumption of the air-conditioning unit depends on the internal and ambient temperature of the base station cabinet. Typically, an internal and ambient temperature of 25 °C is assumed, which results in a constant power consumption for the air conditioning. Typically, this power consumption can account for 10% to 30% of the overall power consumption of a macrocell base station site. Modern site equipment, in particular RRH technology, is migrating to passive cooling approaches with clear benefits to the overall site energy consumption.

A number of base station site power consumption models are available in the open literature to predict the total power consumption of different types of base station with reasonable accuracy. Significant generic models have been developed in Ref. [20–22] based on measurements from specific vendor equipment. In this chapter, we consider the model developed by the GR project, which is similar to the model in Ref. [20] and is based on manufacturers' data. Figure 2.2 depicts a schematic block diagram of the salient power consuming units of a macrocell base station site using the GR model.

The units divide into three categories: those units used per antenna per sector of the base station site, which includes the PA and the transceiver units; those units used per sector, which

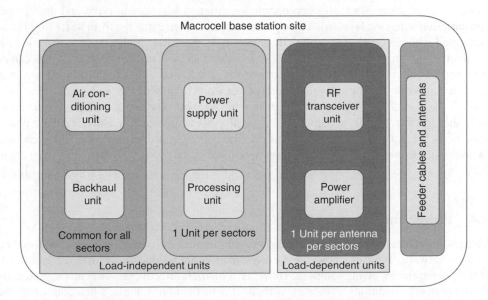

Figure 2.2   Power consuming units of a macrocell base station site

includes the processing and power supply units; and those units used once per base station site, which includes the air-conditioning and backhaul units. A further categorization of the units is based on whether or not there is a dependency on the traffic load. The load-dependent units include the PA and RF transceiver. The power consumption of these components is scaled by a normalized load activity factor α, which ranges from zero to unity. The load-independent units include the power supply, air-conditioning, backhaul, and processing. Though there is some load dependency associated with the backhaul [23] and processing units, its effect is relatively small such that the power consumption of the load-independent units is treated as constant with time. Then, a mathematical expression for the power consumption of a base station site is given by Eq. (2.1).

$$P_{bts} = P_{ac} + P_{bh} + n_s \cdot \{P_{ps} + P_{pu}\} + \alpha \cdot n_s \cdot n_a \cdot \left\{ P_{trx} + \frac{P_{tx}}{\eta_{pa} \cdot \eta_{cl}} \right\} \qquad (2.1)$$

The term P_{bts} denotes the overall base station site power consumption. The terms P_{ac}, P_{bh}, P_{ps}, and P_{pu} denote the power consumption of the air-conditioning, backhaul, power supply, and processing units, respectively, which are load-independent, where n_s is the number of sectors per base station site. The terms P_{trx} and P_{tx} denote the transceiver and peak PA output power consumption, respectively, which are load-dependent and hence scaled by α. The term n_a denotes the number of RF transmission chains, which corresponds to the number of antennas used. This term only scales those units associated with the RF transmission chain. The power consumption of the PA is given by $P_{pa} = P_{tx}/\eta_{pa}$, where η_{pa} denotes the PA efficiency (i.e., output divided by input power at the quiescent operating point). In practice, the PA efficiency has some dependency on α due to the nonlinear characteristic of the PA. The term η_{cl} accounts for the antenna feed cable efficiency. With an appropriate choice of parameters, Eq. (2.1) can be used to model macro-, micro-, pico-, and femtocell base station power consumption with reasonable accuracy as well as account for MIMO antenna configurations. Table 2.2 shows a list of parameter values appropriate for different base station types and their power consumption at full load (i.e., α). These parameters have been extracted from Ref. [20, 21] and combined with data collected in the GR project [24]. An RRH configuration is assumed for the microcell. The estimates in Table 2.2 demonstrate that the significant power savings can be achieved when using a microcell when compared to a macrocell. The combination of reducing the transmit power and the air-conditioning power while improving the feeder cable efficiency reduces the overall base station power consumption by about nine times. Both pico- and femto-cells show further significant reductions in overall power consumption. However, these power reductions are achieved at the expense of cell site coverage as discussed later in the chapter. In addition, the proportion of power consumption attributed to backhaul increases from about 0.5% in a macrocell to 48% in a femtocell, when the backhaul is based on a fiber optic link.

2.4   Power and Energy Metrics

Metrics provide a means of quantifying the consumption of power or energy in a mobile cellular network. Typically, power and energy metrics are used to compare and contrast different power/energy saving solutions. However, a metric's effectiveness depends significantly on the network conditions when a measurement is made. For example, when comparing the energy consumption of two different RAN configurations or architectures, there is an implicit

Table 2.2   LTE BS site power consumption estimates based on 2010 vendor data

Equipment	Macrocell	Microcell	Picocell	Femtocell
Number of sectors (n_s)	1	1	1	1
Number of antennas per sector (n_a)	1	1	1	1
Transmit power (W)	40	6.3	0.13	0.1
Transceiver unit (W)	13	6.5	1.0	0.6
Processing unit (W)	30	27	3	2.5
Power supply unit (W)	50	50	5	2
Backhaul (fiber) (W)	10	10	10	10
Air-conditioning unit (W)	225	60	0	0
PA efficiency (η_{pa})	30%	23%	7%	4%
Antenna feed cable loss (dB)	−3	−1	0	0
BS site power (W)	594	188	21	18

requirement that both networks are compared when transporting the same absolute number of application data bits. This premise suggests that energy per bit, with units of J/bit, provides a useful energy metric. The method developed in the GR project to compare the energy consumption of two distinct RANs builds on the notion of a J/bit metric or Energy Consumption Rating (ECR).

The energy metric framework is developed by considering a RAN, covering an area Λ_{ran}, which must transport in a predetermined period of time a total of M application bits, which are evenly distributed over A_{ran}. The RAN consists of n homogeneous base station sites (BTS) each of approximate circular radius $D/2$, where D is the intersite distance. From Eq. (2.1), the energy consumption of a BTS, E_{BTS}, is given by Eq. (2.2).

$$E_{BTS} = P_{rh}T_{rh} + P_{oh}T_{oh} \qquad (2.2)$$

The terms in P_{rh} and P_{oh} denote the load-dependent and load-independent peak or maximum power consumptions, respectively, where rh refers to a radio-head-only consumption and oh refers to an overhead-only consumption. From Eq. (2.1) these terms are given by $P_{rh} = n_s \cdot n_a \cdot \left\{ P_{trx} + \frac{P_{tx}}{\eta_{pa} \cdot \eta_{cl}} \right\}$ and $P_{oh} = P_{ac} + P_{bh} + n_s \cdot \{P_{ps} + P_{pu}\}$. The terms T_{rh} and T_{oh} denote time durations over which load-dependent and independent power is consumed, respectively. T_{rh} is defined in terms of a site-average capacity $T_{rh} \stackrel{def}{=} \frac{M/n}{C_{BTS}}$, while T_{oh} is defined in terms of a site-average throughput, $T_{oh} \stackrel{def}{=} \frac{M/n}{S_{BTS}}$, both measured in bit/s, and $T_{rh} \leq T_{oh}$. However, $S_{BTS} = \alpha C_{BTS}$ such that the site-average load activity factor $\alpha = T_{rh}/T_{oh} \leq 1$ and $S_{BTS} \leq C_{BTS}$. Then, Eq. (2.2) may be written in terms of C_{BTS}, α, M and n as shown in Eq. (2.3).

$$E_{BTS} = P_{rh} \cdot \frac{M}{n \cdot C_{BTS}} + P_{oh} \cdot \frac{M}{\alpha \cdot n \cdot C_{BTS}} \qquad (2.3)$$

The number of base station sites n can be approximated by $n = A_{RAN}/A_{BTS}$, where the site coverage area $A_{BTS} \cong \pi \left(\frac{D}{2}\right)^2$. Then, an expression for the total RAN energy consumption in terms of the site parameters is given by Eq. (2.4).

$$E_R = n \cdot E_{BTS} = \left(\frac{P_{rh}}{C_{BTS}} + \frac{P_{oh}}{\alpha \cdot C_{BTS}} \right) \cdot M \tag{2.4}$$

Dividing E_R by M in Eq. (2.4) gives an expression for the RAN energy consumption rating ECR_R, as shown in Eq. (2.5).

$$ECR_R = \frac{E_R}{M} = \frac{\alpha \cdot P_{rh} + P_{oh}}{\alpha \cdot C_{BTS}} = \frac{P_{BTS}}{S_{BTS}} = ECR_{BTS} \tag{2.5}$$

Equation (2.5) demonstrates the use of the ECR metric when determining the absolute RAN energy consumption because $E_R = M \cdot ECR_R$. Importantly, both the RAN and BTS ECR are equal for a homogeneous deployment of sites and are given by the ratio of site-average power consumption (P_{BTS}) to site-average throughput (S_{BTS}). Equation (2.5) is formulated on the parameters of a base station site. Without loss of generality, a formulation based on the parameters of a sector (or cell) is straightforward because the total number of cells is equal to $n \times n_s$.

Equation (2.5) can be used as a comparison framework between two or more distinct RAN architectures. Let $E_{R,1}$ and $E_{R,2}$ denote the energy consumption of two distinct RANs (RAN$_1$ and RAN$_2$), where $E_{R,1} \geq E_{R,2}$. A figure of merit termed the RAN Energy Consumption Gain is defined as $ECG_R = \frac{E_{R,1}}{E_{R,2}}$ on the basis that RAN$_2$ consumes less energy than RAN$_1$ owing to more energy-efficient base station sites. An expression for the ECG_R in terms of the base station site parameters is given by Eq. (2.6), where RAN$_1$ and RAN$_2$ transport M_1 and M_2 application bits, respectively.

$$ECG_R = \frac{M_1 \cdot (\alpha_1 \cdot P_{rh,1} + P_{oh,1})/S_1}{M_2 \cdot (\alpha_2 \cdot P_{rh,2} + P_{oh,2})/S_2} = \frac{M_1}{M_2} \cdot \frac{P_1/S_1}{P_2/S_2} = \frac{M_1}{M_2} \cdot \frac{ECR_1}{ECR_2} \tag{2.6}$$

In Eq. (2.6), the ratio M_2/M_1 constitutes a second RAN figure of merit termed the RAN Throughput Gain (TPG) denoted by TPG_R. For clarity, in Eq. (2.6), the subscript BTS denoting base station site has been omitted.

Equation (2.6) forms a basis for comparing the energy efficiency of distinct RANs across a broad range of operating configurations. This is achieved by applying the constraints $T_{oh,2} = T_{oh,1}$ and $T_{rh,2} \leq T_{rh,1}$. From Eq. (2.3), the first constraint gives $\frac{M_2}{\alpha_2 n_2 C_2} = \frac{M_1}{\alpha_1 n_1 C_1}$ while the second constraint gives $C_2 \geq C_1$, where n_i for $i = 1, 2 \ldots$ denotes the number of sites per RAN. As $n_1/n_2 \equiv A_2/A_1$, where A_1 and A_2 are the base station site areas and $A_1 \geq A_2$, then the constraint $\frac{M_2}{\alpha_2 n_2 C_2} = \frac{M_1}{\alpha_1 n_1 C_1}$ gives an expression for the RAN TPG as $TPG_R = \frac{M_2}{M_1} = \frac{S_2/A_2}{S_1/A_1}$. That is, the RAN Throughput Gain is equal to the ratio of the base station site area throughput densities (bit/s/km^2) between RAN$_2$ and RAN$_1$. Substituting $\frac{M_2}{M_1} = \frac{S_2/A_2}{S_1/A_1}$ into Eq. (2.6) gives Eq. (2.7).

$$ECG_R = \frac{1}{TPG_R} \cdot \frac{ECR_1}{ECR_2} = \frac{P_1/A_1}{P_2/A_2} \tag{2.7}$$

Equation (2.7) illustrates a key relationship between the RAN ECG and the base station site area power densities (W/km^2). However, Eq. (2.7) also demonstrates a fundamental relationship between the product $ECG_R \cdot TPG_R$, shown in Eq. (2.8), and the ratio of the base station site Energy Consumption Ratings (J/bit).

$$ECG_R \cdot TPG_R = \frac{ECR_1}{ECR_2} \tag{2.8}$$

The product $\text{ECG}_R \cdot \text{TPG}_R$ forms an overall figure of merit when comparing two RANs, which takes into account the RAN energy efficiency as measured by the ECG and the RAN throughput efficiency as measured by the TPG. Importantly, this composite figure of merit is determined by the site ECR metrics. Equation (2.8) clearly demonstrates the inverse relationship between energy efficiency and throughput efficiency where by increasing one reduces the other while their product is bound by the ratio $\text{ECR}_1/\text{ECR}_2$. Then, the most efficient RAN in terms of least energy consumed and greatest throughput achieved is the one that maximizes the product $\text{ECG}_R \cdot \text{TPG}_R$. In summary, the three defining figures of merit for comparing RANs in terms of the site metrics P/A (W/km^2), S/A (bit/s/km^2), and P/S (J/bit) are shown in Eq. (2.9).

$$\text{ECG}_R = \frac{P_1/A_1}{P_2/A_2}, \quad \text{TPG}_R = \frac{S_2/A_2}{S_1/A_1}, \quad \text{ECG}_R \cdot \text{TPG}_R = \frac{P_1/S_1}{P_2/S_2} \tag{2.9}$$

The energy and throughput efficiencies of various LTE RAN configurations are explored in Section 2.5. However, the formulas in Eq. (2.9) provide a number of key insights into the bounds on RAN ECG and TPG. For example, consider a homogeneous RAN that is signal-to-interference (SIR) limited with load activity factor $\alpha_1 = 1$, intersite distance D_1, and base station site power consumption P_1. If the intersite distance is reduced from D_1 to D_2 using the same base station (i.e., $P_2 = P_1$) with $\alpha_2 = 1$, then $S_2 \cong S_1$ giving $\text{ECG}_R \cdot \text{TPG}_R = 1$. This scenario corresponds to cell densification using the same BTS infrastructure. Equation (2.9) proves that the RAN ECG decreases as D_2 decreases because $\text{ECG}_R = 1/\text{TPG}_R = A_2/A_1 = D_2^2/D_1^2$ while the RAN TPG increases as D_1^2/D_2^2. However, if a lower power base station is used instead (i.e., $P_2 < P_1$), then the expressions in Eq. (2.9) can be used to determine a least intersite distance $D_2 = D_1\sqrt{P_2/P_1}$ with $\text{ECG}_R = 1$ and $\text{TPG}_R = P_1/P_2$. By increasing D_2, it is possible to increase ECG_R while reducing TPG_R subject to the constraint $\text{ECG}_R \cdot \text{TPG}_R = P_1/P_2$. This bound demonstrates that joint energy savings and throughput gains are attainable at reduced intersite distances provided the new base station has lower power. In the next section, a variety of energy saving strategies for an LTE-based RAN are explored building on the metric analysis of this section.

2.5   Energy and Throughput Efficiency in LTE Radio Access Networks

In this section, the ECG and TPG figures of merit developed in Section 2.4 are used to investigate an LTE RAN accounting for intercell interference as well as noise. The analysis is based on a Shannon capacity model of the LTE physical layer. The interference is characterized by an average value related to the load activity factor α. The analysis is carried out for a homogeneous deployment of single sector or cell base station sites using antennas characterized by unity gain in the azimuth plane (i.e., omnidirectional) and gain G in the elevation plane. Each base station transmits with the same RF power P_{tx}. This approach provides vivid insights into the planning of RANs based on energy as well as throughput requirements [25].

A key performance metric in RANs is the average signal-to-interference-and-noise ratio (SINR) defined as $\text{SINR} = \frac{S}{I+N}$, where S is the average received signal power, I is the average received interference, and N is the noise power. The total interference from K neighboring sites is given by Eq. (2.10).

$$I = \sum_{k=1}^{K} \alpha_k I_{\text{max},k} \qquad (2.10)$$

The value of the *SINR* depends on the position of the mobile within a cell. Often, the cell-edge SINR is used as a determining quality of service (QOS) metric. Another important QoS metric is the cell-average SINR obtained by averaging the SINR over the cell area. The analysis assumes that the load activity factors α_k in every cell (i.e., serving and interfering) are equal such that $\alpha_k = \alpha \ \forall k$. Then, $I = \alpha I_{\text{max}}$ where $I_{\text{max}} = \sum_{k=1}^{K} I_k$ is the maximum interference power at the mobile. The SINR can be written as a function of the minimum signal-to-interference ratio (SIR), the signal-to-noise ratio (SNR), and the cell load activity factor α as shown in Eq. (2.11) where $\text{SIR}_{\text{min}} = \frac{S}{I_{\text{max}}}$ and $\text{SNR} = \frac{S}{N}$.

$$\text{SINR} = \cfrac{1}{\cfrac{\alpha}{\text{SIR}_{\text{min}}} + \cfrac{1}{\text{SNR}}} \qquad (2.11)$$

The SIR and SNR are averaged values taken over the small-scale fading of S and I. The magnitude of SIR_{min} at the mobile is determined by the network geometry as well as the antenna configuration. It is common practice to model the interference as a Gaussian noise. Therefore, the SNR is given by $\frac{S}{N} = \frac{\text{EIRP}}{LN}$, where the effective isotropic radiated power $\text{EIRP} = GP_{tx}$ for transmit antenna gain G. The term in L is a linear path loss given by $L = MKd^\gamma$, where M denotes a shadow fading margin, γ is the path loss exponent, K is the median path loss at a distance of one kilometer, and d is the distance in kilometers between the mobile and base station. The dB path loss is modeled by $L_{dB} = 128 + 37.4 log_{10}(d_{km})$, which is based on the 3GPP system simulation specification [25]. A dB shadow-fading margin of 8.7 dB is used throughout the analysis, which corresponds to a 95% single-cell area probability for a shadow-fading standard deviation of 8 dB. The Shannon formula is used to provide a mapping between capacity C and SINR as shown in Eq. (2.12) where the terms in EIRP and L have been substituted.

$$C = B log_2 \left(1 + \cfrac{1}{\cfrac{\alpha}{\text{SIR}_{\text{min}}} + \cfrac{MKd^\gamma N}{GP_{tx}}} \right) \qquad (2.12)$$

Equation (2.12) gives the capacity or peak rate at the mobile's position for a system bandwidth B. The site- or cell-average capacity is obtained by taking the average of C over the cell area (i.e., $C_{\text{BTS}} = \frac{4}{\pi D^2} \int_{\text{cell area}} C \, dA$). This calculation is usually completed numerically or by system level modeling of the RAN. Then, an expression for the cell-average spectral efficiency is given by $\eta_{\text{BTS}} = \frac{C_{\text{BTS}}}{B}$ and the cell-average throughput is given by $S_{\text{BTS}} = \alpha C_{\text{BTS}}$.

The average values for the cell capacity and throughput are combined with the RAN metrics and figures of merit developed in Section 2.4 to evaluate a number of options for saving energy in a RAN. In the following sequel, energy savings obtained by reducing the cell size, deploying small-cell technologies, and using sleep modes are evaluated. Also, the energy efficiency of various HetNet topologies is investigated. The parameters for these investigations are shown in Table 2.3 with power consumption values being extracted from the data presented in Table 2.2

Table 2.3   BTS technology parameters

Cell	Macro	Micro	Pico	Femto	Units
P_{tx}	46	38	21	20	dBm
P_{oh}	315	147	18	14.5	W
P_{rh}	279	41	2.86	3.1	W
G	18	6	2	0	dBi

and applied to Eq. (2.1) for a base station site with one antenna unit ($n_a = 1$) and, therefore, one sector or cell ($n_s = 1$).

2.5.1   Reducing Cell Size

In this subsection, the impact on energy consumption and throughput is evaluated when the size of a cell is reduced without changing the base station site hardware or system bandwidth. This scenario is equivalent to reducing the intersite distance or increasing the number of base station sites per km². The impact of reducing D on ECG_R, TPG_R and their product ETG_R is considered. The RAN with the greater cell size is identified as RAN_1 and, without loss of generality, the load activity factor in RAN_1 is set equal to unity (i.e., $\alpha_1 = 1$). From Eq. (2.9), expressions for the ECG_R and TPG_R in terms of the intersite distances and cell-average spectral efficiencies are given by Eq. (2.13) where the cell area ratio $\frac{A_2}{A_1} = \left(\frac{D_2}{D_1}\right)^2$.

$$ECG_R = \left(\frac{D_2}{D_1}\right)^2 \frac{P_{rh,1} + P_{oh,1}}{\alpha_2 P_{rh,2} + P_{oh,2}} \qquad TPG_R = \left(\frac{D_1}{D_2}\right)^2 \frac{\alpha_2 \eta_2}{\eta_1} \qquad (2.13)$$

Figure 2.3(a), (b), and (c) show plots of ECG_R, TPG_R, and ETG_R as a function of the ratio $\frac{D_2}{D_1}$ with α_2 as a parameter, respectively. The results were obtained for the macrocell parameters shown in Tables 2.2 and 2.3 with a value of $D_1 = 1$ km, which gives a cell-average spectral efficiency $\eta_1 = 3.255$ bit/s/Hz. The results demonstrate a number of important conclusions based on the common practice of operators to increase system capacity by increasing the number of base station sites per km².

From Figure 2.3(a), the ECG_R is always < 1 for $\alpha_2 < 1$ and as D_2 decreases with respect to D_1, the ERG_R decreases rapidly with $ECG_R \ll 1$ for $\frac{D_2}{D_1} < 0.5$. When $\alpha_2 = 1$, the ECG_R increases marginally as D_2 decreases because η_2 increases marginally as D_2 decreases. Figure 2.3(a) also illustrates the impact of cell load on energy efficiency. The ECG_R increases with decreasing load. That is, a lightly loaded RAN consumes less energy than a heavily loaded RAN. However, the difference in ECG_R is relatively small as the load α only affects the radio-head power consumption (P_{rh}), which is typically small compared to the overhead power consumption (P_{oh}). When $T_{rh} < T_{oh}$, an opportunity exists to exploit sleep mode principles such as discontinuous transmission (DTX).

In contrast, Figure 2.3(b) shows that the TPG_R exhibits the opposite characteristics to the ECG_R. That is, the TPG_R increases with decreasing D_2 and is lower when the network is more

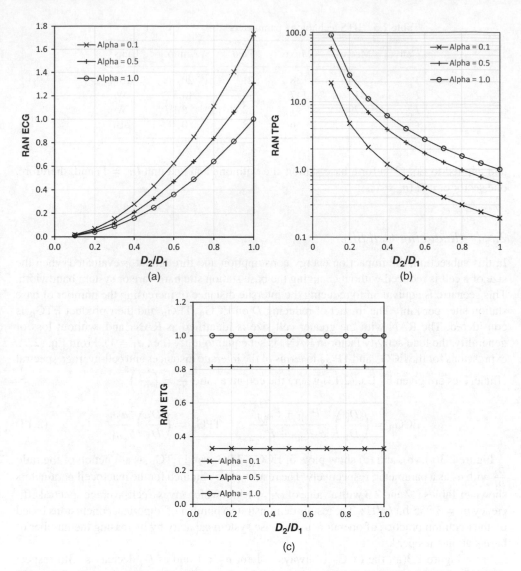

Figure 2.3 Energy and throughput figures of merit for a macrocell RAN. (a) ECG_R, (b) TPG_R, (c) ETG_R

lightly loaded. For values of $D2/D1 < 0.5$, the TPG increases rapidly. A point of interest is when $ECG_R = TPG_R$. This equality indicates that the performance of RAN_2 has achieved the largest possible joint energy consumption and throughput gains with respect to RAN_1. In particular, when $ECG_R = TPG_R = 1$, the joint gain $ETG_R = 1$ corresponds to the case that RAN_2 achieves the same energy and throughput performance as RAN_1 at intersite distance D_2. From Figure 2.2(a) and (b), the only point when $ECG_R = TPG_R = 1$ is when $D_2 = D_1$. This result demonstrates that increasing the RAN throughput by reducing the intersite distance only will always result in an increased RAN energy consumption. Figure 2.3(c) plots $ETG_R = ECG_R \cdot TPG_R$ versus D_2 with α_2 as a parameter. The graphs show that ETG_R remains constant for $0.1 \leq \frac{D_2}{D_1} \leq 1$ but increases with α_2 to a maximum value of unity. The constant characteristic of $\mathrm{ETG_R}$ can be attributed to the RAN performances being SIR limited.

2.5.2 Reducing Cell Size and BTS Power Consumption

The results in (A) showed that in order to increase throughput by reducing intersite distance, the power consumption of the BTS should be reduced. In this section, the impact of replacing macrocells with microcells using the BTS data in Tables 2.2 and 2.3 is examined. Under full load, the macrocell consumes 594 W while the microcell consumes 184 W. Using the ECG, TPG, and ETG metrics the least intersite distance and the least cell bandwidth are defined and evaluated when $\mathrm{ECG_R} = \mathrm{TPG_R} = 1$ (i.e., $\mathrm{ETG_R} = 1$). That is, $\mathrm{ECG_R} = \mathrm{TPG_R} = 1$ corresponds to the condition that $\mathrm{RAN_2}$ using microcells achieves the same throughput and energy consumption as $\mathrm{RAN_1}$ using macrocells.

Figure 2.4(a), (b), and (c) show plots of $\mathrm{ECG_R}$, $\mathrm{TPG_R}$, and $\mathrm{ETG_R}$ as a function of the ratio $\frac{D_2}{D_1}$ with α_2 as a parameter, respectively. As in (A), the results were obtained for a value of $D_1 = 1$ km corresponding to a cell-average spectral efficiency $\eta_1 = 3.255$ bit/s/Hz in $\mathrm{RAN_1}$. The results show similar trends to those obtained in Figure 2.3. However, as $P_2 \ll P_1$, there is an appreciable range of α_2 values when $\mathrm{ETG_R} > 1$ for $0.1 \leq \frac{D_2}{D_1} \leq 1$. That is, the reduced power consumption of a microcell can be exploited to enhance the throughput of $\mathrm{RAN_2}$ without consuming more energy compared to $\mathrm{RAN_1}$ which uses macrocells.

In order to quantify these gains, the least intersite distance when $\mathrm{ECG_R} = \mathrm{TPG_R} = 1$ can be determined from Eq. (2.13) as shown in Eq. (2.14) when the system bandwidths are the same (i.e., $B_2 = B_1$).

$$\frac{D_2}{D_1} = \sqrt{\frac{\alpha_2 P_{\mathrm{rh},2} + P_{\mathrm{oh},2}}{P_{\mathrm{rh},1} + P_{\mathrm{oh},1}}} = \sqrt{\frac{\alpha_2 \eta_2}{\eta_1}} \tag{2.14}$$

From analysis, an exact solution to Eq. (2.14) is obtained when $\alpha_2 = 0.15$ and $\eta_2 = 5.595$ bit/s/Hz giving $\frac{D_2}{D_1} = 0.508$ or $D_2 = 508$ m. Any value of $D_2 < 508$ m will result in $\mathrm{RAN_2}$ consuming more energy than $\mathrm{RAN_1}$ though achieving a higher RAN throughput. However, the energy consumption gains in $\mathrm{RAN_2}$ are achieved by lowering α_2, which reduces the intercell interference, in order to increase the cell-average spectral efficiency η_2. This approach has the undesired effect of operating each cell of $\mathrm{RAN_2}$ with a very low throughput efficiency ($\frac{\alpha \eta}{B}$), which significantly under utilizes the available radio resource. For example, the cell-average throughput efficiencies in $\mathrm{RAN_1}$ and $\mathrm{RAN_2}$ are 3.255 bit/s/Hz and 0.839 bit/s/Hz, respectively, which correspond to absolute cell-average throughputs of 65 Mbit/s and 16.7 Mbit/s, respectively. This ratio is required so that $\mathrm{RAN_2}$, which has almost four times as many cells as $\mathrm{RAN_1}$, achieves the same RAN throughput and energy consumption when the system bandwidths are the same (i.e., $B_2 = B_1 = 20$ MHz).

In order to avoid operating $\mathrm{RAN_2}$ with a very low throughput efficiency, its system bandwidth B_2 can be reduced instead. The amount of this reduction can be calculated from the metric expressions in Eq. (2.10) for the case when $\alpha_2 = \alpha_1 = 1$ and $\eta_2 \cong \eta_1$. These constraints correspond to both RANs being fully loaded and SIR limited such that the cells in both RANs have approximately the same throughput and spectral efficiencies. Equation (2.15) shows expressions for $\mathrm{ECG_R}$ and $\mathrm{TGP_R}$ in terms of the BTS powers, bandwidths, and intersite distance ratio

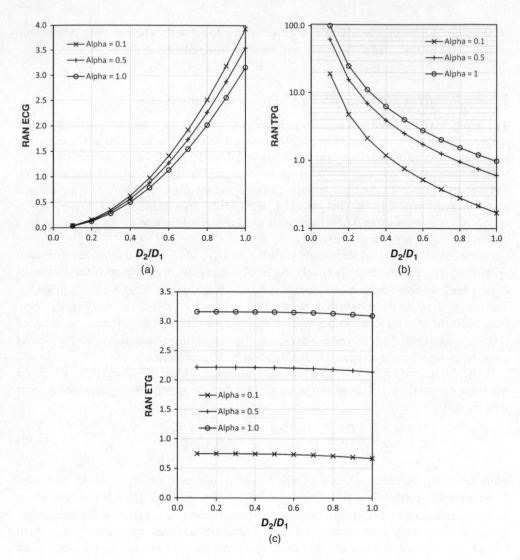

Figure 2.4   Energy and throughput figures of merit for a small-cell RAN. (a) ECG_R, (b) TPG_R, (c) ETG_R

where the cell-average throughputs are given by $S_i = \alpha_i B_i \eta_i, i = 1, 2$.

$$\text{ECG}_R = \left(\frac{D_2}{D_1}\right)^2 \cdot \frac{P_1}{P_2} \qquad \text{TPG}_R = \left(\frac{D_1}{D_2}\right)^2 \cdot \frac{B_2}{B_1} \qquad (2.15)$$

For $\text{ECG}_R = \text{TPG}_R = 1$, the first expression in Eq. (2.15) gives $\frac{D_2}{D_1} = \sqrt{\frac{P_2}{P_1}} = 0.563$ or $D_2 = 563$ m while the second expression gives $\frac{B_2}{B_1} = \left(\frac{D_2}{D_1}\right)^2 = \frac{P_2}{P_1} = 0.317$ or $B_2 = 6.34$ MHz for $B_1 = 20$ MHz. The implication of this result is far reaching–if there is no requirement to increase RAN throughput, then the system bandwidth requirement can be reduced by using smaller, low powered cells. Alternatively, without affecting the RAN throughput of each, several operators sharing a single wideband infrastructure would be more energy-efficient. Though the power consumption attributed to backhaul has been included in the analysis, the provision of backhaul connectivity in small cells would need to be resolved.

As $ETG_R = ECG_R \cdot TPG_R$, then combining the expressions in Eq. (2.15) gives a fundamental equations for the RAN energy throughput gain in terms of BTS powers and system bandwidths as shown in Eq. (2.16).

$$ETG_R = \frac{B_2}{B_1} \cdot \frac{P_1}{P_2} = \frac{P_1/B_1}{P_2/B_2} \tag{2.16}$$

The expression for ETG_R in Eq. (2.16) is independent of the intersite distance ratio $\frac{D_2}{D_1}$ and depends only on the ratio of the power spectral densities between fully loaded RANs that have SIR limited performance. This characteristic is observed in Figures 2.4(c) and 2.3(c) for $\alpha_2 = 1$, where ETG_R versus $\frac{D_2}{D_1}$ is almost constant. As discussed earlier, a point of particular interest is when $ECG_R = TPG_R$, which corresponds to when the performance of RAN_2 achieves the largest possible joint energy consumption and throughput gains with respect to RAN_1. Equating the expressions in Eq. (2.15) gives the expression in Eq. (2.17) for the optimum intersite distance ratio that gives the highest joint energy consumption and throughput gain of RAN_2 over RAN_1. That is, Eq. (2.17) defines the intersite distance ratio for least energy consumption and highest throughput gain.

$$\frac{D_2}{D_1} = \left(\frac{B_2}{B_1} \cdot \frac{P_2}{P_1} \right)^{\frac{1}{4}} \tag{2.17}$$

The 1/4 exponent in Eq. (2.17) demonstrates the stringent restrictions on $\frac{D_2}{D_1}$ when simultaneously constrained by energy and throughput requirements. The macro/microcell scenario considered in Figure 2.3 where $P_1 = 594$ W, $P_2 = 188$ W, and $B_2 = B_1 = 20$ MHz gives $\frac{D_2}{D_1} = 0.75$ or $D_2 = 750$ m for $D_1 = 1000$ m. This result agrees with the plots in Figures 2.4(a) and (b) when $\alpha_2 = 1$. The corresponding values of ETG_R is given by $\frac{P_1}{P_2} = 3.16$ while $ECG_R = TPG_R = \sqrt{ETG_R} = 1.78$. Equations (2.16) and (2.17) illustrate that the key drivers in RAN energy and throughput performance are system bandwidth and BTS power consumption. The former should be maximized and the latter minimized.

2.5.3   BTS Sleep Mode

The impact of sleep mode can be assessed in the context of changing the intersite distance between the base station sites of a regular, homogeneous macrocell deployment. As considered previously, in the analysis, each BTS uses a single antenna with an omnidirectional azimuth beam pattern. Then, a RAN of n_2 operating cells with an intersite distance D_2 can be transformed into a sleep mode RAN of n_1 operating cells with an intersite distance $D_1 = \sqrt{3}D_2$ by powering down $n_2 - n_1$ cells, where $\frac{n_2}{n_1} = \frac{A_1}{A_2} = \left(\frac{D_1}{D_2} \right)^2 = 3$. This configuration equates to one where no cells in the RAN for intersite spacing D_2 have adjacent cells that are operational when in sleep mode. Providing both intersite distances D_2 and D_1 achieve SIR limited performance, then there is no need to increase the transmit power on the downlink, though consideration should be given to the uplink sensitivity to ensure that mobile uplink transmissions are satisfactorily received.

For this configuration, the expressions for ECG_R and TPG_R in Eq. (2.10) can be modified for the sleep mode case as given in Eqs. (2.18) and (2.19). The term in β is a sleep mode factor on the BTS overhead power consumption with a value between 0 and 1. However, the overhead power consumption due to backhaul is not scaled by β as it is unlikely that the backhaul would be placed into sleep mode for signaling reasons. In the analysis next, a value of $\beta = 0.5$ is used throughout.

$$ECG_R = \frac{(P_1 + 2\beta P_{oh,1})/A_1}{P_2/A_2}$$

$$= \frac{1}{3} \cdot \left(\frac{\alpha_1 P_{rh,1} + (1 + 2\beta)P_{oh,1} + 2(1 - \beta)P_{bh,1}}{\alpha_2 P_{rh,2} + P_{oh,2}} \right) \qquad (2.18)$$

$$TPG_R = \frac{S_2/A_2}{S_1/A_1} = 3 \cdot \left(\frac{\alpha_2 B_2 \eta_2}{\alpha_1 B_1 \eta_1} \right) \qquad (2.19)$$

In order to calculate the ECG_R in Eq. (2.18) for a macrocell sleep mode scenario, the following equalities are used: $B_2 = B_1, P_{rh,2} = P_{rh,1}$, and $P_{oh,2} = P_{oh,1}$ with values being taken from Tables 2.2 and 2.3. In this macrocell example, a value of $D_2 = 500$ m is used giving an equivalent intersite distance in sleep mode of $D_1 = 866$ m. The condition for establishing sleep mode is realized when $TPG_R = 1$ for $\alpha_1 = 1$, which corresponds to the sleep mode RAN being fully loaded. Through analysis, the value of η_1 was calculated at 3.26 bit/s/Hz. The equivalent load and spectral efficiency in the RAN without sleep mode activated are given by α_2 and η_2, respectively. Solving for these values when $TPG_R = 1$ using Eq. (2.19) gives $\alpha_2 = 0.21$ and $\eta_2 = 5.19$ bit/s/Hz. That is, when the load in the RAN falls below ~ 0.21 the RAN will enter sleep mode. At this point, the ECG_R is determined using Eq. (2.18) with $\alpha_1 = 1$ and $\alpha_2 = 0.21$. Then, for $P_{rh} = 279$ W and $P_{oh} = 315$ W, an $ECG_R = 0.82$ is obtained. A value of $ECG_R = 0.82$ means that the macrocell RAN in sleep mode consumes 18% less energy compared with normal full load operation. If the exercise is repeated using microcells instead of macrocells for $D_2 = 500$ m, an $ECG_R = 0.74$ is achieved corresponding to 26% less energy being consumed by the microcell RAN in sleep mode. However, repeating the exercise using picocells for $D_2 = 150$ m gives an $ECG_R = 0.88$, which corresponds to only 12% less energy being consumed by the picocell RAN in sleep mode. The higher potential energy savings in a microcell RAN is attributed to the larger percentage overhead power in microcells (78% compared to 53% in macrocells), which can be scaled down by β during sleep mode. However, in a picocell RAN, the backhaul power consumption, which accounts for almost 50% of the BTS power, is not scaled by β and becomes a limiting factor. If the backhaul power in picocells is scalable by β, then the energy savings rise to 30%.

2.5.4   Heterogeneous Networks

In this sequel, the energy efficiency of HetNets is explored by considering various small-cell overlays of a macrocell RAN. ECG_H, TPG_H, and ETG_H denote energy, throughput, and joint energy/throughput figures of merit for the HetNet RAN. The calculation of these figures of merit is performed in a straightforward manner using the fundamental metrics of cell-average area power density and area throughput density as these quantities can be added to represent the superposition of different HetNet layers. The analysis considers the evolution of a homogeneous macrocell RAN with an intersite distance D_1 and an operating bandwidth B_1. The HetNet is realized by adding a homogeneous overlay of small cells with an intersite distance D_2 and operating bandwidth $B_2 = B_1$ MHz. The small-cell overlay is implemented on an adjacent frequency band, thereby eliminating interference between layers. The intersite distance D_2 is selected to give efficient HetNet energy and throughput characteristics. For comparison, the energy/throughput characteristics of homogenous small-cell networks are determined when operating over the same total spectrum bandwidth $(B_1 + B_2)$. Expressions for the ECG_H and TPG_H derived from the area power density and area throughput density are shown in Eqs. (2.20) and (2.21), respectively, where subscript 1 refers to the macrocell RAN underlay and subscript 2 refers to the small-cell RAN overlay.

$$\mathrm{ECG}_H = \frac{P_1/A_1}{P_1/A_1 + P_2/A_2} = \frac{\mathrm{ECG}_2}{1 + \mathrm{ECG}_2} \tag{2.20}$$

$$\mathrm{TPG}_H = \frac{S_1/A_1 + S_2/A_2}{S_1/A_1} = 1 + \mathrm{TPG}_2 \tag{2.21}$$

The terms in ECG_2 and TPG_2 can be evaluated using the formulas in Eq. (2.15) for SIR limited performance when the load factor for both layers is set to unity. The ETG_H is obtained by taking the product of ECG_H and ETG_H. As discussed previously, the joint energy and throughput gain is maximized when $\mathrm{ECG}_H = \mathrm{TPG}_H$. Equating ECG_H and TPG_H, as defined in Eqs. (2.20) and (2.21), gives the expression in Eq. (2.22) in terms of the BTS powers, bandwidths, and intersite distances.

$$\frac{B_2}{B_1} \cdot \left(\frac{D_1}{D_2}\right)^2 + \frac{B_2}{B_1} \cdot \frac{P_1}{B_2} + 1 = 0 \tag{2.22}$$

Substituting $x = D_2/D_1, a = P_1/P_2$ and $b = B_2/B_1$ into Eq. (2.22) gives the general algebraic expression $(1 + ab)x^2 + b = 0$. Evaluation of this function shows that there is no solution for $x = (D_2/D_1)^2$ in the range $0 \leq \left(\frac{D_2}{D_1}\right)^2 \leq 1$. This result illustrates a key characteristic of two tier networks—there is no optimal trade-off between energy savings and throughput gain. The addition of an overlay network always increases the overall energy consumption within the RAN. Equation (2.20) shows that ECG_H is always less than unity. The best way to make a two-tier HetNet energy efficient is to maximize ECG_2, that is, maximize the energy consumption gain in the overlay with respect to the underlay. Therefore, the primary reason for adding a small-cell overlay is to enhance the RAN capacity and this is demonstrated by Eq. (2.21) whereby the overlay throughput gain TPG_2 linearly adds to the underlay throughput TPG_1 to give TPG_H.

The performances of three HetNet topologies are compared and contrasted with various single tier small-cell topologies in Table 2.4. A homogeneous macrocell RAN (i.e., RAN_1) with $D_1 = 500$ m and $B_1 = 10$ MHz is used as a reference baseline for comparing the various topologies explored.

In Table 2.2, the HetNet configurations HN1, HN2, and HN3 correspond to micro/macrocell, pico/macrocell, and pico/microcell RANs, respectively. In the case of HN1&2, D_2 has been selected with respect to D_1 to achieve $\mathrm{ECG}_2 = \mathrm{TPG}_2$ with the original macrocell RAN forming the underlay. In the case of HN3, the underlay has been replaced by a microcell RAN with

Table 2.4 HetNet versus small cell—energy and throughput gains

RAN	Macro	Micro	Micro	Pico	Pico	HN1	HN2	HN3	Units
B_2	20	10	20	10	20	10	10	10	MHz
D_1	500	500	500	500	500	500	500	375	m
D_2	500	375	446	216	257	375	216	216	m
ECG	1	1.78	2.51	5.31	7.52	0.64	0.84	1.33	
TPG	2	1.78	2.51	5.34	7.51	2.78	6.33	7.12	
ETG	2	3.17	6.32	28.35	56.47	1.78	5.33	9.45	

$D_1 = 375$ m, which gives ECG = TPG in the microcell RAN with respect to the original macrocell RAN. Then, a picocell overlay is added with $D_2 = 216$ m, which establishes ECG = TPG in the picocell overlay with respect to the microcell underlay. The results in Table 2.2 show that in HN1&2, the $ECG_H < 1$ with the pico/macro deployment being more energy-efficient than the micro/macro deployment. Both HetNets achieve $TPG_H > 1$ with the pico/macro HetNet showing significantly greater throughput gains than the micro/macro HetNet. A similar trend is observed for the ETG_H figure of merit values. In contrast, HN3 achieves an $ECG_H > 1$, which is attributed to the energy savings achieved by replacing the macrocell underlay with a microcell underlay. The picocell overlay in HN3 helps to boost the throughput significantly in HN3 giving the highest TPG_H and ETG_H of the three HetNet configurations considered.

However, Table 2.2 also shows results for homogeneous small-cell deployments encompassing microcell and picocell RANs only. What is striking about these results is that small-cell RANs offer significantly better energy and throughput gains than HetNets. Even for the same bandwidth of 10 MHz, a picocell RAN can achieve a joint energy/throughput gain of 28.35 compared with the best HetNet, which achieves 9.45. The enhancements are even greater when the small-cell bandwidth is increased to 20 MHz giving an ETG of 56.47 in the picocell RAN. While further increases in the TPG_H and ETG_H for a HetNet can be achieved by allowing both layers to share the total system bandwidth of 20 MHz, the increase in intercell interference limits the gains achievable. An upper bound for a 20 MHz HetNet with under- and overlays sharing the same frequency band can be calculated from the data in Table 2.2. For example, a 20 MHz microcell underlay (Table 2.2, colum 4) with ECG = TPG ~ 2.5 and a 20 MHz picocell overlay (Table 2.2, colum 6) with ECG = TPG ~ 7.51 can be combined to give a 40 MHz HetNet with an ECG $= ((1/2.5) + (1/7.5))^{-1} = 1.875$, a TPG $= 2.5 + 7.5 = 10$, and an ETG $= 18.75$. As these figures of merit represent the HetNet performance when the under- and overlay are on separate 20 MHz bands, the performance when sharing the same 20 MHz band is expected to be significantly less. Therefore, for the prevailing power consumption models used in this analysis, small cells would appear to offer significantly greater potential than HetNets for saving energy while increasing throughput even when backhaul effects are taken into account.

2.6 Conclusions

This chapter has taken a comprehensive look at the energy consumption problem in broadband wireless networks, in particular cellular mobile radio networks. With research on 5G considering techniques to increase throughput such as cell densification, HetNets, and massive MIMO, there remains a significant dearth of the underlying theories and principles needed to deliver the required increase in data throughput without compromising the overall RAN energy efficiency. The issue has been addressed in this chapter by developing metrics that enable different RAN architectures to be fairly compared in terms of relative energy consumption and throughput figures of merit. In particular, a new composite figure of merit has been developed called the RAN Energy Throughput Gain (ETG_R), which enables salient RAN parameters to be set in order to achieve the best possible combination of energy reduction and throughput gain. The ETG_R is formed by taking the product of the ECG_R and the TPG_R and demonstrates the inverse relationship between throughput gain and energy consumption gain. Importantly, the basis of these figures of merit relates to comparing the fundamental metrics of area power density in W/km^2, area throughput density in $bit/s/km^2$, and, in the case of the ETG_R, the energy consumption rating in J/bit. It is the optimization of these figures of merit that drives the design of an energy-efficient RAN. At the centre of such designs is the requirement for power consumption models of the RAN hardware, in particular the BTS. Such models should accurately account for the load-independent as well as the load-dependent causes of power consumption. Furthermore, the backhaul consumption should be taken into account, particularly in small cells such as pico- and femtocells where it can account for approximately 50% of the BTS power consumption. For these cell types, the backhaul can be a limiting factor to achieving an energy-efficient deployment as backhaul power consumption cannot be readily controlled.

The second part of this chapter examines a number of RAN architectures based on the LTE standard. Four key scenarios were considered: reducing intersite distance, deploying small-cell technologies; introducing sleep modes; and deploying HetNets. In all four cases, the RAN ECG, TPG, and ETG figures of merit were used to characterize the RAN energy and throughput efficiencies. The RAN configurations were evaluated for constant transmit power and SIR limited performance, which is characteristic of urban deployments where macrocell intersite distances of ≤ 1 km are common. When intersite distance was reduced for the same BTS hardware, the figures of merit confirmed that the energy consumption increased monotonically. The throughput also increased monotonically demonstrating the benefit of using smaller cells to increase the RAN capacity but the throughput gains were only achieved at the expense of increased energy consumption.

In order to mitigate this behavior, the macrocells were replaced by lower power microcells. The figures of merit were used to identify two salient deployment options. The first option established the least intersite distance between microcells that resulted in the microcell RAN having the same energy consumption and throughput as the macrocell RAN (i.e., $ECG_R = TPG_R = ETG_R = 1$). For the same system bandwidth, this operating point is achieved by lowering the cell load α to a feasible level that limits the intercell interference. The main enhancements experienced by the microcell RAN in this case were higher peak and cell-edge throughputs (or capacities). By increasing the intersite distance at the feasible load, the ECG_R can be increased above unity but the TPG_R will decrease below unity such that their product ETG_R remains almost equal to one. Alternatively, if the least intersite distance is held constant while the load factor is increased, then the ECG_R remains almost constant at unity while the TPG_R increases giving the desired increase in throughput with reduced cell size for no further increase in RAN energy consumption.

However, operating the microcell RAN at the feasible load significantly under-utilizes the available radio resources. If no further increase in area throughput density is required, then an alternative strategy is to reduce the system bandwidth B instead of the cell load factor. For this case, each cell of both the microcell and macrocell RANs can be operated at full load with $\alpha = 1$, and, therefore, each cell achieves approximately the same cell-average spectral efficiency. Then, both microcell and macrocell RANs achieve the same power spectral density giving the fundamental relationship that the bandwidth ratio B_2/B_1 is equal to the BTS power ratio P_2/P_1 when $ECG_R = TPG_R = 1$. The LTE standard allows the air-interface bandwidth to be selected from a range of bandwidth values equal to 1.4, 3, 5, 10, 20 MHz. As each bandwidth mode operates with full transmit power, further reductions in power consumption could be considered by selecting the minimum transmit power needed to achieve SIR limited performance while meeting a cell-edge peak throughput rate.

The second option established the least intersite distance for $ECG_R = TPG_R$ when the load factor $\alpha = 1$. That is, all RANs are compared at maximum load. This condition established an operating point, which jointly maximized the ECG_R and the TPG_R given by Eq. (2.17). In particular, when the systems had the same bandwidths and their performances were SIR limited, the ratio of intersite distances and the ratio of BTS powers followed a fourth power law. This relationship confirmed that reductions in BTS power consumption can be traded for joint improvements in energy and throughput efficiency.

The behavior of a RAN in sleep mode was investigated. Though placing cells into sleep mode offers energy savings by scaling down the overhead power in nonoperational cells excluding backhaul, the savings are realized for low load factors. Over the diurnal day, the opportunity for entering sleep mode may be limited leading to small overall energy savings. The savings from sleeping microcells is potentially greater than that obtained by sleeping macro- or picocells. With respect to macrocells, the microcells exhibit a higher proportion of overhead power that can be scaled down during sleep mode. However, in picocells, the backhaul power consumption, which is not scaled down, becomes a limiting factor.

The energy and throughput efficiency of HetNets were investigated in the context of evolving a macrocell RAN into a large-cell underlay and small-cell overlay using separate frequency bands to eliminate interference between the layers. In addition, the energy and throughput performance of a number of homogeneous small-cell deployments were compared with the HetNet options. The main conclusions drawn were that small-cell deployments achieved considerably greater energy savings and throughput gains than HetNets. In HetNets, the addition of an overlay contributes an extra energy consuming component into the overall RAN while inhibiting the optimum trade-off between energy savings and throughput gains. The main advantage of a small-cell overlay is the enhancement of throughput. However, the gains remain significantly less than those achievable by using a homogeneous deployment of small cells. Even when the two tiers share the same frequency band, allowing interlayer interference, the upper bound on potential energy savings and throughput gains were considerably less than those achieved by small cells.

Overall, the chapter has established a robust evaluation framework for assessing the energy and throughput performance of an LTE-based RAN. The results demonstrate the fundamental roles played by the cell-site power consumption model, including the backhaul, the cell-average area power and throughput densities (efficiencies), and the energy consumption rating in joules/bit. Based on the models and analyses provided, the evolution of a macrocell cell RAN into a high throughput, energy-efficient RAN might be satisfactorily achieved by first evolving into a HetNet of small cells over macrocells followed by removing the macrocell underlayer leaving a small-cell RAN. An important ingredient in future small-cell deployments will be the minimization of the energy consumption in the backhaul. The backhaul power consumption can account for 50% of the overall BTS power consumption in a picocell or femtocell and can become a limiting factor especially when considering sleep mode options. An interesting solution to this problem might be to transform the macrocell underlay into a collection of massive MIMO-based macrocells providing backhaul connectivity to numerous off-grid, small cells. When cloud RAN techniques are taken into the technology mix, the options for energy and throughput efficient RANs become prolific. Hopefully, this chapter can contribute to this rapidly expanding field by establishing a consistent methodology for accurately assessing energy and throughput trade-offs in RAN planning.

References

[1] J. He, P. Loskot, T. O'Farrell, et al., "Energy efficient architectures and techniques for green radio access networks," in Communications and Networking in China (Chinacom), Beijing 2010.

[2] J.H. Tsai, H.W. Lo, and W.C. Chou, "Evaluation of mobile services for the future of 3G operators," IJMC, vol. 7, no. 4, pp. 470–493, 2009.

[3] A. Fehske, G. Fettweis, J. Malmodin, and G. Biczok, "The global footprint of mobile communications: the ecological and economic perspective," IEEE Commun. Mag., vol. 49, no. 8, pp. 55 62, 2011.

[4] EC, "Impacts of information and communication technologies on energy efficiency," *Bio Intelligence Service*, 2008.

[5] Vodafone, "Sustainablility report-for the year ending 31 March 2011," *Vodafone Group plc*, 2011.

[6] W. Riaz, J. Gutierrez, and J. Pedersen, "Strategies for the next generation green ICT infrastructure," in *ISABEL International Symposium on Applied Sciences in Biomedical and Communication Technologies*, Bratislava, 2009.

[7] B. Heile, "Smart grids for green communications," IEEE Wireless Commun., vol. 17, pp. 4–6, 2010.

[8] C. Han, et al., "Green radio: radio techniques to enable energy-efficient wireless networks," IEEE Commun. Mag., vol. 49, no. 6, pp. 46–54, 2011.

[9] B. Badic, T. O'Farrrell, P. Loskot, and J. He, "Energy efficient radio access architectures for green radio: Large versus small cell size deployment," in *IEEE Vehicular Technology Conference Fall (VTC 2009-Fall)*, Anchorage, Alaska, USA, pp. 1–5, 2009.

[10] H.-O. Scheck, "ICT & wireless networks and their impact on global warming," in *Procedings of European Wireless Conference*, Lucca, Italy, pp. 911–915, 2010.

[11] M. Lasanen, et al., "Environmental friendly mobile radio networks: approaches of the European OPERA-Net 2 project," in *20th International Conference on Telecommunications (ICT 2013)*, Casablanca, Morocco, May 6–8, 2013.

[12] G. Auer, et al., "How much energy is needed to run a wireless network?," IEEE Wireless Commun., vol. 49, no. 6, pp. 40–49, 2011.

[13] Cisco, Cisco Visual Networking index:global Mobile, Cisco, 2012.

[14] K. Gerwig, "Tablet devices could change user behaviour and network capacity planning," *SearchTelecom.com*, March 2012.

[15] O. Holland, et al., "Opportunistic load and spectrum management for mobile communications energy efficiency," in *IEEE 22nd International Symposium on Personal, Indoor and Mobile Radio Communications (PIMRC 2011)*, Totonto, 2011.

[16] J. Elling, M. Sorensen, P. Mogensen, and E. Lang, "Mobile broadband network evolution towards 2015-a Copenhagen area case study," Telektronikk, vol. 106, no. 1, pp. 138–148, 2010.

[17] Vodafone, "Performance data summary," 2012. [Online]. Available: http://www.vodafone.co.uk/our -responsibilities/performance-data-summary/. [Accessed 22 March 2014].

[18] Digital Communications Knowledge Transfer Network, "Energy Efficient Wireless Communications (Green Radio Access Networks)," in *Wireless Technology and Spectrum Working Group*, London, 2011.

[19] J. Louhi, "Energy efficiency of modern cellular base stations," in *29th International Telecommunications Energy Conference*, Rome, 2007.

[20] M. Deruyck, W. Joseph, and L. Martens, "Power consumption model for macrocell and microcell base stations," Trans. Emerging Telecommun. Technol., vol. 25, no. 3, pp. 320–333, 2014.

[21] F. Richter, A. Fehske, and G. Fettweis, "Energy efficiency aspects of base station deployment strategies for cellular networks," in *IEEE 70th Vehicular Technology Conference (VTC 2009-Fall)*, Anchorage, Alaska, USA, 2009.

[22] G. Micallef, "*Energy efficient evolution of mobile broadband networks*," PhD Thesis, Aalborg University, 2013.

[23] S. Tombaz, et al., "Impact of backhauling power consumption on the deployment of heterogeneous mobile networks," in *IEEE Global Communications Conference (GLOBECOM 2011)*, Houston, TX, 2011.

[24] W. Guo, C. Turyagyenda, H. Hamdoun, S. Wang, P. Loskot and T. O'Farrell, et al., "Towards a low energy LTE cellular network: Architectures", Proceedings of 19th European Signal Processing Conference, Barcelona, Spain, 29 August - 2 September 2011, Invited Paper.

[25] J. Salo, M. Nur-Alam, and K. Chang, "Practical introduction to LTE radio planning," 2010. [Online]. Available: http://digitus.itk.ppke.hu/ takacsgy/lte_rf_wp_02Nov2010.pdf. [Accessed 22 March 2014].

3

Energy-Efficiency Metrics and Performance Trade-Offs of GREEN Wireless Networks

Marco Di Renzo

Paris-Saclay University, Laboratory of Signals and Systems (UMR-8506), CNRS – CentraleSupelec – University Paris-Sud XI 91192 Gif-sur-Yvette (Paris), France
marco.direnzog@lss.supelec.fr

3.1 Introduction

3.1.1 Ubiquitous Mobility and Connectivity: The Societal Change

Since the turn of the century, there has been a tremendous growth in the mobile data market. The number of subscribers and the demand for wireless services has escalated. Indeed, the penetration of mobile services has exceeded that of the power grid. There are 48 million people in the world who have mobile phones, even though they do not have electricity at home [1]. In this context, mobile communications may be allowed to be an indispensable commodity by most, and mobile data, video, as well as television services are also becoming an essential part of everyday life. With the introduction of the Android operating system and the iPhone, the use of ebook readers such as the iPad, and the success of social networking using Facebook, the demand for (cellular) data traffic has grown significantly in recent years. Thus, communication on the move has proven to be transformational, and mobile operators struggle to satisfy the data traffic demands in wireless (cellular) networks, while keeping their costs at minimum to maintain profitability.

3.1.2 Mobile Data Traffic: The Forecast

A further explosion of mobile data traffic is predicted. According to CISCO's estimates predicted in February 2011 [1], the 2010 mobile data traffic growth rate was higher than

Green Communications: Principles, Concepts and Practice, First Edition.
Edited by Konstantinos Samdanis, Peter Rost, Andreas Maeder, Michela Meo and Christos Verikoukis.
© 2015 John Wiley & Sons, Ltd. Published 2015 by John Wiley & Sons, Ltd.

anticipated. The global mobile data traffic grew 2.6-fold in 2010, nearly tripling for the third year in a row. Furthermore, according to CISCO's estimates published in February 2013 [2], the global mobile data traffic grew 70% in 2012, and it reached 885 petabytes per month at the end of 2012, up from 520 petabytes per month at the end of 2011. These growth rates of mobile data traffic resemble those of the fixed network observed during 1997–2001, when the average yearly growth was 150%. The overall mobile data traffic is expected to grow to 11.2 exabytes per month by 2017, a 13-fold increase over 2012, which corresponds to a compound annual growth rate (CAGR) of 66% from 2012 to 2017. More particularly, the Asia Pacific and North America regions will account for almost two-thirds of the global mobile traffic by 2017. Middle East and Africa will experience the highest CAGR of 77%, increasing 17.3-fold over the forecast period. The Asia Pacific region will have the second highest CAGR of 76%, increasing 16.9-fold over the forecast period. The emerging regions of Latin America as well as Central and Eastern Europe will have CAGRs of 67% and 66%, respectively. When combined with the Middle East and Africa, the above-mentioned emerging market will represent an increasing share of the total mobile data traffic, which is expected to be up from 19% at the end of 2012 to 22% by 2017 [2].

3.1.3   Mobile Data Traffic: The In-Home Scenario

Furthermore, a survey conducted by the CISCO Internet Business Solutions Group (IBSG) indicates that much of the mobile data activity takes place in the home [1, 2]. In particular, it has been estimated that the percentage of time spent using the mobile Internet at home is approximately 40%. The amount of mobile data traffic on the move is approximately 35%, whereas the remaining 25% of mobile Internet use occurs at work. The relatively high percentage of home-based mobile data use suggests that next-generation cellular networks require specific data access points installed by home users to satisfy the huge demand for data traffic, and, at the same time, to get improved indoor voice and data coverage. By using these home access points, the telecommunications operators may be able to offload, in a cost- and energy-effective manner, the data traffic onto a fixed network, either by offering their subscribers dual-mode mobile phones or through the employment of femtocells, which are considered the key enabling technology to handle the growing demands for mobile data traffic in the home [3]. In particular, to meet the demand of massive mobile data growth, IDATE Research & Consulting and Infonetics Research has forecast the employment of 39.4 million femtocell units and a $2.98 billion market by 2015 [4].

3.1.4   Next-Generation Cellular Networks: The Compelling Need
            to be "Green"

The unprecedented surge of mobile data traffic in the cellular industry has motivated telecommunications operators and researchers to develop new transmission technologies, protocols, and network infrastructure solutions for maximizing both the achievable throughput and the spectral efficiency (SE). On the other hand, little or no attention has been devoted to energy consumption and complexity issues. As a result, the Information and Communication Technology (ICT) sector contributes substantially to the global carbon emissions. In particular, at the time of writing the ICT sector represents around 2% of the global carbon emissions already, of which mobile networks contribute about 0.2%. This is comparable to the worldwide carbon emissions of airplanes, and about a quarter of the worldwide carbon emissions of cars. Furthermore, this amount is expected to increase every year at a rapid pace due to the massive increase of the mobile data traffic. Currently, there are more than 5 million base stations (BSs) serving mobile users, each consuming an average of 25 MW/hour/year [5, 6]. In addition to the environmental aspects, the energy costs represent a significant portion of the network operators operating expenditure (OPEX). While each BS connected to the electrical grid may cost approximately $3000 per year to operate, off-grid BSs operating in remote areas generally run on diesel power generators and may cost ten times more [5]. Furthermore, with the advent of data-intensive cellular standards, such as the long-term evolution advanced (LTE-A) system, the energy consumption of each BS can increase up to 1400 W, and the energy cost of each BS may reach $3200 per annum with a carbon footprint of 11 tons of carbon emissions [7]. The radio network itself adds up to 80% of an operator's entire energy consumption. In this context, the development of revolutionary clean-slate wireless communications technologies that are capable of meeting the forecast mobile data traffic growth while reducing the carbon footprint of next-generation cellular networks is a compelling necessity.

3.1.5 Addressing the Energy Efficiency Challenge: Green Heterogeneous Networks

The rising energy cost and carbon footprint of operational cellular networks have motivated both network operators and regulatory bodies, such as the 3rd Generation Partnership Project (3GPP) and the International Telecommunication Union (ITU), to develop innovative solutions for improving the energy efficiency (EE) of cellular systems. This emerging trend has attracted the interest of researchers worldwide to develop "green heterogeneous networks" [8]. In heterogeneous cellular networks, low-power nodes are overlaid within a macrocell hence creating a wireless heterogeneous network. These low-power nodes may be picocells, femtocells, fixed and mobile relays, remote radio heads, distributed antenna elements, and so on. Thanks to the increased heterogeneity, these networks expand the coverage, improve the network capacity, reduce the energy consumption, and enhance the link reliability through a more dense deployment of low-cost and low-power access points. The reason behind all these potential advantages and, in particular, the EE of the heterogeneous network architecture is simple: *the densification of access points inherently reduces the distance between the network elements*. Since, based on electromagnetic laws, the received power falls off with the transmission distance and obeys an inverse power law where the exponent is known as the path loss exponent, this implies that reducing the distance has a beneficial impact on both the capacity and the transmission power. More specifically, the capacity can be increased and the transmission power can be reduced. As a consequence, heterogeneous cellular network architectures are considered as a strong potential enabler for the design of spectral-efficient and energy-efficient cellular networks and for striking a flexible trade-off between these two competing performance indicators.

Numerous collaborative projects have been launched worldwide for addressing the energy efficiency of mobile communications systems. Notable examples are as follows:

- The "Energy Aware Radio and NeTwork TecHnologies" (EARTH) [9] project.
- The "Towards Real Energy-efficient Network Design" (TREND) [10] project.

- The "Cognitive Radio and Cooperative strategies for Power saving in multistandard wireless devices" (C2POWER) [11] project.
- The "GREENET – An early stage training network in enabling technologies for green radio" [12] project.
- The "Green Terminals for next-generation wireless systems" (GREEN–T) [13] project.
- The GreenTouch consortium [14], whose mission is to deliver the architecture, the specifications, and the roadmap to increase, by 2015, the network's energy efficiency by a factor of 1000 compared to the 2010 levels.

Furthermore, in recent press releases (e.g., IP/09/393 [15]), ICT players have been warmly invited to develop innovative technologies in support of a greener world and to make people more aware of how they use energy. In this context, "green heterogeneous networking" constitutes a wide ranging research discipline that intends to cover all layers of the protocol stack and various system architectures, as well as to identify the fundamental trade-offs between spectral efficiency, energy efficiency, system and signal processing complexity, and system-wide performance.

3.1.6   The Emerging Paradigm Shift: From the SE to the SE Versus EE Trade-Off

The conventional response to the surge of mobile data traffic is the proposal of advanced transmission technologies and protocols designed for maximizing the SE. In fact, since the SE is directly linked to the notion of Shannon capacity [16], until recently it has been considered to be the main performance indicator to fueling the design and optimization of wireless communications systems in general and cellular networks in particular. As a result, the vast majority of transmission technologies and protocols used in the operational cellular and mobile networks have been designed by taking into account diverse factors, such as throughput, quality-of-service (QoS), availability, scalability, without paying specific attention to the energy consumption. With this design methodology, the operational cellular systems can only achieve energy savings at the cost of a performance and/or throughput degradation. Explicitly, it is crucial to develop power-efficient, low-complexity solutions that still satisfy the target QoS and throughput requirements.

To this end, transmission technologies and protocols should be designed and optimized for next-generation cellular networks by using more appropriate performance indicators, which explicitly take the energy consumption and the system's complexity into account. This implies that new performance metrics quantifying the EE of mobile networks have to be introduced, in addition to the SE, for the design of emerging green heterogeneous networks. Furthermore, it is clear that these new performance metrics will introduce new fundamental trade-offs that have to be accurately investigated in order to appropriately quantify the performance of emerging green heterogeneous networks. The objective of this chapter is to summarize the main efforts of the research community in the definition of these energy-efficiency metrics as well as to describe the most important and fundamental trade-offs that emerge with these new performance metrics at hand. Energy-efficiency metrics and performance trade-offs are described in the next two sections, respectively.

In order to better understand the importance of these new EE metrics and the new trade-offs that emerge from their adoption, let us provide two simple examples from a physical layer point of view (1) the point-to-point single-input-single-output (SISO) additive white Gaussian noise (AWGN) channel and (2) the point-to-point multiple-input-multiple-output (MIMO) Rayleigh fading channel.

SE versus EE Trade-Off of the SISO-AWGN Channel. A widespread definition of EE metric is the throughput per unit energy [16, 17]. By considering the SISO-AWGN channel, the ratio between the SE (η_{SE}), defined as the throughput per unit bandwidth, and the EE, defined as the throughput per unit energy (η_{EE}), can be formulated as follows [16, 18]:

$$\eta_{EE} = \frac{\eta_{SE}}{N_0(2^{\eta_{SE}} - 1)}$$

where N_0 is the noise power. This simple formula highlights that the EE is monotonically decreasing when increasing the SE.

SE versus EE of MIMO-Rayleigh Fading Channel. MIMO communications constitute promising techniques for the design of future wireless communications systems, including the fifth-generation cellular networks. In simple terms, the capacity of MIMO systems is proportional to $\min\{N_t, N_r\}$, where N_t and N_r represent the number of transmit and receive antennas, respectively [19]. This implies that the throughput may be increased linearly with the number of antennas. As a consequence, MIMO techniques provide high data rates (SE) without increasing the spectrum utilization and the transmit power. However, in practice, MIMO systems need a multiplicity of associated circuits, such as power amplifiers, RF chains, mixers, synthesizers, filters, which substantially increase the circuit power dissipation of the BSs [5, 20–22]. Recent studies have clearly shown that the EE gain of MIMO communications increases with the number of antennas, provided that only the transmit power of the BSs is taken into account and their circuit power dissipation is neglected. On the other hand, the EE gain of MIMO communications remains modest and decreases with the number of active transmit antennas, if realistic power consumption models are considered for the BSs [23]. As a result, while the SE advantages of MIMO communications are widely recognized, the EE potential of MIMO communications for cellular networks is not well understood.

Thus, the important consideration from these two toy examples is that current solutions that are spectral efficient may turn out to be suboptimal in terms of EE. This leads to the conclusion that energy-efficient solutions may operate relatively far from the Shannon capacity. However, improving the EE at the cost of the QoS (SE/throughput) for the end user may be unacceptable in commercial networks. Therefore, an important message emerge: the development of beneficial wireless communications techniques striking an attractive SE versus EE trade-off for next-generation cellular networks is a compelling necessity [18].

3.2   Energy-Efficiency Metrics

The very broad term "green communications" lacks clear scientifically based definitions and quantifiable metrics. Currently, it is more of a marketing term than a standard to strive for. To truly address this problem on a transformational level, high-risk and high-reward research is required that integrates all aspects of communications stack and peripheral interactions. Most importantly, metrics and their associated measurement science that define green communications from combined energy efficiency and network optimization perspectives must be developed. Metrics are essential to providing guidance to manufacturers and service providers to help them make better decisions regarding infrastructure development and purchases. This is where energy-efficiency metrics play an important role. These metrics provide information in order to directly compare and assess the energy consumption of various components and the overall network. In addition, they also help us to set long-term research goals of reducing energy consumption. With the increase in research activities pertaining to green communications and hence in number of diverse energy-efficiency metrics, standards organizations such as the European Technical Standards Institute (ETSI) and the Alliance for Telecommunications Industry Solutions (ATIS) are currently making efforts to define energy-efficiency metrics for wireless networks [24, 25].

Such metrics may be classified as *energy efficiency metrics* or *energy consumption metrics*. An energy-efficiency metric corresponds to the ratio of attained utility (e.g., the transmission distance reached, the area covered, the output power, the bits transmitted) to the consumed power/energy used. On the other hand, an energy consumption metric corresponds to the energy/power consumed per unit of attainable utility. Energy-efficiency metrics of telecommunication systems can be classified into three main categories [26–28]:

- *Facility-level metrics*, which relate to high-level systems where the equipment is deployed, such as data centers and Internet Service Providers (ISP).
- *Equipment-level metrics*, which are defined to evaluate the performance of an individual equipment.
- *Network-level metrics*, which assess the performance of equipments while also considering features and properties related to capacity and coverage of the network.

A comprehensive taxonomy of these metrics is available in [5, 22, 26–32]. In what follows we provide a list of widely used metrics to quantify the energy efficiency of green communications. For simplicity, the analytical formulation of the energy metrics is avoided in the present chapter, but it is available in the references cited provided.

- *Power usage efficiency (PUE)* [28] is used to evaluate the performance of power hogging data centers. The PUE is defined as the ratio of the total facility power consumption to the total equipment power consumption. It is a good metric to quickly assess the performance of data centers at a macro level, but it fails to account for energy efficiency of individual equipments. This is a facility-level metric.
- *Data center efficiency (DCE)* [28] is the reciprocal of the PUE. This is a facility-level metric.
- *Power per user (PPU)* [26], which is measured in W/user, is the ratio of the total facility power to the number of users. This is an equipment-level metric.
- *Energy consumption rating (ECR)* [26], which is measured in W/Gbps, is the ratio of the normalized energy consumption to the effective full-duplex throughput. The ECR provides the manufacturers a better insight into the performance of hardware components. However, it does not account for the network load. This is an equipment-level metric.
- Since even the busiest networks do not always operate at full-load conditions, it would be useful to complement metrics such as the ECR to incorporate the dynamic network conditions, such as energy consumption under different loads. Such metrics are equipment-level metrics and they are as follows [5, 26]:
  - *ECRW* (ECR-weighted), which is the ratio of the energy consumption over the effective system capacity by taking into account full, half, and idle conditions;
  - *ECR-VL* (energy-efficiency metric over a variable-load cycle), which is the average energy rating in a reference network described by an array of utilization weights;

- ○ *ECR-EX* (energy-efficiency metric over extended-idle load cycle), which is the average energy rating in a reference network, where extended energy savings capabilities are enabled;
- ○ *Telecommunications energy efficiency ratio (TEER)* introduced by the ATIS and the *Telecommunication equipment energy efficiency rating (TEEER)* introduced by Verizons Networks and Building Systems, which consider the total energy consumption as the weighted sum of the energy consumption of the equipment at different load conditions.
- *Absolute energy efficiency (AEE)* [5], which accounts not only for the consumed power, the bit rate, but also for the temperature aspect of the system since classical thermodynamics is based on the absolute temperature of the system under analysis. This is an equipment-level metric.
- *Performance indicator in rural areas (PIrural)* [5], which is measured in Km^2/W, is the ratio of the total coverage area and the power consumed at the site. This metric is useful since rural areas are characterized by a low load and the main target is the maximization of the coverage area. This is a network-level metric.
- *Performance indicator in urban areas (PIurban)* [5], which is measured in Users/W, is the ratio of the number of users based on average busy hour traffic demand by users and the power consumed at the site. This metric is useful since urban areas have higher traffic demand than rural areas; hence, capacity is more important than the coverage area. This is a network-level metric.
- *Power per Unit Area (PUA)* [22], which is measured in W/m^2, is equal to the network average power usage divided by the coverage area of the network. The metric focuses on the total network power (or, equivalently, the total energy consumption) and is closely related to the CO_2 emissions and the associated carbon footprint. It is further a very relevant quantity at low traffic loads, as in this case the network is coverage limited rather than capacity limited. This is a network-level metric.
- *Energy per bit (EpB)* [22], which is measured in Joule/bit, is defined as the network energy consumption during the observation period divided by the total number of bits that are correctly delivered in the network during the same time period. Since the network energy consumption is simply the (average) power multiplied with the observation period, this metric may, equivalently, be described as the (average) network power in relation to the (average) data rate. This is both an equipment- and network-level metric, depending on how it is measured.
- *Energy consumption gain (ECG)* [6], which defined as the ratio of the energy consumed by the baseline and the one consumed by the system under test. The ECG is useful for comparing two different systems. This is both an equipment- and network-level metric, depending on how it is measured.
- *Energy efficiency*, which is measured in bit/Joule, is defined as the number of bits that can be transmitted per unit energy. This metric is very often used for the analysis of the energy efficiency of wireless communications systems. This is both an equipment- and network-level metric, depending on how it is measured.

In conclusion, the proper evaluation of the energy efficiency of green communications is a serious matter that the green networking community shall face soon. In that regard, the adoption of effective energy efficiency metrics is instrumental to assess the potential gains. In this section, we have summarized the most common energy-efficiency metrics that can be found in the literature. In general, due to their intrinsic difference and relevance, no single metric can represent the whole state of the system. Nevertheless, choosing a metric rather than another may yield significantly different evaluation results. As a consequence, a large consensus needs to be reached as soon as possible on a reduced set of well-defined performance indicators to promote fair and accurate cross-comparisons among different solutions.

3.3 Performance Trade-Offs

Generally speaking, energy efficiency means using less energy to accomplish the same task. In case of communications systems, the task to be accomplished could be a file transfer, a phone call, and so on. It is clear that improving the energy efficiency of a communication system is expected to have some "costs" in terms of other aspects of their overall performance. In Section 3.1.6, we have provided two simple examples that support this intuition by considering the SISO-AWGN channel and the MIMO-Rayleigh fading channel. We have shown that improving the EE inevitably implies reducing the SE, which for MIMO systems may lead to switching off some antenna elements in order to decrease the overall power dissipation. This section is aimed at summarizing some fundamental trade-offs to be considered when designing an energy-efficient communication system.

The SE versus EE trade-off is one of the main trade-offs that arise in the design of green wireless networks. In Ref. [18], the authors have recently identified four main fundamental trade-offs that are now widely accepted by the research community. These four trade-offs are as follows:

- *SE versus EE trade-off*, which, for a given available bandwidth, provides the balance between the achievable rate and the energy consumption of the system.
- *Deployment efficiency (DE) versus EE trade-off*, which provides the balance between the deployment cost, throughput, and energy consumption in the network as a whole.
- *Bandwidth (BW) versus power (PW) trade-off*, which, for a given transmission rate, provides the balance between the utilized bandwidth and the power needed for transmission.
- *Delay (DL) versus PW trade-off*, which provides the balance between the average end-to-end service delay and the average power consumed for transmission.

3.3.1   The SE Versus EE Trade-Off

As mentioned in Section 3.1.6, the SE is defined as the capacity per unit bandwidth and the EE is defined as the capacity per unit energy. The SE is a widely accepted criterion for wireless network optimization. The peak value of SE is always among the key performance indicators by standardization bodies. On the other hand, the EE has raised the attention of standardization bodies only recently [33]. For example, reducing the number of active RF chains (i.e., active antenna elements in MIMO-aided BSs) for the sake of reducing the power consumption of BSs has been actively discussed within 3GPP standardization bodies [34]. The main idea is to enable the BSs to use only a subset of the available antenna elements in order to reduce the power consumption during low traffic periods. The reason of this choice is that in practical systems the SE versus EE trade-off has a mathematical formulation that is more difficult than that provided in Section 3.1.6, and that the circuit power consumption significantly affects the overall power efficiency of communications networks. The SE versus EE trade-off is also a function of the hardware impairments. For example, the power amplifiers (PAs) play an important role in the power efficiency of wireless communications systems [5]. Their role is to increase the power level of the transmit signal so that the corresponding received signal can be demodulated by the receiver meeting a given error probability requirement. Two important metrics quantifying the performance of PAs are linearity and efficiency. The linearity of the response of a PA is an important factor for wireless communications since the distortion of the signal causes an increase in the required transmit power to meet the same error rate requirement and an irreducible error floor. The drain efficiency is the ratio between the output RF power and the input DC power; therefore, this is a measure of how much DC power is converted to RF power. High efficiencies are required to minimize the needs for thermal dispersion and to decrease the energy consumption. However, high efficiency and high linearity are conflicting requirements in PAs. If PA linearity is required, there must be a direct relationship between the output power and the power supplied to the PA. On the other hand, in order to be power efficient the electronic device/system should use a limited amount of power even when a high output power is needed. This conflicting behavior leads, for example, to find the right modulation scheme in order to balance the SE (amplitude modulations) and the EE (constant-envelope modulations).

3.3.2   The DE Versus EE Trade-Off

The DE is a measure of the system throughput per unit of deployment cost, which is an important network performance indicator for mobile operators. The deployment cost consists of both capital expenditure (CAPEX) and OPEX. For radio access networks, the CAPEX comprises the infrastructure costs and the OPEX includes operational and maintenance costs. The main issue behind this trade-off is that the DE is estimated during the network planning phase, whereas the EE is estimated when the network is operational. A typical example where this trade-off arises is the choice of the cell coverage. Usually, the DE is reduced by increasing the cell coverage as much as possible, since this reduces the expenditure costs. On the other hand, increasing the cell coverage by having bigger cells results in the need of increasing the transmission power of the BSs to reach the users in the cell edges, which in turn reduces the EE averaged over the cell of interest. The heterogeneous cellular network infrastructure described in Section 3.1.5 is considered to be a promising solution for addressing this trade-off. In particular, with the combination of macrocells and micro/pico/femtocells, coverage and capacity provision can be decoupled into different tiers of the network, where the macrocells handle coverage and mobility issues whereas the femtocells focus on local throughput. In addition, relay stations and remote radio heads may further help in improving the DE versus EE trade-off, since they are of much lower cost and smaller coverage compared to macro BSs, hence bringing mobile users closer to the network and making the system deployment more flexible.

3.3.3   The BW Versus PW Trade-Off

Both BW and PW are limited resources in wireless communications. From Shannon's capacity formula for a simple SISO-AWGN channel, the relation between transmit power and signal bandwidth for a given transmission rate, R, can be formulated as follows:

$$PW = WN_0(2^{R/W} - 1)$$

which shows that the transmit power increases by decreasing the signal bandwidth. Furthermore, it shows that as the bandwidth increases without limit, there is a minimum power limit that cannot be overcome. This formulas also highlight that, for a given data transmission rate, the expansion of the signal bandwidth is preferred in order to reduce the transmit power and thus to achieve a better energy efficiency. This trend for bandwidth demand is indeed the core tenet of the different generations of cellular systems. In particular, the second and third generations of wireless communications systems use a fixed BW transmission. On the other hand, future deployments of LTE and LTE-A standards provide a higher flexibility in spectrum usage so that the transmission BW can be tuned for different applications (i.e., by using carrier aggregation). Furthermore, new transmission concepts, such as the cognitive radio (CR) [5], are capable of adjusting the modulation order according to the available BW as function of time, hence supporting a more flexible use of the BW. However, both carrier aggregation and CR incur in some practical overhead, such as multiple RF chains and spectrum sensing, both of them increasing the energy consumption. Thus, it is important to pay attention to how these technologies will be integrated in next-generation green wireless networks.

3.3.4   The DL Versus PW Trade-Off

DE, SE, and BW are physical layer-oriented metrics, which account for either the system efficiency or network resources. The DL (or service latency) is, on the other hand, a measure of the QoS experienced by the user, and it is closely related to the upper layers of the protocol stack and to the traffic statistics. In the second generation of mobile systems, the DL was mostly associated with the signal processing time and with the propagation delay. On the other hand, emerging wireless communications standards, such as the LTE-A, provide a wide plethora of wireless services and applications with heterogeneous DL requirements. Therefore, in this emerging application scenarios, it is important to understand the trade-off between transmit power and delay. By still using the Shannon's capacity formula for a simple SISO-AWGN channel, PW can be formulated as:

$$PW = WN_0T_b(2^{1/T_bW} - 1)$$

where W is the signal bandwidth and $DL = T_b = 1/R$ is the delay for transmitting one bit, given the rate R. This formula shows that the transmission power increases by decreasing DL. In practical systems, however, the DL versus PW trade-off is more complicated and its mathematical analysis needs advanced information and queuing theories. From queuing theory, it is known that the average DL of a packet queue is determined by the statistics of the traffic arrivals and departures. In general, there is no closed form expression available to show the direct relation between DL and PW. Therefore, the investigation of simplified but approximate models is desirable for providing insights for practical system design.

In summary, in this section we have outlined some fundamental trade-offs that originate in green wireless networks. Although the trade-offs have been illustrated by considering idealized toy examples showing monotonic behaviors among the different metrics of interest, it is worth emphasizing that in practical systems the trade-off relations usually deviate from simple monotonic curves, thus making the design and optimization of green communications networks a challenge.

3.4 Conclusion

Future mobile communications networks will not only be optimized for capacity, but also their design will implicitly take energy efficiency into account. This implies that new performance metrics quantifying the energy efficiency of mobile networks are necessary, which inevitably introduce fundamental performance trade-offs. In this chapter, we have outlined the most common and widely used energy efficiency metrics and have summarized the main performance trade-offs arising from their utilization.

Acknowledgments

This work was supported in part by the European Commission under the auspices of the FP7–PEOPLE MITN–GREENET project (Grant 264759) and the FP7–PEOPLE MITN–CROSSFIRE project (Grant 317126).

References

[1] Cisco visual networking index: global mobile data traffic forecast update, 2010–2015, [Online]. Available: http://newsroom.cisco.com/ekits/Cisco VNI Global Mobile Data Traffic Forecast 2010 2015.pdf.

[2] Cisco visual networking index: global mobile data traffic forecast update, 2012–2017, [Online]. Available: http://www.cisco.com/en/US/solutions/collateral/ns341/ns525/ns537/ns705/ns827/white paper c11-520862.pdf.

[3] J. G. Andrews, H. Claussen, M. Dohler, S. Rangan, and M. C. Reed, "Femtocells: past, present, and future", IEEE J. Sel. Areas Commun., 30, 3, 497–508, 2012.

[4] Femtocells: major telecom equipment vendors, operators showing greater interest in femtocell market. [Online]. Available: http://www.idate.org/.

[5] Z. Hasan, H. Boostanimehr, and V. K. Bhargava, "Green cellular networks: a survey, some research issues and challenges", IEEE Commun. Surveys Tuts., 13, 4, 524–540, 2011.

[6] C. Han, T. Harrold, S. Armour, I. Krikidis, S. Videv, P. M. Grant, H. Haas, J. S. Thompson, I. Ku, C.-X. Wang, T. A. Le, M. R. Nakhai, J. Zhang, and L. Hanzo, "Green radio: radio techniques to enable energy-efficient wireless networks", IEEE Commun. Mag. 49, 6, 46–54, 2011.

[7] Telecommunication Predictions 2010, Technology, Media & Telecommunications Industry Group, Deloitte. [Online]. Available: http://www.deloitte.com/assets/Dcom-UnitedStates/Local Assets/Documents/TMT us tmt/us tmt telecompredictions2010.pdf.

[8] J. Hoydis, M. Kobayashi, and M. Debbah, "Green small–cell networks", IEEE Veh. Technol. Mag., 6, 1, 37–43, 2011.

[9] https://www.ict-earth.eu/.

[10] http://www.fp7-trend.eu/.

[11] http://www.ict-c2power.eu/.

[12] http://www.fp7-greenet.eu/.

[13] http://greent.av.it.pt/.

[14] http://www.greentouch.org/.

[15] http://europa.eu/rapid/press-release IP-09-393 en.htm.

[16] G. Y. Li, Z. Xu, C. Xiong, C. Yang, S. Zhang, Y. Chen, and S. Xu, "Energy-efficient wireless communications: tutorial, survey, and open issues", IEEE Trans. Wireless Commun., 18, 6, 28–35, 2011.

[17] X. Wang, A. V. Vasilakos, M. Chen, Y. Liu, and T. Kwon, "A survey of green mobile networks: opportunities and challenges," ACM/Springer Mobile Netw. Appl., 17, 1, 4–20, 2012.

[18] Y. Chen, S. Zhang, S. Xu, and G. Y. Li, "Fundamental tradeoffs on green wireless networks," IEEE Commun. Mag., vol. 49, no. 6, pp. 30–37, 2011.

[19] H. Huang, C. B. Papadias, and S. Venkatesan, MIMO Communication for Cellular Networks, Springer, 2011.

[20] S. D. Gray, "Theoretical and practical considerations for the design of green radio networks," IEEE Vehicular Technology Conference – Spring, Keynote speech, 2011. [Online]. Available: http://www.ieeevtc.org/conf -admin/vtc2011spring/5.pdf.

[21] G. Auer, V. Giannini, I. Godor, P. Skillermark, M. Olsson, M. Imran, D. Sabella, M. Gonzalez, C. Desset, and O. Blume, "Cellular energy efficiency evaluation framework," IEEE Vehicular Technology Conference – Spring, pp. 1–6, 2011.

[22] G. Auer, O. Blume, V. Giannini, I. Godor, M. Imran, Y. Jading, E. Katranaras, M. Olsson, D. Sabella, P. Skillermark, and W. Wajda, "D2.3: energy efficiency analysis of the reference systems, areas of improvements and target breakdown," EARTH: Energy Aware Radio and Network Technologies, 2012. [Online]. Available: https://bscw.ictearth.eu/pub/bscw.cgi/d71252/EARTH WP2 D2.3 v2.pdf.

[23] F. Heliot, M. A. Imran, and R. Tafazolli, "On the energy efficiency–spectral efficiency trade–off over the MIMO Rayleigh fading channel," IEEE Trans. Commun., vol. 60, no. 5, pp. 1345–1356, 2012.

[24] Alliance for Telecommunications Industry Solutions, ATIS Report on Wireless Network Energy Efficiency, ATIS Exploratory Group on Green (EGG), 2010.

[25] European Telecommunications Standards Institute, Environmental engineering (EE) energy efficiency of wireless access network equipment, ETSI TS 102 706, v1.1.1, 2009.

[26] A. P. Bianzino, A. K. Raju, and D. Rossi, "Apple-to-Apple: a framework analysis for energy-efficiency in networks," Proc. of SIGMETRICS, 2nd GreenMetrics Workshop, 2010.

[27] T. Chen, H. Kim, and Y. Yang, "Energy efficiency metrics for green wireless communications," 2010 International Conference on Wireless Communications and Signal Processing (WCSP), pp. 1–6, 2010.

[28] C. Belady, et al., Green Grid Data Center Power Efficiency Metrics: PUE and DCIE, The Green Grid, 2008.

[29] L. A. Suarez, L. Nuaymi, and J.-M. Bonnin, "An overview and classification of research approaches in green wireless networks," EURASIP J. Wireless Commun Netw, vol. 2012, no. 1, p. 142, 2012.

[30] A. Amanna, A. He, T. Tsou, X. Chen, D. Datla, T. R. Newman, J. H. Reed, and T. Bose, "Green communications: a new paradigm for creating cost effective wireless systems," 2013. [Online]. Available: http://filebox.vt .edu/users/aamanna/web%20page/Green%20Communications-draft%20journal%20paper.pdf.

[31] Wei Wang, Zhaoyang Zhang, and Aiping Huang, "Towards green wireless communications: metrics, optimization and tradeoff," Green IT: Technologies and Applications, 1st edn, Berlin Heidelberg: Springer-Verlag, 2011, pp. 26–27.

[32] M. Nasimi, F. Hashim, and C. K. Ng, "Characterizing energy efficiency for heterogeneous cellular networks," IEEE Student Conference on Research and Development, pp. 198–202, 2012.

[33] 3GPP TSG–SA#50 SP–1008883GPP, 3GPP Work Item Description, Study on system enhancements for energy efficiency, 3GPP TSGSA, Istanbul, Turkey, Tech. Rep., 2010, agreed Work Item 500037 (Release 11).

[34] 3GPPTSG–RAN WG2 #67, "eNB power saving by changing antenna number," [Online]. Available: http://www .3gpp.org/ftp/Specs/html-info/TDocExMtg–R2-67–27297.htm, contribution R2–094677 from Huawei.

4

Embodied Energy of Communication Devices

Modeling Embodied Energy for Communication Devices

Iztok Humar[1], Xiaohu Ge[2], Lin Xiang[2], Minho Jo[3], Min Chen[2] and Jing Zhang[2]

[1] *Faculty of Electrical Engineering, University of Ljubljana, Ljubljana, Slovenia*
[2] *Huazhong University of Science and Technology, Wuhan, Hubei, China*
[3] *Korea University, Seoul, Korea*

4.1 Introduction

During past few decades, the Information and Communication Technology (ICT) has played a very important role in a modern society. However, the extensively growing numbers of users, new product and service launches as well as the rising service usage times are putting great demand on energy consumption in ICT infrastructure today.

4.1.1 Energy Consumption of ICT in Figures

From the operating point of view, approximately 600 TWh or 3% of total worldwide electrical energy is being consumed by the ICT [1, 2], causing approximately 2% of the world's CO_2 emissions, and this number is even expected to grow to 1,700 TWh by the end of 2030 [3]. Telecommunication equipment (mobile, fixed, and communication devices, excluding consumer entertainment and media) currently contributes 30% to this figure or roughly 1% of the world's total electricity consumption, resulting in 30 TWh per year in United States (US) only. The major branch of this sector is mobile telephony responsible for a half of this energy

Green Communications: Principles, Concepts and Practice, First Edition.
Edited by Konstantinos Samdanis, Peter Rost, Andreas Maeder, Michela Meo and Christos Verikoukis.
© 2015 John Wiley & Sons, Ltd. Published 2015 by John Wiley & Sons, Ltd.

consumption [2, 4]. It is expected to grow[1] further with a still rising demand for ICT services in developing countries and due to their above-mentioned popularity.

Being a large consumer, the ICT community has increasingly recognized the importance of energy-related topics during last years. There are at least two strong motivating factors driving further research and improvements in this field. The first one is definitely to minimize the environmental impact of this sector on the climate change caused by increased CO_2 and other greenhouse gases concentration levels in the atmosphere emitted mostly due to the use of fossil fuels as a primary source for production of electrical energy. As its usage in ICT sector has grown lately, it is estimated to contribute a 2.7% to the global CO_2 emissions in 2020, resulting in 1.43 Gtons per year. Besides their corporate responsibility regarding the environmental protection, the telecommunication operators are becoming increasingly aware of their energy bills, which are estimated to grow to approximately four billion EUR of their operating expenses (OpEx9 in European Union). Thus, the reduction of energy consumption has also a direct economic effect.

4.1.2 The Approaches to Reduce ICT Energy Consumption

From the two above-mentioned reasons and in the perspective of the energy-efficient society, it is desirable that the ICT sector reduces its energy consumption. Consequently, the researchers are suggesting many different approaches to reduce operating costs and effects on the environment.

There are three main basic concepts aiming at energy consumption of ICT [1] and are as follows:

1. Re-engineering – is based on introduction and design of more energy-efficient elements for network device architectures, rearrangement and optimization of the organization of devices as well as at reduction of their intrinsic complexity levels.
2. The dynamic adaptation of network or device resources – modulates the capacities of network interfaces and processing engines to follow actual traffic loads and requirements. It is achieved by two power aware capabilities: dynamic voltage scaling and idle logic – a dynamic trade-off between packet service performance and power consumption.
3. Sleeping/standby approaches – are used to selectively drive unused network/device portions to low standby modes and to wake them up only if necessary.

4.1.3 The Problem of Past Researches

In previous works, the authors suggested different approaches to reduce the energy consumption and improve energy efficiency of ICT for each of the above-mentioned concepts. However, some of the proposed approaches related to the concept of network re-engineering or sleeping/standby strategies suggest adding network elements in order to decrease operating energy of the existing elements.

An example of such approach that is discussed later in more detail is a cellular network. Its general idea is to find a trade-off between a decrease in operating energy based on reduced

---

[1] The power consumption is growing 16–20% per year [5].

transmission power and a consequently increased operating energy caused by higher number of cells, where sleep-mode strategy can be used for further energy savings. A common suggestion of these research works is to improve the energy efficiency of cellular networks by using base stations (BS) with a reduced transmission power (reduced operating energy) such as micro, pico, or femtocells, but employing a larger number of them to maintain the coverage and capacity.

Another example is the network connectivity proxying scheme [6] which employs external network proxies that virtually maintain presence for network computers, but lets them powered-down and saves energy while idle. In some cases, this approach may require additional hardware that should be further evaluated from the perspective of energy consumption.

Even if we overlook the problem that most of today's widely employed hardware does not necessarily support a sudden total power-off strategy suggested in many of these researches, there is another notable drawback in the proposed approaches. Although all of these researches successfully proved to reduce the operating energy consumption, they paid no attention to the embodied energy[2] which is consumed by the manufacturers to produce the additional high-tech equipment. Some of the analyses avoid the consideration of the embodied energy claiming that there are no publicly available databases to collect the required parameters; others simply ignore it without any remarks. Until recently, it was thought that the embodied energy content of electronic devices is negligibly small compared to the energy used in operating the device over its life; therefore, most previous researches endeavored for energy conservation on this assumption.

In contrast to approaches from previous studies, this chapter shows how important it is to include embodied energy in energy efficiency modeling. The simulation results of such an extended modeling confirm the important trade-off between operating and embodied energy consumption. The results provide some guidelines for manufacturers, operators, and researchers.

4.2   The Extended Energy Model

It is obvious that a lot of energy is needed by the ICT equipment to operate. However, a lot of energy is also needed to manufacture the device. Hence, to analyze the total energy usage, it is not sufficient to only look at the operating energy consumption, the embodied energy of a device must also be included.

4.2.1   The Embodied Energy and Its Meaning in ICT Technology

By the common definition, the embodied energy E_{EM} is the energy consumed by all the processes associated with the production of a device. The initial embodied energy E_{EMinit} comprises the energy used to acquire and process raw materials, transport, manufacture components as well as to assembly and install all products in the initial construction of the device. The maintenance embodied energy $E_{EMmaint}$ includes the energy associated with maintaining, repairing, and replacing materials and components of the device over its lifetime. Some authors prefer an expanded definition associated with a broader field known as life cycle assessment

---

[2] See Section 4.2.1 for further details on embodied energy.

(LCA) which attempts to characterize all environmental impacts from "cradle-to-grave," extending the above-mentioned components with the energy used for extraction of resources following by the energy for sales, use, demolition, and disposal of equipment.

Even though it seems that the consideration of the embodied energy in the process of total energy evaluation in the field of ICT is a somehow new approach, it has a long tradition in other disciplines, such as building construction [7] and car production [8]. Embodied energy has already been taken into account in some fields of electronics engineering, such as photovoltaic [9] and even with ICT more closely related areas, such as computer industry [10] and mobile phone production [11]. Figure 4.1 shows the relation between the embodied and operating energy for the representative technologies/products of three selected societies (urban, industrial, and information) during their life cycles (a) buildings/house, (b) vehicles/car, and (c) ICT/electronic equipment (a BS[3] which will later be discussed in more detail). Although

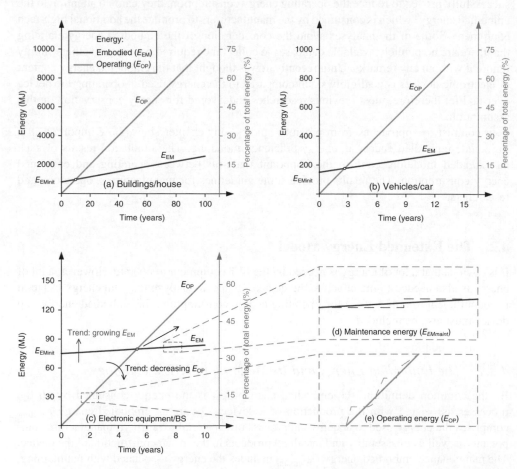

Figure 4.1 Embodied and operating energies of house, car, and electronic/ICT device during their lifetimes

---

[3] Based on our estimation for a general BS, see Section 4.2: Embodied energy estimation for a general BS.

the numbers used in these figures are rough estimations for averages of greatly varying values[4] given by the literature, an obvious difference among them enables us to draw some important conclusions that directly influence ICT area.

One of the most remarkable facts that should be noted in these graphs is the shift of the intersection point between the curves of embodied and operating energy to the upper-right position, explaining that the ratio between embodied and operating energy during the life cycle of modern (electronic or ICT) equipment is much higher in comparison to the house or car, mainly reasoning in a sophisticated and energy-insensitive production of semiconductors, minimized operating energy consumption of these devices, and their relatively short life cycles due to frequent replacements caused by the upcoming technologies.

The scale on the right explains the share (percentage) of initial embodied energy relative to the total energy used in devices lifetime. While the embodied energy represents a relatively small part of total energy in houses (7–10%) or cars (10–15%) in comparison to electronic or ICT devices (30–40%), it is ironic that the problem of embodied energy is studied in depth in the aforementioned disciplines, while research work in the area of ICT simply neglects it.

4.2.2 Embodied Energy Assessment of an ICT Equipment

Several different methodologies to assess the device's initial embodied energy and its environmental impacts exist and are designed to be conducted by the manufacturing companies and their component suppliers. They range from the well-known LCA [12], employing very extensive studies, requiring data that are not easily available and taking a great deal of time to be completed, to the ecological footprint analysis (EFA) [13] and key environmental performance indicators (KEPI) [14] approach which is more easy to use and requires less time and input data. Although some of manufacturers (i.e., [4, 11, 15]) have already conducted at least some parts of LCA, the collected data is usually kept confidentially or provided with a great delay, sometimes also because of the extensive space of time spent to perform assessment (e.g., a report in year 2004 discusses the equipment from the year 2000 or earlier), which is of little use due to quickly changing component materials and developing processes of manufacturing. The data is often provided merely in an aggregated form – that is, the embodied energy for the entire equipment of an operator's network, including radio access networks, data centers, and office equipment. It is thus not possible to distinguish the embodied energy of a single ICT equipment. It should also be noted that most of the publicly available data is given for commercial purposes, which makes them questionable in reliability.

Since no detailed, trustworthy, and up-to-date data is provided by the manufacturers, it is necessary to employ a different approach enabling the external observer to estimate the initial embodied energy. As an illustration, we provide an example of embodied energy estimation for a BS of a cellular network in the next subsection.

In the past, researchers suggested many different methods, but the work [16] does not offer just a versatile approach for estimation of embodied energy but a suitable tool as well, referring to many studies for the embodied energy parameters of different materials and processes. The basic principle of this methodology is to make assumptions on the most important materials and manufacturing processes of electronic equipment and to then use the data from available high-quality studies for individual parts and extrapolate the results to get representative integrative figures.

---

[4] The embodied energy of buildings in different counties depends on traditionally used construction materials, whereas the operating energy depends on the climate conditions.

ICT equipment is generally made of several thousands of components made up of a large variety of materials and substances. Mechanical housing and electrical components are usually made up of metals and plastic polymers. The hart is a set of printed wiring boards with active electronic components such as integrated circuits and passive electronic components such as resistors, capacitors, conductors, and connectors. The power supply may include batteries, but for the sake of their diversity they were excluded from our study.

A great majority of a state-of-the-art electronic device's embodied energy is used for semi-conductors' processing and manufacturing. This is due to very complex processes of wafer manufacturing that may include up to several hundred distinct process steps. The share of energy used for semiconductor devices in our estimation is in agreement with previous research [17] revealing that in computer industry, almost 95% of the energy content for electronics manufacturing goes into the wafer manufacturing/chip packaging. In Ref. [16], they have also noted that in comparison with this numbers, the energy of other phases of electronic manu-facturing (printed board manufacture, board-level assembly, display manufacturing, and final assembly) may be safe to neglect. As ICT equipment consists not only of electronics, but also a large power supply, climate (cooling) equipment, and housing, the share of embodied energy of semiconductors in general telecommunication equipment is not expected to be as high as with computers. The passive electronic components usually account for only 1% of total ini-tial embodied energy [11]. The rest of embodied energy can be summarized in the following groups: bulk and metal materials, conventional manufacturing, components manufacturing and assembly, and cables. The energies of supply chain, transportation, and setup of the equipment also have to be considered.

It should be noted that given approach focuses strictly on materials and manufacturing pro-cesses, consequently some parameters, such as research and engineering activities or software developments (that are also energy consuming), are disregarded, since they are hard to relate to specific processes or products. This is also an interesting direction for further research in the field of total LCA that have to be carried out in cooperation with equipment manufacturers.

4.2.3 Maintenance

The maintenance embodied energy is associated with maintaining, repairing, and replacing materials and components of the ICT equipment over its lifetime. The dashed line in Figure 4.1 shows that maintenance activities are performed consecutively at certain time periods, which can be interpreted as a linear function of time illustrated with the solid line on the same picture. The estimations of maintenance embodied energy slightly differs by the sort of ICT equipment, but in general, it can be estimated by 1% of initial embodied energy per year (maintenance power $P_{EMmaint} = E_{EMinit}/(100 \text{ years})$).

4.2.4 Importance of Lifetime

To be able to compare the manufacturing phase with the operation phase of the ICT equipment within its life cycle, it is essential to consider the time aspect. To do that, the lifetime for the studied products must be reflected. The lifetime of ICT equipment can vary, but it is usually in the interval between 5–15 years. Unfortunately, this corresponds to anticipated commercial lifetime which is in many cases sufficiently shorter than the technical lifetime. Since the new technologies supersede the previous one very expeditiously, the equipment is usually replaced before the end of it technical lifetime, what virtually extends the share of embodied energy in total energy during lifetime. This high-energy intensity of manufacturing combined with rapid turnover in equipment further encourages the need to rethink the suggested energy models. Assumption about lifetime of ICT device has a significant impact on the results of the share of embodied energy in total devices energy.

4.2.5 The Operating Energy

As it has already been noted in the introduction, ICT consumes approximately 600 TWh per year, corresponding to roughly 3% of global electricity consumption [5].

For further analysis, it should be noted that the operating energy E_{OP} can be divided into two parts: a constant part $E_{OPconst}$ representing the fixed energy consumed by ICT equipment that does not depend on the traffic load (i.e., in idle state it is the same as in case of fully loaded system): energy, used to power up a device, battery backup, and so on, and E_{OPlin}, that is linearly scaled with the transmission energy, representing the power needed to process the traffic, perform signal processing, cooling or to cover amplifier and feeder losses in cellular networks. ($E_{OP} = E_{OPlin} + E_{OPconst}$). Consequently, the linear relation of the operating power (P_{OP}) to its transmission power (P_{TX}) can be applied as $P_{OP} = a \cdot P_{TX} + b$.

4.2.6 The Total Energy Consumption Model

With regard to the conclusions of previous chapters, we propose a total energy consumption model of ICT device that includes both the embodied energy and the operating one as follows:

$$E = E_{EM} + E_{OP} = E_{EMinit} + E_{EMmaint} + E_{OPlin} + E_{OPconst} \tag{4.1}$$

The initial embodied energy is expended once in the initial production of the device, whereas maintenance embodied and operation energy accrues over the effective lifetime of the BS and can be expressed as $E_{OP} = P_{OP} \cdot T_{lifetime}$, $E_{EMmaint} = P_{EMmaint} \cdot T_{lifetime}$.

Sleep-mode or power-off strategies, suggested by some of past researches, will further decrease the operating energy in comparison with embodied energy. Similarly, shortening the commercial lifetime of the equipment will additionally alter this relation, strengthening the part of embodied energy.

The ratio between the embodied and operating energy is also given in Figure 4.1. Comparing with total energy, the share of embodied energy of information society devices is much higher (32%) as opposed to the representative products of urban (7%) and industrial (13%) societies.

4.3 Embodied/Operating Energy of a BS in Cellular Network – A Case Study

As we already noted, mobile telephony is the major branch in telecommunications sector, responsible for a half of its energy consumption. A detailed look inside the energy consumption of operating mobile telephony services reveals that only up to 10% is associated with the users equipment, whereas the remaining share of 90% is consumed by the network components [18], more specifically, around two-thirds are used by the BSs.

4.3.1 Overview of Past Studies in BSs Energy Modeling

For this reason, the cellular networks have attracted many researchers with the aim to improve their energy efficiency. We will draw attention to just a few of them that are related to the system-level architectures and features of cellular networks. The main idea of these studies is to save energy with power-off strategy.

The study [19] suggests to power off the underutilized BSs during the period of low traffic. To maintain the coverage with the reduced number of BSs, the transmission operating power of the active BSs has to been increased. The authors search for an optimal energy saving in cellular access network, neglecting the increment of transmission operating power, avoiding the optimization problem of finding the lowest operating energy with different number of BSs and their coverage. Using trapezoidal and real-environment daily traffic patterns, they have proved that the savings of the order of 25–30% in the operating energy can be achieved with an optimal power-off scheme employed.

Similarly, the work [20] strives to find the optimal cell size between large and small cell deployments. The optimization process between increased cell radius (increases transmission operating power) and reduced cell radius (higher number of required cells increases fixed operating power) was upgraded by using sleep-mode strategy, powering off the cells without active users, but maintaining the capacity of the system. Using a simulation, they have shown that the reduction of cell sizes improves the operating energy consumption ratio (the energy, needed to transfer a bit of information), whereas the energy consumption gain (the ratio between the energies of large and small cell deployments) remains constant or increases linearly with the number of cells when using the sleep-mode strategy.

The second main research stream (such as the work provided in Ref. [21]) improves the energy efficiency by employment of two-tier cellular access network. The main idea of two-tier networks is to extend the conventional macro sites with the deployment of femto/pico/microcells covering much smaller area with cell radius between several meters to several hundred meters. They have addressed the optimization problem between the energy consumption by the macrosites (high transmission operating energy) and the additional operating energy required by the employment of femto/pico/microcells. A BS sleep-mode strategy was once more proved to be a promising mechanism to eliminate the energy consumption of underutilized BSs.

The general idea of all the above-mentioned and other similar studies is to find a trade-off between a decrease in operating energy, based on reduced transmission power, and a consequently increased operating energy, caused by higher number of cells, where sleep-mode strategy can be used for further energy savings. A common suggestion of these researches is to improve the energy efficiency of cellular networks by using BSs with a reduced transmission power (reduced operating energy), but employing a larger number of them to maintain the coverage and capacity. On the other hand, the main limitation of the above-mentioned studies is that they neglect the embodied energy of cellular network's equipment.

Due to the continuous efforts of communication systems manufacturers to improve [22, 23] the operating energy efficiency of their systems on one side and the increasingly complex processes of semiconductor manufacturing on the other side, the embodied energy of cellular networks' equipment is obviously far from being neglected. The embodied energy can be as high as half of operating energy in the equipment's lifetime. The previously described suggestions for saving the operating energy can obviously be very boomeranging: to save some operating energy, much additional equipment has to be installed, requiring much more embodied energy. And it is also clearly evident that the embodied energy cannot be avoided by using the power-off strategies.

4.3.2 The Need to Rethink Previous Models

Although none of the previous researches have yet investigated the energy efficiency of cellular network including the embodied energy, many authors have proposed it as an important further research direction. In the introduction of their paper [20], the authors have suggested that "both operating as well as embodied energy need to be considered in the evaluation", but the latter was neglected in the further analysis. The same opinion is shared in Ref. [2], where authors discuss the introduction of femtocells as an enhancement to the architecture of cellular networks. Evaluating the environmental effects of this approach, they noted that the "impact of equipment manufacturing should also be taken into account." It should also not be neglected that this topic has recently attracted some research groups, standardization bodies, as well as research departments of cellular network equipment manufacturers, who have started to work toward the analysis of environmental impacts associated with delivering of their product, which will hopefully result in environmentally friendly outputs and solutions. All these facts urge for a revision of energy-saving models in cellular network [24, 25] – with the consideration of the embodied energy, further investigated in next subsection.

4.3.3 The Embodied Energy of a BS

With the methodology, given in Section 4.2.2 we have made a rough estimation of embodied energy for a general BS. The results are summarized in Table 4.1.

Table 4.1 The constituents of initial embodied energy for materials and processes during the production of a general BS

Material/process	Energy requirement (MJ/kg)	Energy (GJ)
Semiconductor devices (silicon wafers and integrated circuitry)	60,000–120,000	37.2
Printed circuit board manufacturing and assembly	300–500	4.3
Bulk materials (plastic, glass, and rubber)	20–140	5.3
Metals (aluminum, copper, steel, lead, and zinc)	100–400	7.0
Conventional manufacturing (cutting, welding, finish machining, and injection molding)	1–30	1.5
Telecom cables	50	2.3
Supply chain		5.8
Components manufacturing and node assembly		8.1
Transportation, setup		3.5
Total embodied energy of a base station		75.0

To further position these results, we have compared the estimated energy of BS with a powerful state-of-the-art computer server [26]. As expected, the embodied energy of BS precedes the embodied energy of computer server, but the results are in the same order of magnitude. The results are also in concordance with Ref. [4], claiming that approximately 20% of mobile equipment's energy is used by raw materials.

The embodied energy content of BS varies greatly with different construction types. This study is focused on macrocell BS only, albeit the equipment manufacturers provide heterogeneous assortment of micro/femto/picocell BSs varying in their power, dimensions, and the embodied energy. However, lower radiation power usually does not mean lower embodied energy; its share in total lifetime energy is usually even higher with the small cells' BSs comparing with the large ones.

4.3.4   The Operating Energy of a BS

The past analysis [27] has decomposed this energy among power amplifier, idling transceiver, power supply, cooling fans[5], transceiver power conversion, combining/duplexing, central equipment, transmit power, and cabling (see Figure 4.2 for the detail). The authors of previous works have suggested many different approaches to reduce the energy consumption of BSs which have been brought into use by the manufacturers with the aim to improve cellular networks' energy efficiency (the details and estimated reductions of energy consumption are given in brackets): improvements of BS hardware design (improvement of power amplifier and signal processing efficiencies, RF techniques and reduction of feeder losses, and application of free cooling, 40%), usage of several system level architectures (optimal cell sizes, two-tier networks: 30%) and features (resource allocations, sleep-mode strategies: 10–30%), differentiation of BS site solutions (indoor/outdoor solutions, modular solutions: 20%), and the usage of renewable energy sources.

The operating energy of a BS is easier to assess, since we can rely on the previously published data. The values for operating and peak power consumptions provided by BS manufacturers[6] are a good starting point. For a typical macrocell[7] installation, the figures for the net power of a BS can vary from 0.4 to 3.0 kW. In Ref. [18], the authors have estimated the average power consumption of equipment dating before the year 2000 as 1.1 kW. From this time, using the above-mentioned approaches the power consumption of BSs is claimed to be reduced substantially [28], to less than 500 W [22, 31]. Thus, we presume this number as average power consumption of state-of-the-art BS, used for modeling in our study. As the radio access in cellular network is intended to work 24/7, the annual operating energy consumption is roughly 15 GJ, resulting in 150 GJ in the estimated device's lifetime (estimated to 10 years).

The power consumption of every BS depends on traffic load and statistical spread. As it is detailed in Figure 4.1(e) with the dashed line the power consumption is a nonlinear function of time with some high-consumptive busy periods as well as periods with no power consumption

---

[5] This portion can vary due to different climate conditions.

[6] The data was of Ericsson, Nokia, Motorola, and Huawei were collected on their websites.

[7] For the sake of simplicity, this study evaluates single-tier (macrocell) architecture only. However, the employment of smaller (femto/pico/micro) cells is a subject of an ongoing research.

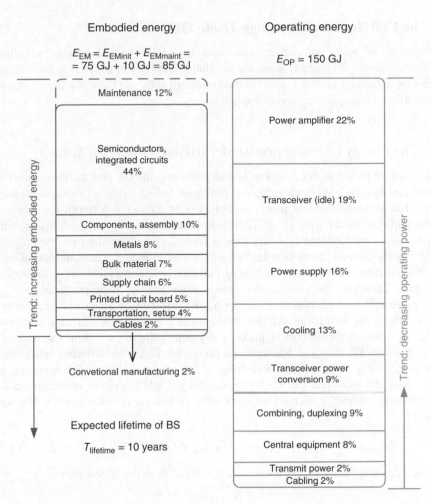

Embodied energy

$E_{EM} = E_{EMinit} + E_{EMmaint} =$
$= 75 \text{ GJ} + 10 \text{ GJ} = 85 \text{ GJ}$

Maintenance 12%

Semiconductors,
integrated circuits
44%

Components, assembly 10%

Metals 8%

Bulk material 7%

Supply chain 6%

Printed circuit board 5%

Transportation, setup 4%

Cables 2%

Trend: increasing embodied energy

Convetional manufacturing 2%

Expected lifetime of BS

$T_{lifetime} = 10 \text{ years}$

Operating energy

$E_{OP} = 150 \text{ GJ}$

Power amplifier 22%

Transceiver (idle) 19%

Power supply 16%

Cooling 13%

Transceiver power
conversion 9%

Combining, duplexing 9%

Central equipment 8%

Transmit power 2%

Cabling 2%

Trend: decreasing operating power

Figure 4.2   The proportion of embodied and operating energy during BS's lifetime, breakdowns, and trends

representing the possible sleep-mode/power-off intervals during low traffic. The average value used in our study can be represented as a linear function of time illustrated with the solid line.

The ratio between embodied and operating energies during the BS's life cycle is evident from Figure 4.2. The results of our estimation can successfully reflect in the aggregated data given by [4] where the authors found that the share of embodied energy in the mobile network's total energy was approximately 15% in 2001, 25% in 2005, and has grown up to 43% by the end of 2006. It also confirms our claim that the operating energy is being reduced on account of embodied energy. However, all these evidences ground an existence of relationship between the embodied and operating energy $E_{OP} = f(E_{EM})$, meaning that the higher embodied energy (such as improvements in BS hardware design) can result in reduced operating energy. This relation was not explored in our study, but it is definitely an interesting question for further research.

4.4   The Cell Number/Coverage Trade-Off

In this section, we apply our energy consumption model to two scenarios, excluding and including the power-off strategy. A simple simulation has been conducted to provide energy consumption optimization with respect to the number of cells and their coverage, considering the trade-off between operating and embodied energy.

4.4.1   The Energy Consumption Model Without Power-Off Strategy

The transmission power of BS antennas is dissipated into the air, and the transmitted signal is deteriorated by path loss, shadowing, and multipath fading effects in wireless propagation channels. The average receiving power decreases with distance r between the receiver and transmitter approximately with $1/(r)^\gamma$ [29], mainly accounting for the path losses, where γ is the path loss exponent that typically ranges from 2 to 5, depending on the propagation environment. Besides, all the mobile devices within the cell require a certain level of received signal-to-noise ratio (SNR), asserting at least minimum receiving power P_{min} for acceptable performance. Therefore, the transmission power of the BS P_{TX} is proportional to $P_{min}(r)^\gamma$, where r represents the cell radius. Considering the link budget for noise, shadowing, and other loss effects (as well as gains), the transmission power can be further scaled with the term $P_{min}(r)^\gamma$. For simplification, suppose a BS with radius $r_0 = 1$ km, the transmission power is $P_0 = 40$ W, then the transmission power for BSs with different cell radius can be expressed as $P_{TX} = P_0 \ (r/r_0)^\gamma$. Referring to the previous section, the operating power $P_{OP} = a \cdot P_0 (r/r_0)^\gamma + b$; and the energy model for the whole system consumption E_{system} for the scenario without power-off strategy of a certain area, covered by n BSs can be expressed as:

$$E_{system} = n \ (E_{EM} + E_{OP}) = n \ (E_{EMinit} + E_{EMmaint} + P_{OP} \cdot T_{lifetime}) \qquad (4.2)$$

where the operating power P_{OP} is a function of cell radius as described above.

4.4.2   The Number/Coverage Trade-Off

The deployment of a larger number of BSs with smaller size (cell radius), such as femto/pico/micro BSs [20, 30] will enable decreased transmit power of BSs as well as reduced operating power and electromagnetic radiation. However, the total power consumption of the system is the multiplication of the number of BSs and a single BS power consumption. As the number of BSs becomes large, the power consumption rises, since the embodied energy of a newly deployed BS adds to the total energy. In theory, there exists a trade-off between the number and the coverage of a single BS, translating into a trade-off between the embodied energy and the operating energy. In practical deployments, the position of the BS, the capacity, and the traffic requirements are always taken into account. However, to explore the number/coverage problem on a simple scenario, we assume the total coverage to be the multiplication of the number of active BSs and a single BS coverage. Under this coverage constraint, the optimal energy consumption is explored in simulation by considering the trade-off between operating and embodied energy consumptions.

4.4.3   The Energy Consumption Model with the Power-Off Strategy

The power-off strategy is an important energy-saving technique by adapting BS activities to traffic dynamics. When the traffic of a certain cell remains at low level, the BS can be shut down for operating energy-saving purposes. The other active BSs should increase their transmission range, thus with larger transmission power, to cover the entire area. When the traffic in a cell increases, the shutdown BS will be activated. Thus, when power-off strategy is applied, the operating power varies with the number of shutdown BSs or depends on the number of active BSs. The average traffic intensity in a day or year varies periodically between its peaks, sometimes assumed to have a trapezoidal or sinusoidal pattern [19]. Consider uniform distribution of users in the cells, and the random use of mobile devices, then the traffic patterns among different BSs are the same except that the traffic peaks are uniformly distributed during the period. Each BS in an area has a power-off probability p, proportional to the share of the low traffic period in the day. At a particular time, there may be M sleeping BSs, where M is a random variable satisfying binomial distribution: $\text{Prob}(M = m) = \binom{m}{n} p^m (1 - p)^{n-m} / (1 - p^n), 0 \leq m \leq n - 1$, and n again is the total number of BSs, covering the area. The energy model for the entire system consumption E_{system} for the scenario with sleeping strategy is expressed as follows:

$$E_{\text{system}} = n \cdot (E_{\text{EMInit}} + E_{\text{EMmaint}}) + \sum_{m=0}^{n-1} (n - m) \cdot \text{Prob}(M = m) \cdot P_{\text{OP}}(n - m) \cdot T_{\text{active}} \quad (4.3)$$

where, as distinguished from Eq. (4.2), the operating energy consumption is the probabilistic average over the distribution of random variable M; and the active time T_{active} of BSs can be estimated by $(1 - p) \cdot T_{\text{lifetime}}$.

When considering the power-off strategy, the operating power of BSs can be reduced by around 25% [2, 19]. The operating energy savings are significant, especially when more BSs are deployed. However, our study argues that the embodied energy consumption will be in larger proportion of the total energy consumption than in the case without power-off strategy. Reconsidering the number/coverage problem, if a large number of smaller BSs are deployed, the embodied energy consumption increase will be dominant in the total energy consumption. That means, the BS is manufactured with great energy cost while it is powered-off during most of its lifetime. Therefore, the consideration of power-off strategy is most necessary for the number/coverage trade-off problem resolution.

4.4.4   Simulation Results

In this section, the presented energy consumption models are evaluated in simple simulation scenarios in order to optimize the total energy consumption by exploring the trade-off between the operating energy and embodied energy. Consider a typical urban area of radius $R = 5$ km

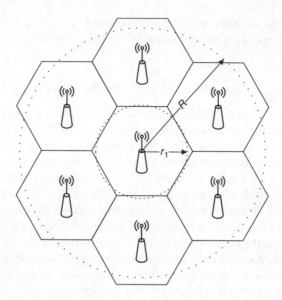

Figure 4.3 The simulation scenario topology

covered by n BSs with the same cell radius r as illustrated in Figure 4.3. During the simulation, the embodied energy is calculated using the estimated data in Section 4.3. The parameters used to evaluate the operating energy are given in Table 4.1 [21].

Parameter	Value
Path loss exponent γ	3.2
Power-off probability p	1/4
Operating power consumption model – parameter a	7.84
Operating power consumption model – parameter b	71.50
Embodied power model parameters	see Section 4.2
Transmission power P_0	40 W

Under this simulation scenario, the energy consumptions with different number of BSs or cell sizes are evaluated in Matlab, both including and excluding power-off strategy, to find the optimal number of BSs or cell size, as depicted in Figure 4.4. It is evident that an optimal number of BSs or cell size with minimal energy consumption exists. When only a small number of BSs with large cell sizes are deployed, the energy consumption is high; this is due to the increased operating energy with cell size. However, when a large number of BSs with small cell sizes are deployed, the embodied energy consumption dominants and leads to the increase in total energy consumption. The optimum is achieved with the trade-off between operating energy and the embodied energy. Compared with past research [32], where the optimal cell size/number of BSs is affected only by the fixed operating energy consumption part,

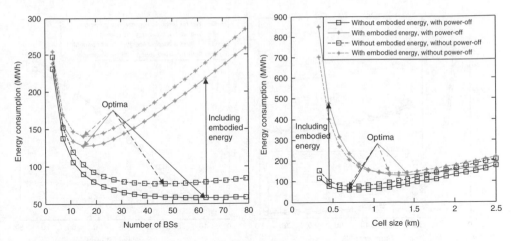

Figure 4.4 The optimal energy consumption of cellular network with respect to number of BSs (a) or cell sizes (b)

the embodied energy in our case has a much stronger effects when the cell size becomes small or the number of BSs becomes large. This finding averts from the suggestions that energy savings can be achieved with a larger number of cells with reduced transmission power. Besides, when sleeping strategy is applied, the operating energy savings will only slightly shift the optimum to higher number of cells due to the decreased operating energy consumption. However, in practical deployments, the power-off strategy is only possible, when the number of BSs is relatively large. This emphasizes the problem of the embodied energy and even raises its share in the total energy, when power-off strategy is applied.

Results on energy efficiency versus the number of BSs are provided in Figure 4.5. Since the system capacity of increases linearly with the number of active BSs [20], the energy consumption ratio (ECR) [33] will decrease with increasing number of BSs. The results ignoring the embodied energy are in accordance with the previous research [20]. However, the energy efficiency of both models, with or without consideration of the embodied energy, is almost the same, when the number of cells is small, but the consideration of the embodied energy reveals that several times more energy is required for the same capacity when the number of cells is large (detailed on the scaled part of Figure 4.5). This again calls for reconsideration of past suggestions in energy saving solutions.

4.5 Discussion and Future Challenges

This chapter extends the energy model of ICT device with the embodied energy constituent. It consists of the energy for production and maintenance of ICT devices, which as shown for the majority of complex electronic ICT devices it is far from being neglected. We have provided and explained a case study of such modeling for a Base station in Cellular network and have shown the important trade-off between operating and embodied energy consumption that provides guidelines for manufacturers, operators, and researchers.

The results given above provide an insight into optimizations considering both operating and embodied energy. It is evident, that the latter plays a very important role and should not simply

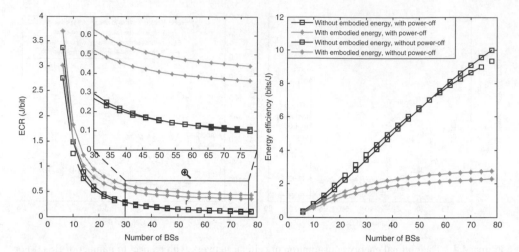

Figure 4.5 The ECR (a) and energy efficiency (b) of cellular network, covering the same area with different number of BSs

be neglected. However, there are still many open questions that manufacturers, operators, and researchers should address in future work.

The ICT equipment manufacturers should advocate total LCA, including the energy consuming activities of research, engineering, and software development. Cooperating closely with component suppliers, they should perform cradle-to-grave assessments of embodied energy for at least a selected set of their radio network product portfolio. This will be of great interest to operators and researchers in relation to network optimization and analysis, respectively. Furthermore, the up-to-date embodied energy estimations should be made publicly available, and simple to interpret and use, even for nonexperts in the field of LCA.

Moreover, the manufacturers should strengthen the awareness of embodied energy and the environmental impacts for the nonmass produced ICT equipment, as they do for mobile phones for example [11]. The manufacturers should inform the customers not only about the improvements in operating energy consumption but also the assessed embodied energy of the equipment should be provided. To ensure this outcome, the standards of embodied energy measurement and estimation should be made a matter of regulation and enforcement by the regulatory authorities.

Operators should perform a cost analysis for building new sites in terms of energy, expenditure, and environmental impacts. Also, the impact of different traffic intensive application on the energy efficiency should also be addressed.

Although there exist some obstacles for public and creditable embodied energy data, this is not an excuse to neglect its influence in energy efficiency modeling and optimizations. We hope this chapter provides further motivation in this perspective, although there remain many open research issues in the embodied energy modeling of ICT equipment.

There is also some evidence that operating energy is being reduced on account of embodied energy. To explore the relationship between the embodied and operating energy is an interesting question for further research.

Although the provided example focuses mainly on cellular networks as they are one of biggest consumers and attract much attention in the recent ongoing research, results of our study are general and can be easily applied to other fields of ICT. For instance, the proposed models are directly applicable to the wireless local area networks. The provided approach of embodied energy consideration may also target other energy efficiency concepts, requiring the deployment of additional hardware to save energy. The network connectivity proxying scheme [6], for example, employs external network proxies that virtually maintain presence for network computers but let them sleep and save energy while idle. This approach may require additional hardware and should be further evaluated by considering the embodied energy limitations. To avoid wasting with the embodied energy of immense proxies employed in thousands of small offices around the world, the implementations using the existing network equipment [6] (either on one of the network servers, implemented within the network interface controller or residing in the already-employed network devices, such as switches) should be greatly encouraged.

Acknowledgments

Dr. Humar would like to acknowledge the support of the research programme P2-0246 – Algorithms and optimization procedures in telecommunications, financed by the Slovenian Research Agency. Dr. Ge would like to acknowledge the support from the National Natural Science Foundation of China, contract/grant number: 60872007; National 863 High Technology Program of China, grant number: 2009AA01Z239; The Ministry of Science and Technology (MOST), International Science and Technology Collaboration Program, grant number: 0903. This research project was also partially supported by the Brain Korea 21 program under NRF, the Ministry of Education, Science, and Technology, the Korean government.

References

[1] W. K. Roth, F. Goldstein, J. Kleinman, Energy Consumption by Office and Telecommunications Equipment in Commercial Buildings, Arthur D. Little, International Energy Agency, Reference No. 72895-00, 2002.
[2] C. Forster, I. Dickie, G. Maile, H. Smith, M. Crisp "Understanding the environmental impact of communication systems", Ofcom Final Report, 2009, http://stakeholders.ofcom.org.uk/binaries/research/technology-research/environ.pdf.
[3] Gadgets and Gigawatts: "Policies for Energy Efficient Electronics", Report, IEA Publications Bookshop, OECD/IEA, 2009.
[4] Ericsson, "Sustainable energy use in mobile communications," White paper, 2007.
[5] G. Fettweis, E. Zimmermann, "ICT energy consumption – trends and challenges," in Proceedings of the 11th International Symposium on Wireless Personal Multimedia Communications, 2008.
[6] B. Nordman, K. Christensen, "Proxying: the next step in reducing IT energy use," Computer, vol. 43, no. 1, pp. 91–93, 2010. DOI: 10.1109/MC.2010.21.
[7] R. J. Cole, P. C. Kernan, "Life-cycle energy use in office buildings," Build. Environ., vol. 31, no. 4, 1996, pp. 307–317.
[8] M. Castro, J. Remmerswaal, M. Reuter, "Life cycle impact assessment of the average passenger vehicle in the Netherlands," Int. J. LCA 2003, vol. 8, no. 5, pp. 297–304.
[9] I. Nawaza, G.N. Tiwarib, "Embodied energy analysis of photovoltaic (PV) system based on macro- and micro-level," Energy Policy, vol. 34, no. 17, 2006, pp. 3144–3152.
[10] E. D. Williams, "Energy intensity of computer manufacturing: hybrid assessment combining process and economic input-output methods," Environ. Sci. Technol., vol. 38, no. 22, 2004, pp. 6166–6174.

[11] P. Singhal, "Life cycle environmental issues of mobile phones," Stage I Final Report, Integrated Product Policy Pilot Project, Nokia, Espoo, Finland, 2005.

[12] ISO 14040:2006, "Environmental management-life cycle assessment-principles and framework," http://www.iso .org/iso/catalogue_detail?csnumber=37456 [Accessed: 11.7.2013].

[13] S. Frey, Development of New Ecological Footprint Techniques Applicable to Consumer Electronics, Brunel University, Middlesex, 2002.

[14] P. Singhal, S. Ahonen, G. Rice, M. Stutz, M. Terho, H. Wel, "Key environmental performance indicators (KEPIs): a new approach to environmental assessment," in Proceedings of Electronics Goes Green 2004+, Berlin, Germany, p. 697–702.

[15] J. Malmodin, L. Oliv, P. Bergmark, "Life cycle assessment of third generation (3G) wireless telecommunication systems at Ericsson," in Proceedings of Second International Symposium on Environmentally Conscious Design and Inverse Manufacturing, Tokyo, pp. 328–334, 2001.

[16] N. Duque Ciceri, T. G. Gutowski, M. Garetti, A Tool to Estimate Materials and Manufacturing Energy for a Product, IEEE/International Symposium on Sustainable Systems and Technology, Washington DC, 2010.

[17] J. Li, Z. Wu, H. C. Zhang, "Application of neural network on environmental impact assessment tools," Int. J. Sustainable Manuf., vol. 1, 2008, pp. 100–121.

[18] C. Schaefer, C. Weber, A. Voss, "Energy usage of mobile telephone services in Germany," Energy, vol. 28, no. 5, 2003, pp. 411–420.

[19] M. A. Marsan, L. Chiaraviglio, D. Ciullo, M. Meo, "Optimal energy savings in cellular access networks," IEEE International Conference on Communications Workshops, Dresden, Germany, 14–18 June 2009.

[20] B. Badic, T. O'Farrrell, P. Loskot, J. He, "Energy efficient radio access architectures for green radio: large versus small cell size deployment," IEEE 70th Vehicular Technology Conference Fall, Anchorage, AK, 20–23 September, 2009.

[21] F. Richter, A. J. Fehske, G. P. Fettweis, "Energy efficiency aspects of base station deployment strategies for cellular networks," IEEE 70th Vehicular Technology Conference Fall, Anchorage, AK, 20–23 September, 2009.

[22] "Huawei Launches Solution to Cut Base Station Power Consumption", 23.1.2008, http://www.3g.co.uk /PR/Jan2008/5656.htm.

[23] A. Larilahti, "Energiatehokkuus ja yritysvastuu kilpailutekijänä", Nokia Siemens Networks, 7.5.2009, http: //www.wwf.fi/wwf/www/uploads/pdf/anne_larilahti.pdf.

[24] I. Humar, X. Ge, L. Xiang, M. Jo, M. Chen, J. Zhang "Rethinking energy efficiency models of cellular networks with embodied energy". IEEE Netw., 2011, vol. 25, no. 2, pp. 40–49.

[25] X. Ge, C. Cao, M. Jo, M. Chen, J. Hu, I. Humar "Energy efficiency modelling and analyzing based on multi-cell and multi-antenna cellular networks". Trans. internet inf. syst. 2010, vol. 4, no. 4, pp. 560–574.

[26] "Computer consumption facts," http://legacy.wattzon.com/stuff/items/kh3hje0e2whxst5s1hkfuri8i /k5x8v0advdf662bzv7odrwolbb.

[27] H. Karl, "An overview of energy-efficiency techniques for mobile communication systems," Telecommunication Networks Group, Technical University Berlin, Tech. Rep. TKN-03-XXX, 2003.

[28] A. Larilahti, "Energiatehokkuus ja yritysvastuu kilpailutekijänä," Nokia Siemens Networks, Ilmasto & yritykset-seminaari, Helsinki, 2009.

[29] X. Cheng, C.-X. Wang, H. Wang, X. Gao, et al., "Cooperative MIMO channel modeling and multi-link spatial correlation properties," IEEE J. Sel. Areas Commun., vol. 30, no. 2, pp. 388–396, 2012.

[30] D. Mavrakis, "Do we really need femto cells?" Vision Mobile, http://www.visionmobile.com/blog/2007/12 /do-we-really-need-femto-cells/.

[31] P.-J. Chung, "Green radio-the case for more efficient cellular base-stations," UK-Taiwan ICT Workshop, Smart Grid & Green Communications, Taipei, Taiwan, 2010.

[32] O. Arnold, F. Richter, G. Fettweis, O. Blume, "Power consumption modeling of different base station types in heterogeneous cellular networks," in Proc. of 19th Future Network & Mobile Summit 2010, Florence, Italy, 2010.

[33] ECR Initiative: "Network and telecom equipment — energy and performance assessment, test procedure and measurement methodology," http://www.ecrinitiative.org.

5

Energy-Efficient Base Stations

Alberto Conte

Alcatel-Lucent Bell Labs, Centre de Villarceaux, Nozay, France

5.1 Introduction

With the explosion of mobile Internet applications and the subsequent exponential increase of wireless data traffic, the energy consumption of cellular networks has rapidly caught the attention of the entire telecommunication community: industrials, operators, academics and government institutions. One of the first actions taken has been to monitor and understand where and by which cellular equipments the energy is consumed. Several studies have been conducted in parallel (e.g. [1 3]), and while the figures may slightly differs, all come to the same conclusion: whatever the technology is used (UMTS, HSPA and LTE), the major part of the energy ($\sim$50–60%) of a mobile network is consumed by the radio access network (RAN), and in particular by the set of base stations, followed by the core network ($\sim$30%), and data centres ($\sim$10%). The impact of the base stations comes from the combination of the power consumption of the equipment itself (up to 1500 W for a nowadays macro base station) multiplied by the number of deployed sites in a commercial network (e.g. more than 12000 in the United Kingdom for a single operator [3]). In order to effectively improve the energy efficiency of the future mobile networks, it is thus important to focus the attention on the base station. This chapter aims a providing a survey on the base stations functions and architectures, their energy consumption at component level, their possible improvements and the major problems that must be faced in order to make such improvement effective. When possible, examples from the 3GPP LTE technology are provided, in order to help the reader to map generic concepts to real-life networks. However, it is not the intention of this chapter to present in detail the LTE technology. The reader interested in an extensive presentation of this technology can refer to Ref. [4].

Green Communications: Principles, Concepts and Practice, First Edition.
Edited by Konstantinos Samdanis, Peter Rost, Andreas Maeder, Michela Meo and Christos Verikoukis.
© 2015 John Wiley & Sons, Ltd. Published 2015 by John Wiley & Sons, Ltd.

5.2   BS Architecture

5.2.1   *Generic Cellular Network Architecture*

The mobile networks are generically decomposed into core network (CN) and RAN. As concrete example, Figure 5.1 shows the functional architecture of LTE networks as specified by 3GPP; here, RAN is called evolved universal terrestrial radio access network (eUTRAN, [5]) and the core network is called Evolved Packet Core (EPC, [6, 7]).

The core network interfaces the mobile network with the "external" world, allowing the mobile terminals to access the public Internet, specific applications (e.g. VoIP), and private networks (e.g. corporate networks). At transport layer, the CN is in charge of providing the IP anchoring for the user equipment (and thus its IP address) such that the UE peers (e.g. a video server) are not impacted by the terminal mobility; all the data received from or destined to the UE goes through the anchor point (PDN Gateway in 3GPP LTE), whatever the position and speed of the UE. It is the job of the core network to track the movements of the UE and to forward its traffic to right position, in general using some type of IP tunnelling (e.g. 3GPP uses GTP [8]). Thanks to the anchoring function, the core network is also well suited to perform traffic control and shaping, allowing or denying the access to specific services in accordance with the subscriber profile. At control layer (MME, HSS and PCRF functions in 3GPP LTE), the core network in charge of managing the subscriber profiles and to perform authentication, authorization and accounting (AAA) of each user.

The core network does not handle any RF aspects specifics to the radio channel, its action stops at IP level (layer 3). It is the role of the RAN to take in charge all the radio specificities and to hide them from the core network. It handles the signalling and data exchanges with the mobile terminal (or User Equipment, UE) over the radio channel on one side, and interfaces with the core network, through the backhaul, on the other side. In modern architectures (e.g. LTE), the RAN is composed essentially by the base stations, as single equipment; other architectures, based on multiple elements, are also possible and have been used in the previous generation of cellular networks, for example in 3G/UMTS networks (with NodeB and RNC)

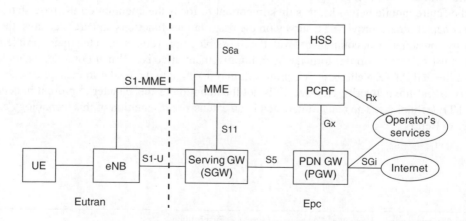

Figure 5.1   3GPP LTE architecture

or in 2G/GSM networks (with BTS and BSC). In the following only the single-element case is considered, the extension to the multi-elements case being straightforward.

5.2.2 Base Station Functions

The identification of the functions implemented by the base station is important in order to understand its hardware and software architecture, introduced in the next sections.

The role of the base station is primarily to take in charge the connection with the user equipments through the last wireless hop, providing all the necessary functions which include the data delivery, the mobility management, the overall radio resource management and the assistance to the UEs for a wide range of operations, such as network discovery, attachment and paging.

In the user plane, the role of the BS is to forward to the right destination (UE) the data received from the core network in the downlink direction (DL: BS → UE) and to perform the reverse operation on the uplink (UL: UE → BS). More specifically, the base station user plane comprises of the following:

- L1 or PHY (physical) layer, including both digital and analogue signal processing, such as symbol and carrier modulation and demodulation, signal amplification and transmission and reception through the antenna system – in 3GPP LTE, this corresponds to implementing the OFDMA (downlink) and SC-FDMA (uplink) PHY layers, whereas in 3G/UMTS this corresponds to implementing the WCDMA PHY layer.
- The medium access control (MAC) layer, in charge of regulating (scheduling) the access to the radio resources (time and frequency spectrum) among the multiple users, QoS levels and flows, also including signalling and control messages – in 3GPP LTE, this layer also includes the hybrid-ARQ (HARQ) functionality and is described in Ref. [9].
- The (radio) link control, including segmentation and reassembly, to adapt the upper-layer packet size to the PHY layer, and retransmissions for error-free transmissions – in 3GPP LTE, this is called the radio link control (RLC) [10].
- Additional PHY agnostic functions, such as security (integrity and ciphering) and header compression – in 3GPP LTE, this is performed by the Packet Data Convergence Protocol (PDCP) [11].
- In single-element RANs, there is a need for packet buffering, forwarding and reordering between peer BSs during mobility – in 3GPP LTE, this is also performed by the PDCP [11].
- On the network side, the backhaul interface (e.g. Ethernet or microwave, depending on the deployment scenario) and the IP (tunnelling) stack (GTP in 3GPP LTE).

In the control plane, the base station relies on the user plane stack (L1, MAC, RLC and PDCP in 3GPP LTE) for the transport of signalling and control messages over the air interface. The BS control plane is in charge of a wide range of tasks, among which broadcast of system information, assistance to network detection and selection (e.g. synchronization), connection control (including security key generation), channel quality measurements configuration and reporting (of serving and neighbour cells), radio admission control, mobility control (of both idle and active UEs), paging, radio link failure detection and recovery, and so on. In addition, the base station is also in charge of transparently transporting the control plane exchanges

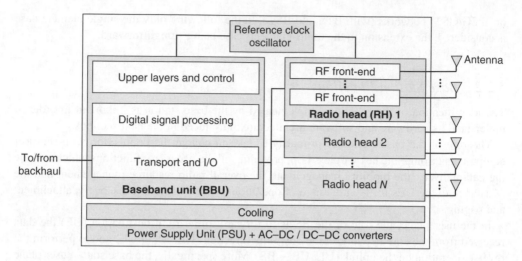

Figure 5.2   Schematic BS architecture

that take place between the UEs and the core network, such as the UE authentication and authorization, or the IP address assignment.

5.2.3 Generic BS Internal Architecture

A very large set of base stations exist in the market, differing by the size, form and internal architecture. This is due by the (co)existence of several cellular technologies, form factors (e.g. macros, micros and femtos; see Section 5.2.4) and telecom equipment makers, each one using its own design. It is thus impossible to provide a unique yet detailed description of a cellular base station; the technical choices to build them are too varied. In order to have a tractable analysis, this section proposes a simplified BS internal architecture, generic enough to be applicable (i.e. adaptable) to describe all kind of base station (Figure 5.2). It will be used in the following of this chapter to analyze the possible choices at component level, the obtained power consumption, and to study the possible evolutions towards more energy-efficient equipments.

The proposed architecture describes the base station in terms of the following five main building blocks:

- The baseband unit (BBU), in charge of all the digital operations
- One or more radio heads, in charge of the analogue operations and working on one or more antennas (e.g. for multi-antenna schemes, such as MIMO)
- A controlled oscillator locked on a GPS source or equivalent in charge of providing a very precise reference clock to the full system, BBU and RHs. For indoor small-size BS deployments (like femtos), where GPS signal is not available and backhaul connectivity not reliable enough to transport a stable clock reference, a stabilized oscillator (e.g. a temperature compensated crystal oscillator, TCXO) can be used
- The power supply unit, also including the different AC/DC and DC/DC converters (base stations normally use an input voltage of 48 V)

- The cooling system, which may vary from air conditioning to fan and even passive cooling in case of very low power base stations (such as femtocells).

The *Baseband Unit* (BBU) is the main block of the digital part of the base station. Referring to the previous section, the BBU is in charge of all the functions from L1/PHY (excluding the analogue part) up to the MAC and network layer, in both user and control plane. It is also in charge of the overall base station control, including O&M. The BBU elementary building blocks are given as follows:

- Transport and I/O layer supporting the different transport protocol stacks (such as IP and IP-tunnelling, GTP in LTE) and including:
  - The backhaul/backbone interface, which can vary from a simple Ethernet (e.g. for femtocells) to optical or microwave link. It is in charge of receiving/sending the IP packets (data and control) from/to the core network and also between the base stations in case of peering (such as in LTE, through the X2 interface).
  - An internal switch in charge of all the I/O that take place during the data/signal processing. Modern base stations abandon the legacy bus architectures to move towards more performing and flexible switches.
  - The in-phase and quadrature (I/Q) interface, used to exchange the digital samples of the baseband signal between the BBU and the radio head(s). This interface can be fully internal (e.g. directly exploiting the I/O switch) or external, e.g. in case of remote radio Head architectures (see Section 5.2.4). Open Base Station Architecture Initiative (OBSAI, [12]) and Common Public Radio Interface (CPRI, [13]) over an optical link are widely used standards for this interface.
- The digital signal processing (or baseband processing), performing the L1/PHY (digital) operations. In the downlink direction, it modulates the MAC-scheduled IP packets into the final baseband signal, described as I/Q digital samples. The digital samples of the baseband signal are then passed to the RH in order to create the final analogue signal that will be transmitted over the air. In the uplink direction, it demodulates the digital samples of the received baseband signal (passed by the RH) in order to extract all the transported information (data and control). More specifically, the most common operations of this block are (non-exhaustive list): FFT/iFFT, channel coding/decoding (turbo/Viterbi), channel estimation, synchronization, MIMO pre-coding, signal detection, digital filtering, up/down sampling and equalization. These operations are very demanding in terms of processing power, and specific hardware accelerators are commonly used for the most critical of them, such as turbo coding/decoding and FTT/iFFT.
- The upper layers and control, including:
  - Shared Memory (and controllers, e.g. DDR 2/3), used by the other blocks to temporary store the data during the different BBU operations. This also includes the buffering of the received data (IP packets) waiting to be sent to the UEs.
  - The "MAC and above" processing (upper layers), including link control (RLC in LTE), security and header compression (PDCP in LTE). In particular, the MAC is in charge of scheduling, that is, selecting which of the packets waiting in the internal buffer(s) will be transmitted in the next transmission slot (scheduling). Packet handling, scheduling and encryption are the most processing greedy operations of this block, with encryption often requiring a specific hardware accelerator.

○ The control and O&M part is in charge of the overall platform control, as well as the support to the remote base station management systems, allowing monitoring and configuring the equipment from the operators O&M platform.

The *Radio Head* (RH) is the analogue part of the base station. Each BS can have one or more radio heads, depending on the number of supported sectors (typically one radio head per sector), carriers and technologies. Each radio head in turn can have one or more antennas, depending on the supported MIMO modes. For example in LTE, MIMO 2×2, requiring two antennas per RH, is currently the default configuration, but higher antenna diversity schemes are supported. Figure 5.3 depicts a generic internal architecture of a radio head, generically composed of the following:

• I/Q interface, used to exchange I/Q digital samples with the BBU
• D↔A Converters, also including radio-head-related digital processing, such as the digital pre-distortion (DPD)
• RF transceiver (RF up/down conversion)
• RF amplifiers (power amplifier, PA, in downlink and low-noise amplifier, LNA, in uplink)
• Antenna interface, including duplexer and high-frequency filters (to split and isolate transmitted and received signals over the same antenna and RF cable), and the RF cable connection to the antenna (feeder or jumper)
• Antenna(s)

In the downlink direction, the digital-to-analogue converter (DAC) is in charge of constructing the analogue baseband signal starting from the digital I/Q samples received from the BBU. The baseband analogue signal is then passed to the transceiver, in charge of translating (up-converting) it into the passband signal, centred around the targeted high-frequency carrier (e.g. 2.6 GHz). The high-frequency passband signal is then amplified by the power amplifier and transmitted to the antenna via the specific interface. From energy consumption point of view, the power amplifier is the most consuming element of a radio head.

The analogue DL chain is not exempt from introducing signal distortions. In particular, high-power amplifiers can have non-linear in–out transfer. For this reason, a loopback (dashed

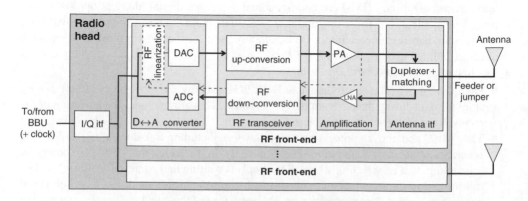

Figure 5.3   Radio head schematic architecture

line in Figure 5.3) is sometime introduced (especially in high-end / high-power macro-base stations); the PA output signal is sent back (through the uplink chain) in order to detect and correct unpredictable distortions. This is done by a (optional) RF-specific digital processing (RF linearization), functionally just before the DAC conversion, and usually including crest factor reduction for decreasing the input signal peak-to-average power ratio (PAPR) and DPD for increasing the PA linearity.

In the uplink direction, the received high-frequency passband signal is filtered and amplified by the low-noise amplifier, before being passed to the transceiver for down-conversion to baseband signal. Several down-conversion techniques can be used, the most common being the superheterodyne and the direct-conversion.

In superheterodyne architectures, the high-frequency signal is converted to a lower intermediate frequency (several successive IF stages are also possible) before obtaining the final baseband signal. Superheterodyne is the preferred choice for high/mid-power base stations (macro and micro BSs). Direct-conversion [14] (or zero-IF) is a simpler technique where the baseband signal is immediately obtained with a single step. Direct-conversion is commonly used in low-power base stations (pico and femto BSs) [15].

In the last stage of UL analogue processing, the baseband signal is passed to the analogue-to-digital converter (ADC), where it is sampled and quantized in order to obtain I/Q samples describing it in the digital domain. The I/Q samples are then sent to the BBU for information extraction (demodulation).

5.2.4   Types of Base Station

As said, there exist a wide set of base stations on the market, differing by the size (i.e. emitted power), spatial-reuse strategy (sectorization), supported carriers and technologies and internal architecture.

Regarding the emitted RF power, base stations are often classified into (see Table 5.1) *large cells* (macro and micro base stations), usually deployed at high heights, such as on top of building roofs or telecom towers, and *small cells* (pico and femto base stations), deployed at ground level (e.g. at home or on top of street lamps).

Macro base stations are deployed to provide coverage over wide areas, several hundreds of metres to few kilometres, depending on the deployment environment (e.g. urban vs. rural), whereas micros are typically used in urban environments to fill smaller spots (few hundreds of meters) or to provide additional capacity in crowded environments (e.g. stadium and large shopping malls). Pico base stations can serve a very limited amount of users (10–20

Table 5.1   Macro/micro/pico/femto.

BS type	N_{sec}	P_{max} (W)	P_{max} (dBm)
Macro	3	40.0	46
Micro	1	5.0	37
Pico	1	1.0	30
	1	0.25	24
Femto	1	0.1	20

at maximum) over a short range area (few tens of meters, maximum ~150 m), whereas femtocells are usually deployed in indoor/in-house environments, serving less than 10 users.

The architecture of large cells is driven by the technical performances (e.g. emitted power, number of parallel active users, multi-antenna schemes), resulting in relatively big and expensive equipments. Due to the high emitted RF power, the requirements on power amplifiers are very challenging to meet; high power over a large linearity region (especially in the case of orthogonal frequency-division multiplexing (OFDM) systems, presenting a high crest factor or PAPR) requires top-class components assisted by DPD and crest factor reduction processing. Even if some alternatives exist, (multi-stage) superheterodyne architectures are the consensual choice for the transceiver. In current large cells, the digital part (BBU) is usually build around several components or even cards, tailored to the specific tasks, such as hardware accelerators and DSPs for digital processing and ciphering, and more generic processors and cards for control and upper layers. However, new architectures tend to move towards system-on-chip solutions, integrating all the BBU functions into a single component. High emitted power translates also into high consumed power and thus high temperatures to dissipate, requiring expensive (and energy consuming) cooling systems, such as air conditioning or at least fan-based free-air circulation (in micro cells). Large cells can support complex multi-antenna diversity schemes, such as MIMO 2×2, 4×4 and in future even more. This implies 2, 4 or more antennas per radio head. Since the antennas of large cells are commonly mounted at high heights, a feeder cable is usually used to connect them to the BS site (e.g. in the basement of the building).

The architecture of small cells is primarily driven by size and cost; since the working conditions are easier (limited RF power ranges), the power amplifiers are often simpler, cheaper and less efficient than in the large cells case and usually do not require any linearization pre-processing (and thus, no loopback in the RH). Transceivers are based on simple direct-conversion solutions, such as zero-IF. System-on-chip solutions are commonly used for the digital part, integrating all the components of the BBU. Cooling is performed by small free-air fans, or by simple passive dissipation in the smallest equipments (femtocells). Small cells have a very limited number of antennas: one or maximum two for MIMO 2×2 schemes. Since the base station site and the antenna are co-located, the antenna connection is usually a loss-free jumper, or even a direct connector.

Base stations also differ by the number of supported sectors per site (Figure 5.4). Macro cells are typically multi-sectors, mostly 3-sectors with roughly 120° angular width. This implies three radio heads for each base station, each one serving a single sector with directional antennas. Micro, pico and femto base stations are typically single sector with omnidirectional antennas.

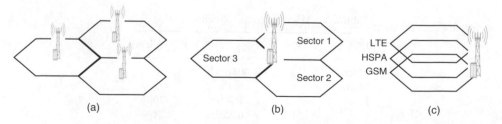

Figure 5.4 Types of base stations (BS) (sectors and carriers). Single-sector (a) versus tri-sector (b) versus multi-carrier or multi-technology (c) BS

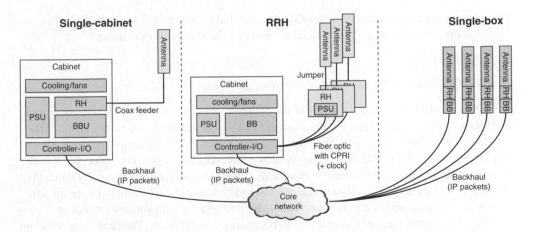

Figure 5.5 Types of BS (single cabinet versus RRH versus in-a-box)

Base stations (and in particular macros and micros) can also support multiple carriers, as well as multiple standards and technologies, such as GSM + HSPA + LTE (Figure 5.4(c)). This allows BS site reuse, reducing the need of acquiring new sites when deploying a new technology (e.g. LTE). Modern base stations are able to support multiple carriers and technologies on a single radio head (per sector).

Finally, three main base station architectures can be identified nowadays (Figure 5.5): the classic single-cabinet architecture, the remote radio head (RRH) and the single-box (or all-in-one) small architecture.

The *single-cabinet* architecture is the classic large base station architecture: all the elements of the BS (except the antennas) are packed into a single cabinet, on one or several blades, at the BS site (e.g. the basement of building). The antennas in turn are mounted at high height (e.g. rooftop) and connected to the base station site through a feeder cable. The drawback is that long feeder cables introduce important signal attenuation, commonly approximated at −3 dB, meaning that half of the signal power is lost in the cable.

Remote Radio Head (RRH) architectures have been introduced to mitigate the problem. In RRH, the radio head is separated from the BBU and mounted close to the antenna. This drastically reduces the distance between RRH and antenna and thus the signal attenuation. The connection between BBU and RRH is based on loss-free optical fibres and CPRI standard [13]. Remote radio head is a first example of "green" base station architecture, since its primary goal is to reduce energy wasting. Its draw-back is a diminished accessibility for maintenance and higher wind load for the antenna tower or mast.

Single-box (or all-in-one) architecture is the natural choice for small base stations. In this architecture, all the BS components, including the antennas are collocated in the same limited-size box (in a way very similar to WiFi access points).

5.3 Base Station Energy Consumption

Several studies have been recently carried on in order to better understand the power consumption of base stations at component level and the impact of traffic load variations on it [14–16].

This section summarizes the main results of these studies and proposes an analytical power model, useful when studying and simulating energy-related aspects of cellular BS.

5.3.1 Analysis of Energy Consumption at Component Level

All the detailed analyses (e.g. [14, 15]) of the power consumed by a base station at component level show similar results: the most consuming blocks (see Section 5.2.3) are the power amplifier, the signal processing (digital and analogue), the power supply unit (especially the AD/DC and DC/DC converters) and the cooling system. However, there exist huge energy consumption differences across the different components. Moreover, such differences vary depending on the type (size) of the base station (macro, micro and femto/pico, as shown in Figure 5.6.

In macro base station, the power amplifier is by far the most consuming component. It participates for approximately 60% to the total BS energy consumption. The remaining 40% are shared between the signal processing (with the digital part consuming two times more than the analogue one), the cooling system (requiring air conditioning) and the power supply (including AC/DC and DC/DC conversion). It must be noted that in some cases, the air conditioning can account for up to 18% of the total BS power consumption. The reason why the power amplifier is the most consuming component in macro BS comes from the need of high RF output power combined with the intrinsic low efficiency of power conversion. Moreover, since macro base

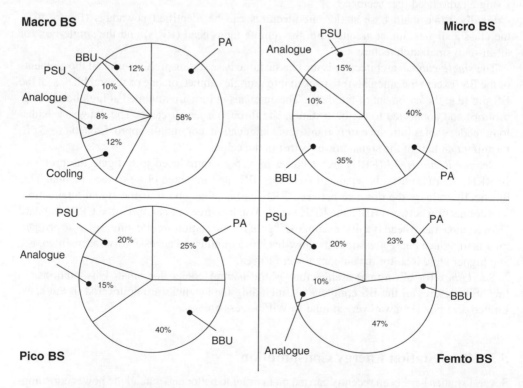

Figure 5.6 Per component energy consumption breakdown for different types of BSs

stations commonly have several sectors (and antennas per sector), several power amplifiers are necessary, one per RF chain (RF front end).

In micro BSs, the required output power and the number of power amplifiers is lower than in the macro case. As a consequence, the power amplifier still accounts for the largest share of energy consumption ($\sim$40%), but the digital processing (BBU) is now at an equivalent level. Since a simple fan is usually sufficient, the cooling impact is largely reduced. The remaining share ($\sim$25%) is shared between the power supply unit ($\sim$15%) and the analogue part ($\sim$10%).

It appears evident that to reduce the energy consumption of large cells (macro and micro) the improvement efforts must be focused on the power amplifier and digital signal processing. For macro base stations efforts in improving heating dissipation and cooling techniques are also very important, with the ambitious objective of replacing highly-consuming air conditioning with much greener fan-based systems.

In small base stations (pico and femto), the power consumption sharing is radically different. The digital processing (BBU) becomes the most consuming block (40–50%), followed by the power amplifier (20–25%). The PSU (power supply unit) accounts for around 20%, with the remaining 10–15% consumed in the analogue part. Cooling (passive) does not impact the power consumption repartition. To reduce the energy consumption of small cell, the efforts must be focused on reducing the impact of BBU and, on a minor part, of PA. Power supply also requires an attentive engineering. The major challenge however comes from the fact that small cells are subject to strong price pressure, making impossible to use high-end high-efficient components.

Finally, it must be noted that in LTE base stations (whatever their size, large and small) the operating conditions of the power amplifier are worsened by the OFDM modulation scheme which has a high PAPR. This forces the amplifier to operate in a linear region between 6 and 12 dB lower than the saturation point (which is in general the optimal efficiency point of a PA) [5]. This reduces the adjacent channel interference (ACI), but increases the PA inefficiency, and thus its power consumption.

5.3.2   Impact of Load Variations

It has been shown that the data traffic load in a cellular network varies during the day, depending on the time [17–20]. For example, in a business (office) area, the data traffic is very high during the day, but very low during the evening and night period. On the contrary, mobile traffic in residential areas tends to be concentrated in the evening and early night hours. However, whatever the deployment area (residential, business, urban, rural, etc.), it is possible to approximate the daily load variation with simple yet realistic models, summarized in Figure 5.7 (100% = average load).

It is thus important to analyze the power consumption of a base station depending on its non-constant load. The best behaviour (from energy consumption point of view) would be a perfect linear one, with zero energy consumption at 0% load (i.e. when the base station is neither sending nor receiving any data from users) and maximum at 100% load. However, the reality is far from this. As an example, Figure 5.8 shows the power consumption of a real LTE macro base station (without any energy-efficiency feature), as function of its load, defined as the number of used resource elements (RE) with respect to the maximum REs available (600) per TTI (here: 600 REs per 1 ms). Since the load can vary per sector, the figure shows the energy consumption for one single sector. In order to keep the result generic, the power

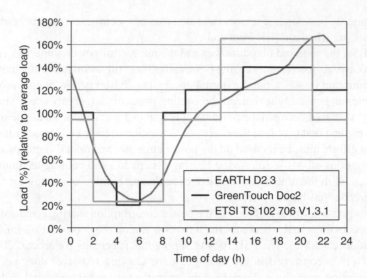

Figure 5.7   Mobile traffic variation profiles from [17] (label EARTH D2.3), [18, 20] (label GreenTouch Doc2), and [19] (label ETSI TS 102 706 V1.3.1)

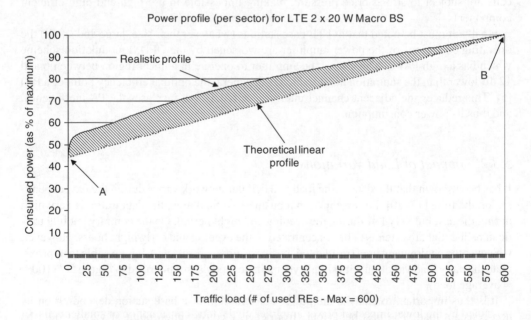

Figure 5.8   Example of BS power consumption as function of traffic load

consumption is normalized with respect to the maximum consumption (per sector). As an example, the consumption of a tri-sector, 2×20 W (MIMO) LTE macro BS is approximately 1500 W, with consumption close to 500 W per sector.

By analyzing Figure 5.8, several important observations can be done on the power consumption of a macro cell (note that the following does not apply in the same way to micro, pico and femto cells):

- The energy consumption of the macro BS effectively grows with the traffic load.
- The growth is not linear; there is an initial step when moving out from 0%. In addition, the slope is steeper at low load than at high loads.
- There is a huge offset at zero load; the energy consumed by a macro BS not serving any user is approximately half of the energy consumed at full load. This translates into very low-energy efficiency at low loads.

The non-linearity of the of the slope implies that it is better to either transmit at full load/power (point A on Figure 5.8) or to not transmit any data (point B). In fact, any other operating point (i.e. the real curve) is always higher (more consuming) than the theoretical linear curve (dashed line). This observation is (one of) the reason at the base of energy-aware scheduling techniques presented in Section 5.4

Even more important, from an energy consumption point of view, is the offset at 0% load (point A). There are several causes to this offset. First, as any real electronic equipment, the base station consumes energy just for the reason of being switched-on: circuits must be power supplied and thus energy is consumed. It is impossible to have ideal equipments that consume absolutely no power when turned-on, even if they *do nothing*. In addition, the BS is always active, even a 0% load. In fact, cellular technologies require that the BSs always broadcast signalling and control channels to assist mobile terminals in important operations, such as detection of service availability, acquisition of system characteristics (e.g. used MIMO scheme), mobility and paging. The mandatory emission of signalling (even when no users are present) is often called *coverage tax*, as it can be seen as the cost of constantly advertising the service availability. It is estimated that approximately 10% of the total RF power is constantly spent in signalling. Figure 5.9 shows the problem on an LTE frame structure; the resource element (squares) occupied by signalling and control channels (even when no user is present) is greyed. It can be observed that an LTE base station is obliged to transmit signalling (e.g. pilots) four times per millisecond (i.e. every 250 us) at least.

The power consumption profile as function of the traffic load drastically varies depending on the type of base station [17, 21]. This is summarized in Figure 5.10, where it can be observed that the dependency from the traffic load is high on large cells but almost non-existent in small BSs (the femtocell power consumption varies from 11 to 9 W approximately). This can be explained by the lower impact of power amplifier on the total consumption of smaller cells. In addition, as the power amplifiers on small cells must be cheap, their performances are often poor, with a flat response in term of power consumption. On the other hand, and whatever the type of BS, it can be noted that the energy consumption of all the other components poorly scale with load. Only the BBU is slightly impacted by the load, but this impact is today very limited. This provides an additional investigation direction for improving the energy efficiency of future base stations. Finally, a (small) non-linearity of the power profile is visible in macro- (as expected) and micro base station. In small cell the profile is linear (and almost flat).

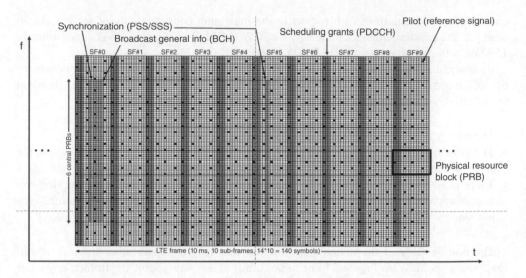

Figure 5.9 Coverage tax: impact of signalling and control channels on LTE frame

5.3.3 Power Models

The goal of the BS power model is to provide a simple and easy-to-use mathematical model of the power consumption of a BS, as function of the traffic load. This is often useful when analyzing and comparing different green wireless solutions. In this chapter, the power model used is a simplified version of the one described in [22], which is an evolution of the model proposed in Ref. [17, 21]. This model allows estimating the base station consumption (P_{in}) as function of the output RF power (P_{out}), the traffic load of the sector (λ) and the type of BS (Table 5.2): number of sectors (N_{sec}), maximum RF power (P_{max}), consumed power at zero load (P_0) and slope (Δ_p). It also models a sleep mode (see Section 5.4), with a specific power consumption (P_{sleep}), that can be activated only at zero load. The model does not distinguish the number of antennas per sector; a 2×20 W macro BS corresponds to a 40 W macro.

$$P_{in} = \begin{cases} N_{sec} \cdot P_{out}, & 0 < \lambda \le 1 \\ N_{sec} \cdot P_{sleep}, & \lambda = 0 \end{cases}$$

$$P_{out} = P_0 + \Delta_p \cdot P_{max} \cdot \lambda$$

Figure 5.11 present the traffic profiles issued by the proposed model in three cases: macro (3 sectors, 2×20 W per sector), micro (5 W) and femto (100 mW).

To conclude this section, it is worth to be noted that other power models have been proposed in the literature in the recent years, as for example [23].

5.4 Evolutions Towards Green Base Stations

The recent attraction drawn by the telecom research and industrial world on green aspects has allowed identifying several promising techniques to reduce the energy consumption of

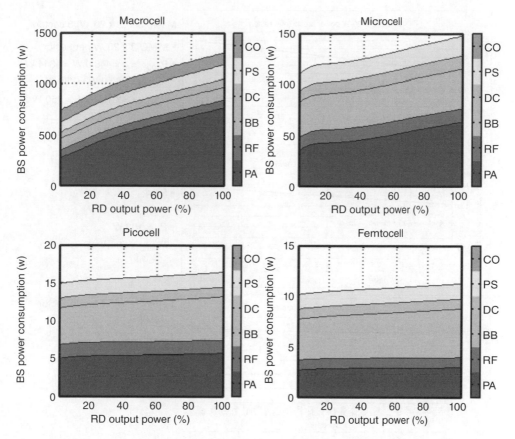

Figure 5.10 Power consumption profiles for different types of base station (year 2010 values) – Courtesy of European Community's 7th Framework Program FP7 project EARTH – Legend: CO = Cooling, PS = power supply, DC = DC–DC converters, BB = baseband processing, RF = RF transceiver (w/o PA), PA = power amplifier

Table 5.2 BS power profile parameters.

BS type	N_{sec}	P_{max} (W)	P_{max} (dBm)	P_0 (W)	Δ_p	P_{sleep} (W)
Macro single cabinet	3	40.0	46	260.0	4.7	150.0
Macro RRH	3	40.0	46	168.0	2.8	112.0
Micro	1	5.0	37	103.0	6.5	69.0
Pico	1	1.0	30	96.2	1.5	62.0
	1	0.25	24	13.6	4.0	8.6
Femto	1	0.1	20	9.6	8.0	5.8

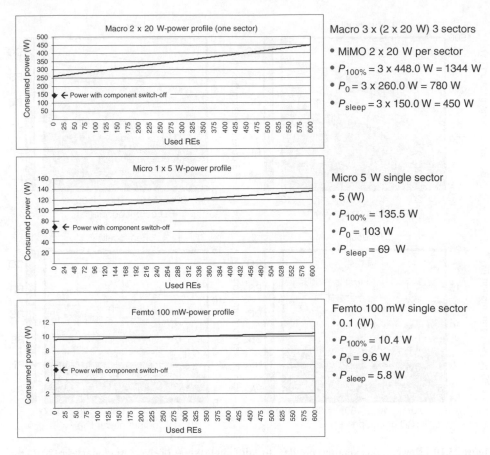

Figure 5.11   Power model profiles for different BS types (macro, micro and femto)

cellular base stations. This section focuses on the evolutions at base station level (component, architecture and operation) from which the higher energy savings are expected, namely, improvements at components level (both hardware and operating-point), implementation of BS stand-by modes, traffic-load adaptation and evolution of BBU architecture, including the cloud approach. When possible, this section also points out the possible limitation and blocking points of the described evolutions.

5.4.1   Component Level Evolutions

New materials and new components architectures constitute the most basic level how wireless BS can effectively benefit of the improvements developed for the worldwide electronic industry. They have an impact on all the BS components, and particularly on the most consuming ones, such as power amplifier, and baseband processing. Indirectly, cooling can also benefit of such improvements; the improved efficiency of the other components usually implies a reduction of the heat to be dissipated.

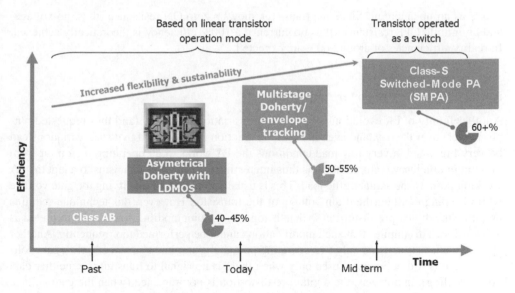

Figure 5.12   Principle of envelope tracking power amplifier (ETPA)

5.4.1.1   New Power Amplifiers architectures

As shown in Section 5.3.1, the PA is responsible for a very significant part of the BS power consumptions, especially in the case of large macro cells. Therefore, high-efficiency PAs have become the subject of intensive research resulting in continuous improvement of performances ([24, 25]).

Several solutions focused on improving the power amplifier efficiency under high PAPR (and backed off) conditions, for example in LTE. Signal conditioning algorithms, such as crest factor reduction for decreasing the PAPR and digital pre-distortion for increasing the PA linearity, enable the PA operation closer to saturation, corresponding to the maximum efficiency working point. These techniques usually come as complement to more radical PA architecture evolutions (Figure 5.12).

The current state-of-the-art is represented by Doherty power amplifiers (DPA), which contain one main (carrier) amplifier always active and an auxiliary one (peaking) active only when signal peaks occurs ([26, 27]). This aspect makes DPA particularly adapted to signals with high PAPR. DPAs show an efficiency of around 40–45%.

Efficiency can be further improved by two new architectures currently under intensive study: envelope tracking PAs (ETPA, [28]) and switch mode power amplifier (SMPA, [29]). In ETPA, the DC supply voltage is constantly adjusted so that to track the variations of the transmitted input. Since the difference between the DC supply voltage and the transmitted envelope voltage is dissipated as heat, envelope tracking PAs allow to drastically reduce this dissipation, and thus to increase the overall efficiency to around 50–55%. EPTA is better view for small cells systems where the instantaneous required pass band is lower.

Class-S SMPA architecture promises very high theoretical efficiency. This comes from the fact that the active component (transistor) is operated in an ON/OFF way: when ON, the transistor acts as a very low resistance (ideally short circuit, such as a closed switch), when OFF it acts as an open circuit. Since the transistor shows a zero ON resistance (thus, no voltage) and an infinite OFF resistance (thus no current), a 100% efficiency is theoretically achieved. In reality, efficiencies of around 60% are expected.

5.4.1.2 Signal-Aware Power Amplifiers

Another family of PA evolutions deals with the variations of traffic (and thus requested output power) over the day, and even in real time. Section 5.3.2 showed that such variations can be very high, and in very low-load conditions the PA works in an operating region far from maximum efficiency (saturation). It is thus interesting to defined mechanisms to adapt the PA working point to the actual traffic load. This is usually obtained by modifying the gate voltage while keeping constant the drain voltage of the transistor. However, this technique requires updating the digital-pre-distortion so that the operating point modification is not interpreted as a distortion. This implies that such modification cannot be performed too frequently. Another approach consists in using only two operating points: the nominal one and one with gate voltage to zero (PA deactivation), used only when there is no signal to transmit (i.e. neither data nor signalling). In this way, the digital-pre-distortion is not impacted (when the gate voltage is at zero there is no signal to pre-distort), allowing very rapid activation/deactivation, to track to real-time (less than 1 ms) traffic variations. During the deactivation period (gain voltage to zero), the power consumption of the PA can be reduced by a factor approaching 10 times with respect to the power consumed at maximum power.

5.4.1.3 Improvements of BBU

The Baseband Unit is currently undergoing important evolutions, brought in particular by important improvements from components and architecture. From an energy point of view, the use of low-consumption DSP and ASICs architectures must be preferred to more flexible but also more energy greedy FPGA architectures. This is today in line with the mainstream trend of moving away from multiple-cards architectures and to adopt integrated solutions, often based on system-on-chip (SoC) architectures. For large base stations, multiple-SoC architectures are hitting the market, where several generic cores are integrated with memory facilities and specific hardware accelerators for computationally expensive operations (such as turbo coding/decoding and ciphering). This integration brings a natural reduction of power consumption of the BBU, and the opportunity for BBU processing scaling, where the internal cores are activated/deactivated according the BS load.

BBU processing scaling requires specific design in terms of both hardware and software. It becomes extremely important to design the BBU parallel processes in a way that processing flexibility can be achieved. As an example, dedicating each processing unit to different parallel operations is not the good approach, because it makes impossible to turn off some processing unit at low load. A good design is in turn to concentrate always-on processes on some processing units and to execute all the load-depending operations together on a pool of processors; at low load, a single processing unit is needed whilst new units can be activated when the load increases.

5.4.2 BS Operation Improvements

The improvements of the base station operation target to improve the proportionality of the power consumed with the traffic load, in particular by exploiting the periods of time when the base station is poorly load (or even not serving any user). Two main techniques have been identified: real-time adaptation to traffic variation and BS stand-by modes.

5.4.2.1 Smart Load Adaptation to Traffic Load Variations

Smart load adaptation techniques apply to active base stations and aim at adapting the BS operating mode to the variation of traffic load. Such solutions combine component-level improvements of the power amplifiers (i.e. signal-aware PAs presented in Section 5.4.1.2 with smart scheduling strategies exploiting such improvements. Two main approaches can be identified [30]: scheduling policies adapting the actual used bandwidth, combined with power amplifiers with adaptive operating point, and scheduling policies creating micro-sleeps periods, using power amplifiers with deactivation mode [31, 32]. The second technique is also known as micro- or cell-discontinuous transmission (DTX) because during the micro-sleep periods the BS suspends its RF transmission. It must be noted that these solutions apply only to OFDM-based networks, such as LTE and LTE-A, and cannot be used in, for example, WCDMA base stations.

Smart load adaptation techniques gains have been estimated [33] to approximately 30% reduction of BS energy consumption.

5.4.2.2 Activation/Deactivation of RF Resources

In this case, the adaptation to the traffic load conditions is obtained by simply activating/deactivating the BS radio resources.

It is possible, for example, to adapt of the number of transmit antennas depending on the traffic load. For example, modern base stations (BSs) use up to eight antennas. When the traffic demand is low, it is possible to switch off some of the RF front ends and use a less spectral efficient transmit mode.

Another approach is to reduce the transmit band. For example, in LTE by using narrower carriers at low loads (e.g. fall-back mode from 20 to 5 MHz). When several carriers are active in the same sector (e.g. a 3-carrier per sector WCDMA Base Station), it is possible to implement a fallback mode which uses only one carrier. A similar approach is possible with multi-technology BS (multi-RAT, e.g. GSM + WCDMA + LTE), by switching off some of the technologies.

In all these cases, a particular attention must be brought to the internal hardware BS architecture. For example, modern multi-carrier and multi-technology base stations (recently hitting the market) tend to use the same RF head for several carriers or several technologies (e.g. one RF head for one sector, offering two 3G carriers and one LTE carrier). In that case, algorithms based on carrier or RAT switch off do not bring significant savings, as the power amplifier remains active for serving the remaining carriers/technologies.

5.4.2.3   Base Station Sleep Modes

During low-load periods, it can happen that a base station becomes unnecessary, either because no users are using it or because the remaining users can be absorbed by neighbour base stations. In these cases, in order to avoid energy wasting, it is interesting to move the base station into some low-consumption mode (sleep or stand-by). Unfortunately currently, deployed base stations (whatever technology) do not implement such sleep modes and can usually work into only two states: full active or complete switch off. Algorithms based on long timescales (i.e. tolerating off–on transients of the order of tens of minutes and above) are not affected by this problem; they can be implemented by turning completely off the base station. This can be achieved even on already deployed base station, using the OAM interface. Algorithms requiring shorter wakes-up delays cannot be based on a complete switch off of the base station. In fact, the delay necessary to move from a power-saving mode to the full active mode depends on the number (and type) of components that are switched off (see, e.g. [34] for an analysis on 3G femtocell). A cold-restart requires several minutes to complete, with the delay increasing with the size of the base station.

New sleep modes must thus be defined. These modes are characterized by the BS wake-up delay (i.e. the time required by the BS to move from the stand-by to the full active mode): the longer is the acceptable wake-up delay, the higher is the number of internal components that can be turned off, and thus the higher are the energy savings. Two main cases can be identified (note that the ultra-fast real-time on/off case corresponds to the cell-DTX presented in Section 5.4.2.1: fast wake-up (few seconds) and longer wake-up time (several tens of seconds to few minutes).

For fast wake-up, very few components can be turned off: the radio head (some care must be brought to digital-pre-distortion stabilization delay), and possible some cooling (in case of single-cabinet, otherwise the RH switch off does not lead to temperature reduction in the BBU cabinet). The other components must remain on, especially the BBU that requires a considerable amount of time to reactivate, incompatible with fast wake-up. It can be concluded that theoretically, a well-designed fast-wake-up sleep mode can target to save the energy consumed by the RH (e.g. ~60% for a macro cell, see Figure 5.6).

If longer wake-up time can be accepted, additional components can be turned off, including the oscillator, the digital signal processing and the upper layers. The backhaul interface and some control process must be kept in order to be able to receive the wake-up request from the neighbour cells or the management system. Cooling can be turned off, while power supply shall be designed to be partially turned off, otherwise it must remain on to supply the backhaul interface and the remaining control processes. A rough estimation of potential energy gains is around 80–90%.

5.4.3   BS Architecture Evolutions

With respect to energy efficiency, two major architecture evolutions are currently being analyzed and defined by the telecom industry (Figure 5.13): massive-MIMO (i.e. the massive increase of the number of active antennas on the radio side) and cloud-RAN (i.e. execution of baseband processes in remotely located farms of generic servers). The two evolutions are not antagonistic and will probably combine together: the centralized high-processing power

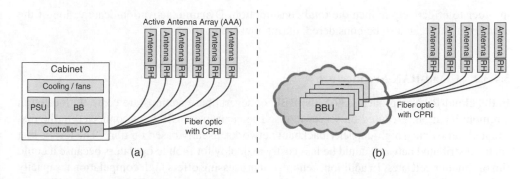

Figure 5.13 BS architecture evolutions: massive-MIMO and Cloud-RAN

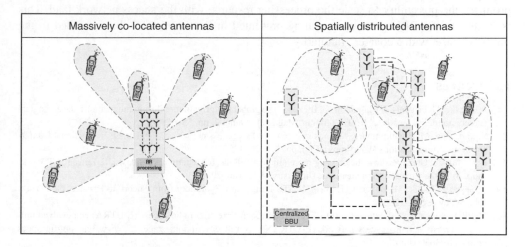

Figure 5.14 Distributed versus centralized massive-MIMO

of cloud-RAN is in fact well adapted to the huge amount of signal processing required to synchronize the emission of massive-MIMO systems.

5.4.3.1 Massive-MIMO Architecture

Massive-MIMO aims at increasing the spectral efficiency through spatial multiplexing ([35–37]). The impact at BS level comes from the multiplication of radio heads and RF front end (hundreds of antennas), centralized on a single site or distributed over a specific region (Figure 5.14). Energy gains come from the fact that the RF power transmitted per antenna is much less than in traditional architecture because the radiated energy can be focused on the receiver. However, such gains risk to be diminished when the high-spectral efficiency of massive-MIMO is not fully exploited, i.e. at low traffic loads. The engineering of such systems is thus challenging ([38]); the used front end must be low power yet efficient enough in order to effectively reduce the total consumption. Dynamic activation/deactivation of the used front ends can also be considered, during low-load periods.

5.4.3.2 Cloud-RAN Architecture

In the cloud-RAN approach ([39]), the BBU of several base stations are centrally located in a remote location, executed on server farms. The centralized BBU are connected to the distributed radios via a high-capacity backhaul networks, usually based on fibre. The idea is that such a distributed network could be less costly to deploy for mobile operators because it could run on smaller cell sites. In addition, centralized processing offers high computational capacity and ease multi base stations cooperation, for example, distributed MIMO (e.g. CoMP, coordinated multi-point transmission/reception defined by 3GPP) or massive MIMO. From the energy efficiency point of view, the centralization of BBUs (also known as BBU-pooling) promises the possibility to scale the processing resource with the access network load. This is similar to the BBU processing scaling presented in Section 5.4.1.3, but extended to the multi-BS case, with a coarser granularity.

References

[1] A. Fehske; G. Fettweis; J. Malmodin; G. Biczok; "The global footprint of mobile communications: the ecological and economic perspective," IEEE Commun. Mag., vol. 49, no. 8, pp. 55–62, 2011.

[2] S. Vadgama; M. Hunukumbure; "Trends in green wireless access networks," 2011 IEEE International Conference on Communications Workshops (ICC), pp. 1–5, 2011.

[3] P. J. Chung; "Green Radio—the case for more efficient cellular base stations," (slides) UK-Taiwan ICT Workshop: Smart & Green Communications, University of Taiwan, 2010).

[4] S. Sesia; I. Toufik; M. Baker; LTE – The UMTS Long Term Evolution: from Theory to Practice, New York: John Wiley & Sons, 2009.

[5] 3GPP Technical Specification 36.300, "Evolved universal terrestrial radio access (E-UTRA) and evolved universal terrestrial radio access network (E-UTRAN); overall description; stage 2," www.3gpp.org/ftp/Specs/html-info/36300.htm.

[6] 3GPP Technical Specification 23.401; "General packet radio service (GPRS) enhancements for evolved universal terrestrial radio access network (E-UTRAN) access," www.3gpp.org/ftp/Specs/html-info/23401.htm.

[7] 3GPP Technical Specification 23.402, "Architecture enhancements for non-3GPP accesses," www.3gpp.org/ftp/Specs/html-info/23402.htm.

[8] 3GPP Technical Specification 29.060, "General packet radio service (GPRS); GPRS tunnelling protocol (GTP) across the Gn and Gp interface," www.3gpp.org/ftp/Specs/html-info/29060.htm.

[9] 3GPP Technical Specification 36.321, "Evolved universal terrestrial radio access (E-UTRA); medium access control (MAC) protocol specification," www.3gpp.org/ftp/Specs/html-info/36321.htm.

[10] 3GPP Technical Specification 36.322, "Evolved universal terrestrial radio access (E-UTRA); radio link control (RLC) protocol specification," www.3gpp.org/ftp/Specs/html-info/36322.htm.

[11] 3GPP Technical Specification 36.323, "Evolved universal terrestrial radio access (E-UTRA); packet data convergence protocol (PDCP) specification," www.3gpp.org/ftp/Specs/html-info/36323.htm.

[12] OBSAI, Open Base Station Architecture Initiative, www.obsai.com.

[13] CIPRI, Common Public Radio Interface, www.cpri.info.

[14] Abidi, A.A.; "Direct-conversion radio transceivers for digital communications," IEEE J. Solid-State Circuits, vol. 30, no. 12, pp. 1399–1410, 1995.

[15] B. Debaillie; A. Giry; M. J. Gonzalez; L. Dussopt; M. Li; D. Ferling; V. Giannini; "Opportunities for energy savings in pico/femto-cell base-stations," Future Netw. Mobile Summit (FutureNetw), 15–17, pp. 1–8, 2011.

[16] Desset, C.; Debaillie, B.; Giannini, V.; Fehske, A.; Auer, G.; Holtkamp, H.; Wajda, W.; Sabella, D.; Richter, F.; Gonzalez, M.J.; Gódor, I.; Olsson, M.; Imran, M.; Ambrosy, A.; Blume, O.; "Flexible power modeling of LTE base stations" IEEE WCNC 2012, Paris, 2012.

[17] EARTH Project, "Deliverable D2.3 - energy efficiency analysis of the reference systems, areas of improvements and target breakdown," available at: www.ict-earth.eu/publications/deliverables/deliverables.html, 2012.

[18] GreenTouch, www.greentouch.org.

[19] ETSI TS 102 706 V1.3.1 (2013-07): "Environmental Engineering (EE) Measurement Method for Energy Efficiency of Wireless Access Network Equipment"

[20] GreenTouch, "GreenTouch Green Meter Research Study: Reducing the Net Energy Consumption in Communications Networks by up to 90% by 2020," A GreenTouch White Paper, Version 1.0, June 26, 2013

[21] G. Auer; V. Giannini; C. Desset; I. Godor; P. Skillermark; M. Olsson; M. A. Imran; D. Sabella; M. J. Gonzalez; O. Blume; A. Fehske; "How much energy is needed to run a wireless network?," IEEE Wireless Commun., vol. 18, no. 5, pp. 40–49, 2011.

[22] NTT DOCOMO, Alcatel-Lucent, Alcatel-Lucent Shanghai Bell, Ericsson, Telecom Italia, "[r1-113495], base station power model." [Online]. Available: http://www.3gpp.org/ftp/tsgran/wg1rl1/TSGR166b/Docs/R1-113495.zip.

[23] Debaillie B.; Desset, C.; Louagie F.; "A flexible and future-proof power model for cellular base stations," in Vehicular Technology Conference (VTC), Scotland, May 2015

[24] S. C. Cripps; RF Power Amplifiers For Wireless Communications, 2nd edn, Artech House Microwave Library, 2006.

[25] J. Joung; C. K. Ho; S. Sun; "Green wireless communications: a power amplifier perspective," 2012 Asia-Pacific Signal & Information Processing Association Annual Summit and Conference (APSIPA ASC), pp. 1–8, 2012.

[26] W. H. Doherty; "A new high efficiency power amplifier for modulated waves," Proc. Inst. Radio Eng., vol. 24, no. 9, pp. 1163–1182, 1936.

[27] B. Kim; J. Kim; I. Kim; J. Cha; "The Doherty power amplifier," IEEE Microw. Mag., vol. 7, no. 5, pp. 42–50, 2006.

[28] J. Kim; J. Son; S. Jee; S. Kim; B. Kim; "Optimization of envelope tracking power amplifier for base-station applications," IEEE Trans. Microwave Theory Tech., vol. 61, no. 4, pp. 1620–1627, 2013.

[29] A. Wentzel; C. Meliani; W. Heinrich; "RF class-S power amplifiers: state-of-the-art results and potential," 2010 IEEE MTT-S International Microwave Symposium Digest (MTT), pp. 812–815, 2010.

[30] A. Ambrosy; O. Blume; H. Klessig; and W. Wajda; "Energy saving potential of integrated hardware and resource management solutions for wireless base stations," in 2011 IEEE 22nd International Symposium on Personal Indoor and Mobile Radio Communications (PIMRC), pp. 2418–2423, 2011.

[31] P. Frenger; P. Moberg; J. Malmodin; Y. Jading; and I. Godor; "Reducing energy consumption in LTE with cell DTX," in 2011 IEEE 73rd Vehicular Technology Conference (VTC Spring), pp. 1–5, 2011.

[32] R. Gupta; E. Calvanese Strinati; and D. Ktenas; "Energy efficient joint DTX and MIMO in cloud radio access networks," in 2012 IEEE 1st International Conference on Cloud Networking (CLOUDNET), pp. 191–196, 2012.

[33] Ambrosy, A.; Blume, O.; Ferling, D.; Jueschke, P.; Wilhelm, M.; Xin Yu; "Energy savings in LTE macro base stations," 7th IFIP Wireless and Mobile Networking Conference (WMNC), 2014, pp. 1–8, 2014.

[34] I. Haratcherev and A. Conte; "Practical energy-saving in 3G femtocells," in IEEE Green Broadband Access (GBA) workshop, in conjunction with ICC 2013, 2013.

[35] T. L. Marzetta; "Noncooperative Cellular Wireless with Unlimited Numbers of Base Station Antennas," IEEE Trans. Wireless Commun., vol. 9, no. 11, pp. 3590–3600, 2010.

[36] H. Q. Ngo; E. G. Larsson; T. L. Marzetta; "Energy and spectral efficiency of very large multiuser MIMO systems," IEEE Trans. Commun., vol. 61, no. 4, pp. 1436–1449, 2013.

[37] J. Hoydis; S. ten Brink; M. Debbah; "Massive MIMO in the UL/DL of cellular networks: how many antennas do we need?," IEEE J. Sel. Areas Commun., vol. 31, no. 2, pp. 160–171, 2013.

[38] E. G. Larsson; F. Tufvesson; O. Edfors; T. L. Marzetta; "Massive MIMO for next generation wireless systems," eprint arXiv:1304.6690, 2013.

[39] Lin, Y.; Shao, L.; Zhu, Z.; Wang, Q.; Sabhikhi, R. K.; "Wireless network cloud: architecture and system requirements," IBM J. Res. Dev., vol. 54, no. 1, pp. 4:1–4:12, 2010.

6

Energy-Efficient Mobile Network Design and Planning

Yinan Qi, Muhammad Ali Imran and Rahim Tafazolli
Institute for Communication Systems (ICS), University of Surrey, Guildford, Surrey, UK

6.1 Introduction

The exponential increase of user data often requires planning and design of a data communication system and the operation and management of such a system. The latter issues are discussed in the next several chapters and this chapter mainly deals with planning and design of a wireless network. Network design and planning consists of three steps: network topology design, network synthesis and network realization [1–3]. In the first step, the locations and connecting patterns of the network nodes are determined. In the second step, the network parameters, such as size, are optimized subject to certain quality of service (QoS) requirements. In the last step, the network is realized with designed configuration parameters. These steps might be conducted in an iterative manner to gradually optimize the network. For a wireless network, network design and planning principles can be applied to both the core network and the radio access network (RAN). The main focus of this chapter is on the RAN. In the development of today's second and third generation cellular mobile radio networks, such as GSM and WCDMA, the main optimization objectives in network design and planning are focused on providing optimum throughput subject to coverage requirements and a required level of QoS. The "Green" aspect does not draw much attention and therefore is rarely addressed. However, the rapid evolution of telecoms industry is accompanied by a huge increase of energy consumption of mobile networks in recent years [4, 5]. Coupled with sharp rising cost of the energy resources in the past few years, the operators are forced to further exploit the potential of network design and planning to reduce both capital expenditure (CAPEX) and operational expenditure (OPEX), mainly arising from the energy consumption. In the efficient design and deployment of the fourth-generation and beyond fourth-generation cellular networks, for example 3GPP LTE/LTE-Advanced, it is proposed by industry consortium that the

Green Communications: Principles, Concepts and Practice, First Edition.
Edited by Konstantinos Samdanis, Peter Rost, Andreas Maeder, Michela Meo and Christos Verikoukis.
© 2015 John Wiley & Sons, Ltd. Published 2015 by John Wiley & Sons, Ltd.

energy consumption will be taken into consideration as an additional optimization target for all the operators [6].

The aim of this chapter is to draw a fundamental and comprehensive picture of the state-of-the-art "green" network design and planning schemes, algorithms and methodologies, mainly focused on the first two steps aforementioned. The organization of the chapter is as follows: the potential of cell parameter optimization to achieve energy saving is introduced in the next section. Sections 6.3 and 6.4 discuss the most promising green solutions for urban and rural scenarios, respectively. The last section concludes this chapter and provides some insight into the future network architecture evolution path.

6.2 Deployment: Optimization of Cell Size

Studies have identified that the increasing energy consumption is mostly related with the higher BS site density to provide huge amount of data transmission capability and high level of coverage [7]. The transmission power can be significantly reduced with a smaller cell size due to the fact that the transmitted signals suffer lower path loss in the air. Meanwhile, considering the transmission power can merely be counted as a part of overall energy expenditure, smaller cells with higher cell deployment density clearly cause additional static power consumption in terms of circuitry power, site cooling power, and so on, which are almost irrelevant with transmission power. The saving potentials should be carefully designed to balance the trade-off between throughput and energy expenditure. Another important consideration when planning the network is the traffic, in particular, the spatial traffic variations. In a high traffic area, it is expected to deploy more cells to allocate more resources for satisfactory user experience. Otherwise, smaller number of large cells can be deployed to meet the traffic requirement. Some of the previous works [8, 9] have made a common assumption that the spatial traffic distribution is uniform to guarantee that the peak traffic can be handled with the required QoS throughout the entire region covered by the cellular system. However, the spatial traffic distribution is not uniform in realistic scenarios. The uniform traffic assumption causes inefficiency because the number of BSs deployed in the low traffic area is more than needed. Hence, a part of the energy consumed by those BSs is wasted. When designing the cellular network parameters, all the aforementioned aspects should be considered to offer realistic solutions.

6.2.1 System Model

Considering a hexagon cell with radius R as shown in Figure 6.1, the area of the cell is denoted as S and we assume that the entire cellular system is able to cover a large region with area U, where the spatial traffic density varies according to the locations. This model is commonly used in system level analysis for cellular networks.

6.2.1.1 Traffic Model Within a Cell

The multi-class MMPP/M/1/D-PS queue model (a single server processor sharing queue, with Markov-modulated Poisson arrival process, Markovian service time and finite capacity) has been widely used to study and dimension the telecommunication systems for more than a

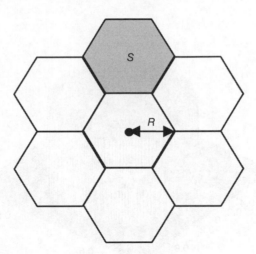

Figure 6.1   Cell definition

century [10, 11]. Surprisingly, the simplicity of the model does not impair the evaluation accuracy. The assumptions made in this model are as follows:

- No more than S sessions and D data connections can be admitted in a cell at the same time. The session arrival rate is λ_S and with s sessions and the data connection arrival rate is $s\lambda_D$.
- The user service time is exponentially distributed with mean value T; hence, the user service rate is exponentially distributed with mean value $\mu_s = 1/T$.
- In each data connection, the amount of information transferred is exponentially distributed with mean value G; hence, the data connection service time is exponentially distributed with mean value $\mu_d = B/G$, where B is the throughput.

In this traffic model, one of the parameters, that is, session arrival rate, is controlled by a Poisson process. It is typical in modelling the traffic types where time-varying arrival rates capture some of the important correlations between inter-arrival times and is applicable in services such as VoIP and FTP. However, this traffic model is not suitable to model the continuous traffic types, for example, video streaming.

6.2.1.2   Spatial Traffic Variation Model

Realistic traffic is envisaged to vary according to the locations, which implies that the session generating rate λ_S as well as the data connection generating rate λ_D are functions of the BS locations and can be expressed as $\lambda(r, \theta)$ in a polar coordinate system. For simplicity of the analysis, we assume that the spatial traffic is a step function as shown in Figure 6.2. The realistic traffic model might be far more complicated than this simple model. However, any realistic traffic model with traffic hotspots can be decoupled into similar step functions. Hence, it is adequate for the principle estimation of the energy savings and the optimization of network design and the planning parameters can be easily extended from this simple model. Note that

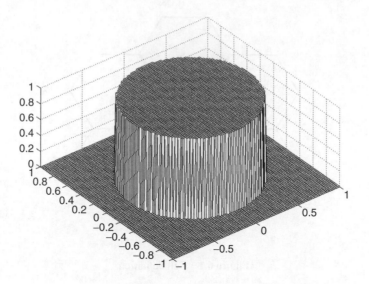

Figure 6.2 Normalized spatial traffic variation model

this model is a peak traffic model, which implies that although the traffic varies with time, it will never exceed the values in this model, guaranteeing required QoS at any time.

6.2.1.3 Propagation Model and Coverage

Normally, the received signal suffers from path loss, slow fading and fast fading. A simple model taking path loss and slow fading into consideration is given as:

$$P_{rx}(r) = K\left(\frac{r}{r_{ref}}\right)^{-\gamma}\varphi P_{tx} \qquad (6.1)$$

where P_{tx}, P_{rx}, r_{ref}, γ and φ represent transmit and receive power, reference distance and propagation path loss exponent and a random variable modelling shadowing effects, respectively, and K is a constant which can be obtained through measurement and its value is given in Ref. [12].

Based on this propagation model, coverage, as another critical performance metric is defined as the fraction of cell area where the received power is able certain threshold P_{min}:

$$C = \frac{1}{S}\int_S r\Pr(P_{rx}(r) \geq P_{min})drd\theta \qquad (6.2)$$

6.2.1.4 Quality of Service (QoS)

The QoS can be referred to several related metrics such as latency, jitter, packet dropping probability, and so on. In this chapter, QoS is mainly determined by the blocking probability that a new generated request is rejected because there are no available resources to be allocated to the new user. According to Ref. [13], the blocking probability p_{blk} can be easily calculated by applying the infinitesimal generator matrix based on the steady state probability as:

$$p_{blk} = \frac{\sum_{d=0}^{D} \pi_{d,S} \lambda_s + \sum_{s=0}^{S} \pi_{D,s} s \lambda_d}{\sum_{s=0}^{S} \sum_{d=0}^{D} \pi_{d,s}(s\lambda_d + \lambda_s)} \tag{6.3}$$

where $\pi_{d,s}$ is the steady state probability for state (d, s) which indicates that the dth data connection request of the sth session arrives. The numerator represents the probability that the incoming session/data connection requests are rejected and the denominator stands for the summarised probability of all steady states.

6.2.2 Optimization of Cell Parameters

In the previous deployment strategies, the cell size is designed to satisfy the maximal traffic in the entire network, which is obviously not an energy-efficient solution. The basic idea of the new deployment strategy is to separate the entire service region according to the traffic level. For the normalized spatial traffic variation depicted in Figure 6.2, we divide the covered area into two parts: a dense zone where the traffic is intensive and a sparse zone where the traffic is relatively low. In the dense zone, the target blocking probability must be met, that is, $p_{blk} \leq p_k^T$. However, in the sparse zone, if we maintain the same cell size, the blocking probability achieved might be significantly smaller than the target one due to the lower traffic demand, thus wasting a large amount of energy. Therefore, we can enlarge the cell size within the sparse zone so that the BSs receive more traffic requests but the coverage and QoS requirements will not be impaired. By doing this, the number of BSs used in the entire network is reduced as well as the overall consumed energy.

Within the dense zone, the radius of the cell is assumed to be R_d and the area of each cell is S_d; whereas in the sparse zone, the radius of the cell is $R_s = \mu R_d$ and the area $S_s = \mu^2 S_d$. A heuristic scheme can be employed to optimize the cell sizes in two zones with different traffic demands:

- **Step1**: For the dense zone, determine $R_{d, BP}$ according to the blocking probability requirement.
- **Step2**: Calculate the average number of data connections within a cell and $R_{h, C}$ according to the coverage requirement.
- **Step3**: Choose $R_d = \min (R_{d, BP}, R_{d, C})$.
- **Step4**: Choose a proper $R_s = \mu_0 R_d$ to satisfy the target blocking probability in the sparse zone.
- **Step5**: Calculate coverage of the cells within the sparse zone.
- **Step6**: If the coverage requirement is met, the algorithm is ended; else reduce $\mu_{n+1} = \mu_n - \Delta\mu$ and goes back to step 5.

In order to validate the feasibility of this scheme, we set up a simulation scenario with LTE link budget and use the FTP service model in Ref. [14]. The target blocking probability is set as 1% and we assume the entire covered area is a circle with radius 100 km and the dense zone is a circle with radius 20 km. The new scheme will be compared with conventional scheme where no separation is conducted and the network planning is merely based on the maximum

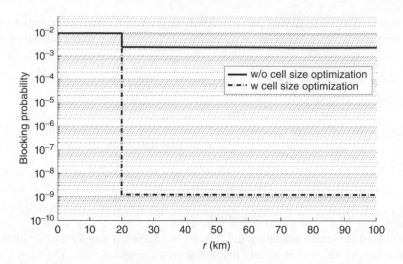

Figure 6.3   Blocking probability

traffic, tailoring to a uniform cell size. We define the energy saving power ratio Φ_{sav} to exploit the energy saving as:

$$\Phi_{sav} = \frac{P_{new}}{P_{uni}} \qquad (6.4)$$

where P_{uni} is the overall power consumption for the uniform cell size.

We use the aforementioned scheme to find the optimal $\mu = 1.4$ and the corresponding $\Phi_{sav} = 0.82$, which means 18% energy consumption saving can be achieved because less number of BSs are deployed in the low traffic zone. The blocking probability is shown in Figure 6.3. If the cell size is unchanged in the sparse zone, the blocking probability sharply decreases due to the very low traffic demand. On the contrary, if the cell size is enlarged, more traffic is generated within one cell and the achieved blocking probability is closer to the target.

Now we come to the conclusion that by separating the whole network into two zones, the cell size can be optimized based on the spatial traffic distribution. With smaller number of BSs deployed in the sparse zone, the overall energy consumption is successfully reduced. Actually, the whole area can be divided into more zones according to the spatial traffic variation. In such a case, the network planning strategy is able to adapt itself to the spatial traffic variation in a more efficiently manner.

6.3   Network Design and Planning for Urban Areas

The planning of a network cannot be easily modified once the rollout is complete. It is difficult to relocate the already deployed base station sites of a legacy network. However, it does not mean that the network is not able to adapt itself to the change of the operation conditions. In one aspect, the network layout does not depend on the deployed but on the active BSs. Hence, it can be dynamically changed by adaptively switching on/off some BSs under different traffic circumstances. In another aspect, the densification of the current network can be tailored with additional macro BSs and/or micro BSs to reduce the inter-site distance (ISD) when the required traffic demand is growing. In the latter case, it is more important to leverage the deployment options for savings of energy by dynamic adaptation when the already deployed and the additionally deployed base station sites are fixed.

6.3.1 Adaptive On/Off Strategies to Change the Network Layout

The daily traffic profile has revealed that there are long periods in which the traffic load is low and the BSs are unnecessarily activated [14]. Different on/off schemes have been investigated to save energy, where the number of active BSs are adjusted dynamically adapting to the real-time traffic load. Therefore, unnecessary network nodes are switched off and the network layout is changed to save energy. It should be noted that the required QoS should not be compromised in the procedure of minimization of energy expenditure.

The on/off scheme can be simply categorized into two approaches: static and dynamic [15]. While the static approach only applies on/off scheme for a fixed potion of all network nodes in the period of one day, the dynamic approach is more adaptive in the sense that the on/off operations can be applied to the BSs depending on the traffic variations. With a simplified linear daily traffic variation pattern assumed as shown in Figure 6.4(a), the energy savings of two approaches are demonstrated in Figure 6.4(b) [15].

Two static on/off schemes, denoted as ½ and ¼, are demonstrated, where $1/x$ means that 1 out of x BSs are activated during the off operation period. As shown in the figure, the dynamic approach benefits from more flexibility and stronger adaptation capability and achieves higher percentage of energy savings. In addition, the presented results also clearly deliver an important message: the achieved energy saving is rather depending on the variation of the traffic, that is, the slope from peak to off-peak d, than one the absolute values of the traffic, that is, the peak traffic and the off-peak traffic.

6.3.2 Adaptive (De)sectorization

It has been pointed out that switching off the entire BS site might not be realistic because the transmission power of the BSs is assumed to be adjustable in a large dynamic range to meet the coverage requirement [16], which is normally impractical in real systems because of the power amplifier and RF link constraints. A more effective approach is adaptive (de)sectorization, where the BSs are able to adapt themselves to the temporal traffic variation by switching off some sectors and changing the beamwidth of the remaining sectors.

As shown in Figure 6.5, each BS consists of three sectors and each sector is defined as a hexagon with radius R. Sectorized antenna is employed in each sector.

Here, we use another commonly used metric of interest G-factor to evaluate coverage:

$$G_{factor} = \frac{P(BS_i \rightarrow UE)}{\sum_{j \neq i} P(BS_j \rightarrow UE) + P_{therm}} \quad (6.5)$$

where $P(BS_k \rightarrow UE)$ is the received power from BS k to the user equipment (UE), given in mW and P_{therm} is the thermal noise power given in mW. The long-term coverage can defined as:

$$C = \frac{1}{S_a} \int_{S_a} \Pr\{G_{factor}(x, y) \geq G_{factor,min}\} dx dy \quad (6.6)$$

where S_a is the area of a sector.

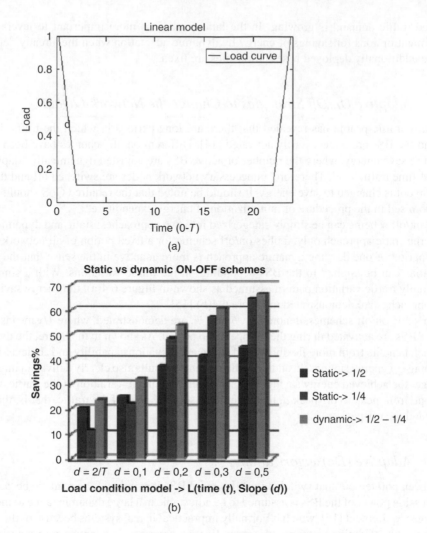

Figure 6.4   Energy savings and daily traffic profile (a) daily traffic profile (b) energy savings

When traffic demand is at high level, all three sectors within one BS are activated. When traffic demand is lower, one sector out of three sectors is switched off. Here for simplicity, we assume sector 1 is switched off. Handover is performed throughout the system. The users in the silent sector, that is sector 1, will be assigned to other sectors or even other BSs based on the long-term received signal strength, that is summation of path loss and shadowing loss. Apparently, those users in the silent sectors will suffer from lower coverage because of the weaker receive power. To maintain the coverage level of the silent sector, we need to change the antenna pattern of the other two remaining sectors.

Considering the 3D antenna pattern in Ref. [17], given by horizontal (azimuth) and vertical (elevation) orientations, the azimuth antenna pattern used in LTE system is given by:

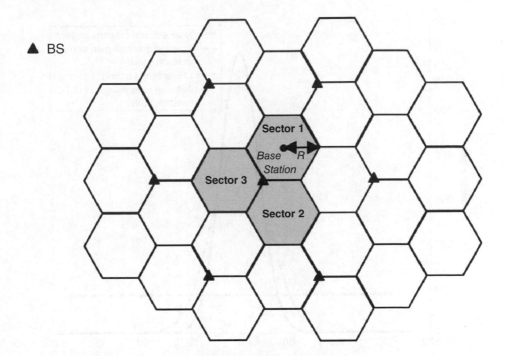

Figure 6.5 BS deployment

$$G_{az}(\Theta) = G_{max} - \min\left\{ 12\left(\frac{\Theta}{\Theta_{3\,dB}}\right)^2, G_{f2b} \right\}$$

$$G_{max} = 14dBi, \Theta_{3\,dB} = 65°, G_{f2b} = 20\,dB$$

$$-180° \leq \Theta \leq 180° \tag{6.7}$$

where G_{max} is the boresight antenna gain, Θ is the angle between the sector and the mobile (BS-UE) line of sight and the sector boresight, $\Theta_{3\,dB}$ is the 3 dB angle, also defined as beamwidth and G_{f2b} is the antenna front to back ratio. The elevation pattern is:

$$G_{el}(\Phi) = G_{max} - \min\left\{ 12\left(\frac{\Phi}{\Delta_\Phi}\right)^2, G_{f2b} \right\}$$

$$\Delta_\Phi = 10°, -180° \leq \Phi \leq 180° \tag{6.8}$$

where Δ_Φ is the elevation width and Φ is the downtilting angle. This downtilting angle is decided by the antenna height H_{ant} (typical value is 20–40 m, here is 20 m) and the distance D between the BS and the crossing point of the horizon and the antenna main lope orientation, that is $\Phi = \mathrm{argtan}(H_{ant}/D)$. It is a key factor to the intra- and inter-cell interferences. It should be optimized to minimize the interference and maximize the long-term throughput, which

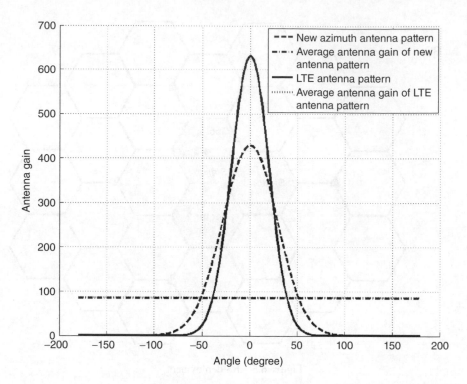

Figure 6.6   Azimuth antenna patterns

is a function of G_{factor}. For the sake of simplicity, we assume the downtilting angle keeps unchanged after sector 1 is switched off.

Even with optimized downtilting angle, turning off one sector will cause coverage problem. In order to cover the silent sector, it is natural to change the direction of other sections to offer a satisfactory coverage. For example, sector 3 antenna can be redirected from 180° to 120° and the beamwidth of two remaining sectors can be expanded from 65° to 95°. After adjustment, the new azimuth antenna pattern is shown in Figure 6.6.

In Ref. [18], a well-defined Macro cell power model based on measurement is presented, where the power consumption in watts is estimated individually for each subsystem. One BS consists of power amplifier (PA), main supply, DC part, RF link, Base band and cooling equipment. The power consumption of each component is shown in Figure 6.7. If one sector is switched off, the energy saving is not straightforward 1/3. PA and RF link consumption can be reduced by approximately 1/3. The other power consumption associated with DC, BB processing and cooling will not be significantly reduced. In a nutshell, the overall power consumption can be reduced by approximately 21%.

As we mentioned before, when minimizing the energy consumption the QoS should not be impaired. Turning off one sector not only brings the coverage problem, but it also increases the traffic demand of the remaining active sectors due to the new users handed over to them and might increase the blocking probability. The same MMPP/M/1/D-PS queue model aforementioned in the cell size optimization section can also be used here to analyze the blocking probability.

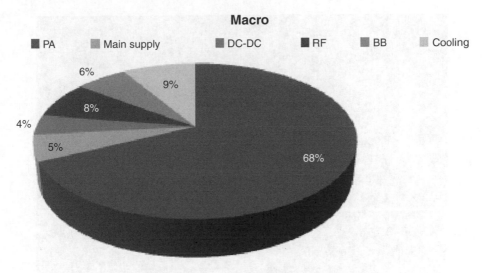

Figure 6.7   Power consumption breakdown

How the energy efficiency of an LTE system can be impacted by adaptive (de)sectorization is depicted in Figures 6.8–6.10 with following assumptions:

- Two tiers of BSs (19 BSs and 57 sectors) with central one as the reference one and inter-cell distance is 500m.
- UEs are placed randomly following a uniform distribution. One sector allows up to 50 UEs and 20 data connections.
- We assume OFDMA system without any cooperation among sites or any fractional frequency reuse patterns.
- LTE based link budget and propagation model.
- There are two main QoS constraints: the target blocking probability is 0.01 and the target coverage is 99%.

The G_{factor} map before and after switching off sector 1 is shown in Figure 6.8.

When sector 1 is switched off, the steering direction of sector 3 is changed and the beamwidth of the remaining two sectors is increased. The direct consequence is that one sector is receiving more interference because of the larger beamwidth, leading to G_{factor} degradation as well as the long-term system throughput. Figure 6.9 depicts the cumulative distribution function (CDF) of G_{factor} and it is indicated that the average value of G_{factor} is smaller when sector 1 is switched off. An interesting observation is that some users (the part in the circle) might actually benefit from this adaptive sectorization. Those users are previously strongly interfered by sector 1. Since sector 1 is off now, the interference is reduced and their performance is improved.

It should be noted that when traffic is further decreasing, we can switch off two sectors out of three sectors and change the antenna pattern to the omnidirectional to maintain the coverage. It will be shown in the following results that switching off two out of three sectors is beneficial when the traffic load is extremely low. The user arrival rate per km^2 is defined in Ref. [18]

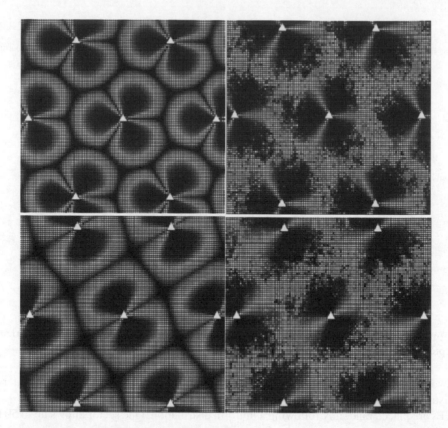

Figure 6.8 G_{factor} map (upper left: 3 sectors, pathloss only; upper right: 3 sectors, pathloss + shadowing; lower left: sector 1 is switched off, pathloss only; lower right: sector 1 is switched off, pathloss + shadowing)

and changes in 24 hours as shown in the upper part of Figure 6.10(a). Note that since the data connection arrival rate is constant, the temporal traffic variation is purely demonstrated by the fluctuation of the user arrival rate per km². The data connection arrival rate is set as 10 per UE per second per km². When the traffic demand in terms of user arrival rate is low, one sector is switched off to save energy. The coverage area of the remaining two sectors is then increased to 1.5 times of its original size, which means that the user arrival rate for each active sector is 1.5 times of its original value now. In the mean time, the average system throughput is degraded as shown in Figure 6.8 (the amount of reddish areas is decreased). As a consequence, the blocking probability increases.

However, if we carefully choose the switching off point, we can guarantee that the blocking probability is still below the target value (peak value of the reference/benchmark system). Obviously, the longer the sector is silent, the better energy efficiency we can achieve. Actually, the switching point is chosen where the blocking probability is exactly the same as the target value to maximize the energy saving. We define the energy saving ratio μ_{sav} as the ratio of the overall system energy consumption with sectors switching off over that without switching off. Figure 6.10(b) indicates that if we do not switch off sectors when the traffic demand is low,

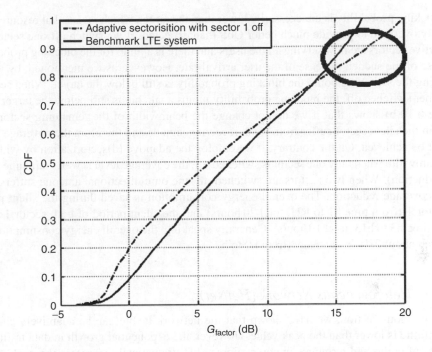

Figure 6.9 Cumulative distribution function (CDF) of G_{factor}

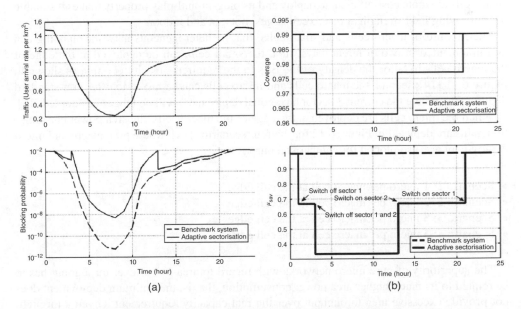

Figure 6.10 (a) Traffic demand and blocking probability, (b) coverage and saving ratio

the blocking probability is unnecessarily lower than the target, which means that resources are actually overpaid to provide much better QoS than expected. If we switch off one sector, the two active sectors are now serving all the users and there is a jump for the blocking probability because of the sudden increase of the user arrival rate. However, as we mentioned, by wisely selecting the switching point, the blocking probability is still below the target. When sector 2 is further switched off, the blocking probability jumps again but is still below the target level. Figure 6.10(b) shows that if we do not change the beamwidth of the remaining sectors, the users in the silent sector suffer from weak receiving signals and therefore the coverage target cannot be achieved. On the contrary, if we employ the adaptive (d)sectorization by widening the beamwidth when only one sector is switched off, the coverage level is well maintained (just 1% reduction). When two sectors are switched off, the omnidirectional antenna suffers from a 2% coverage reduction. The overall energy consumption is saved during the silent period of sector 1 (from time h1 to h21, total 20 hours) and the silent period of both sector 1 and 2 (from time h3 to h13, total 10 hours). Generally speaking, the overall energy consumption can be saved about 28% in each day (from h0 to h24).

6.3.3   Heterogeneous Network (HetNet)

In the previous section, we have shown that the network layout can be adaptively changed when traffic is lower than the peak value. However, the exponential growth in data traffic and number of connected terminals in contemporary life dramatically increases the peak traffic, which can only be tackled with densified planning with additional network nodes. Installation of new macro cells are neither convenient nor efficient. On the contrary, compact, low-power small cells are more cost-efficient to deploy and its plug-in-and-play property makes it suitable for improving field strength at macro cell edges, hotspot and indoor.

The mixture of macro cells and small cells, that is, heterogeneous network planning can be implemented in two different ways: first, the small cells can be deployed at the edge of the macro cells to tailor the density of the BSs to the required capacity per area as shown in Figure 6.11 [14]; second, the macro cells and micro cells can be mixed to offer non-standard deployment, where the current 3-sector hexagonal deployment will no longer be adopted [19].

When the micro cells are deployed at the edge of the macro cells, the area power consumption gains are depicted in Figure 6.12 for the four scenarios [14], where pure macro and micro deployment are also plotted for comparison purpose:

- Scenario 1: 1 micro per macro cell at the cell edge
- Scenario 2: 2 micro per macro cell at the cell edge
- Scenario 3: 3 micro per macro cell at the cell edge
- Scenario 4: 5 micro per macro cell at the cell edge.

The superiority of pure micro networks with regard to area power over throughput has to be related to its much higher area power consumption, that is, the optimum deployment does not provide excessive area throughput over the real capacity requirement. Given a throughput request, deployment of additional micro cells can only be efficient when this throughput request if relatively high. If the traffic request is low, the additional capacity gains brought by the micro cells are overwhelmed by the extra power consumption.

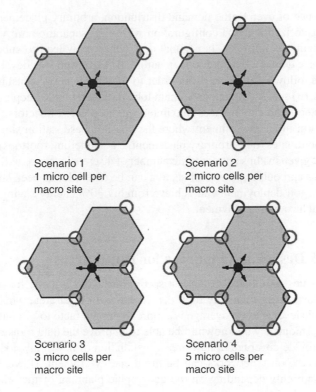

Scenario 1
1 micro cell per
macro site

Scenario 2
2 micro cells per
macro site

Scenario 3
3 micro cells per
macro site

Scenario 4
5 micro cells per
macro site

Figure 6.11   HetNet with micro cells at the edge of macro cells

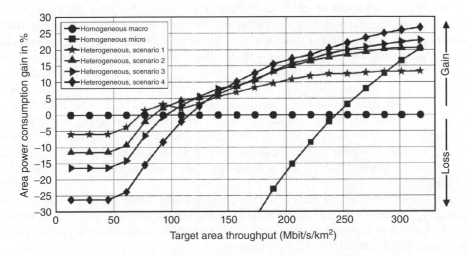

Figure 6.12   Area power consumption gain

As a consequence of even traffic demand distribution, arbitrary placement of base station sites and three-sectorized antenna configuration with 120° separation are valid. However, if the traffic distribution is not even, hexagonal deployment may lead to suboptimal network. In order to optimize deployment solution, an automatic algorithm was developed in Ref. [18] that performs the following main steps, in order to minimize energy consumption: (i) places base station sites, (ii) sets sector antennae main lobe direction (while keeping 120° separation for the three sectors) and (iii) reduces maximum output power of sectors. The algorithm is assumed to perform in an environment where the possible base station site locations form a discrete set of coordinates (non-arbitrary placement, e.g. modelling rooftops); traffic demands are assumed to be given in kbps units at finite number of discrete locations and it is assumed that antenna directions and output powers of sectors can be tuned. It has been shown in Ref. [19] that this non-hexagon deployment can achieve roughly 30% energy saving when compared with conventional hexagon deployment.

6.4   Network Design and Planning for Rural Areas

The green techniques introduced in the last section are more suitable for urban area where the BSs are deployed with relatively high densification. In rural area, where the cell size is large and the traffic level is low, coverage is a more important facto to be considered. In such a case, relaying technique is well known to be able to improve the data transmission in the cell edge and/or to provide coverage in new areas. Installation of new relay nodes (RNs) instead densifying the macro only networks can be more energy efficient. However, the efficiency improvements are highly depending on the geographic planning of the relay nodes and the capabilities of the relay nodes.

The basic idea behind relaying is to receive help from some radio nodes, called relays, to perform more spectrum and energy-efficient communications [20, 21]. RNs can be specifically devoted network nodes or other user devices in the vicinity. The relay node can either be used as a simple repeater which just amplifies and forwards its reception without any further processing, namely amplify-and-forward (AF) [20], or be equipped with more sophisticated baseband processing capability to be able to decode, re-encode and forwards received messages [21]. The latter one is called decode-and-forward (DF). These two schemes have already been part of LTE-Advanced to offer the possibility to extend coverage and increase capacity, allowing more flexible and cost-effective deployment options [22]. Other than these two schemes, compress-and-forward has drawn considerable attentions recently, where the relay node compresses and forwards its observation to the destination, benefiting from receive diversity [23]. In order to further exploit the uncertainty of wireless media, relaying schemes can be combined in a hybrid fashion and achieve improved spectrum efficiency because of the flexibility, [24, 26].

The link level results reveal that the S–D distance and the position of RN play an important role in energy efficiency (EE) performance [19]. When extended to a cellular network scenario, multiple RNs can be deployed in each sector. In such a case, we need to find the optimal number of deployed relay nodes (RNs) and their locations.

Figure 6.13 shows the SINR map with 2 and 4 relay nodes in each sector, respectively. The SINRs of the vicinity areas of the relay nodes are enhanced. When the number of relay nodes increases, a larger area is covered by the relay nodes, which means UEs are more likely to receive help from the relay nodes.

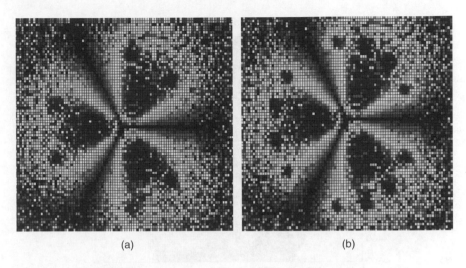

Figure 6.13 SINR map with 2 and 4 relay nodes in each sector

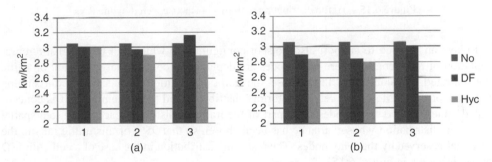

Figure 6.14 Energy efficiency: (a) one RN, (b) 4 RNs in (1) short, (2) middle, and (3) long BS-RN distance

As shown in Figure 6.14, hybrid relay achieves better EE performance than DF and the best EE performance is achieved at the cell edge. With four relays, although the overall energy consumption is increased compared to the one relay case during the time of transmission, the required transmission time for UEs served by RNs is reduced, increasing the idle time of the BSs. In this regard, more UEs can obtain help from the RNs and lead to lower energy consumption. However, this does not mean the number of relays can be increased without limit. At some point, the disadvantage of more offset energy consumption, which is the energy consumption even when transmission power is zero, will outweigh the benefit of improved throughput and transmission power, thus causes more energy consumption. Based on these results, we can conclude that the hybrid relay system that enables a pair of terminals (BS and RNs) to exploit spatial diversity shows significant improvement in energy efficiency performance in terms of consumed energy per km². However, compared with direct transmission, the cooperative strategy only shows significantly improved energy efficiency when the RN is not deployed in the very close area of the BS.

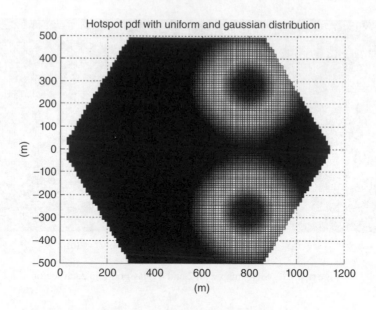

Figure 6.15   Traffic distribution with two hotspots covered by the RNs

Relay can be used to support also hotspot configurations which can be a realistic representation of the scenario where relay nodes are deployed to collect a large amount of traffic concentrated in a small area. In this case, the traffic distribution has been shaped assuming that half of the traffic of the cell is uniformly distributed and the remaining half is concentrated under the two relay nodes. This part of the traffic is distributed according to a spatial Gaussian distribution whose variance has been chosen so that 95% of this traffic falls in the area that is served by the relay nodes. The resulting distribution in the case of a cell with ISD 500 m is shown in Figure 6.15.

The results are shown in Figure 6.16 [25]. The column relative to BS SoTA uses the power model provided in Ref. [18] for state-of-the-art 2010 Base stations. It can be seen that RNs serve the hotspot of traffic offers a significant improvement, lowering the overall energy consumption index of 20.8% in the case of 2-hop relay nodes, and of 5.6% in the case of multicast relay nodes. Even higher gains can be achieved when also base stations are upgraded with a more efficient power model. When newer base stations are deployed alone, a good increase in energy efficiency is reached, but the improvements granted by relay nodes still stack with this higher performance, lowering the corresponding energy efficiency index of an additional 28.4% in the case of 2-hop relays and of 2.5% in the case of multicast relays.

6.5   Conclusions and Future Works

In this chapter, the detailed designing choices as well as the network planning solutions have been explained based on theoretical derivations, performance evaluations and system level studies. The discussed solutions address the high-energy efficiency targets of the future mobile cellular networks which feature very high traffic demanding, highly fluctuating daily traffic profile, and so on.

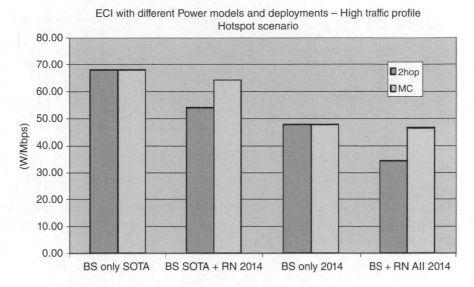

Figure 6.16   Overall ECI in the hotspot scenario with different deployments]

We have shown that the network design parameters, such as cell size, can be optimized according to the spatial traffic variations to save energy. After rollout, the network layout is difficult to change, but there are opportunities to adaptively adjust the network layout by applying some on/off schemes according to daily traffic variations. In particular, the (de)sectorization scheme, where the BS can switch off some of sectors to save energy during low traffic periods, is shown to be promising.

In the literature, it is often stated that the usage of small cells reduces transmit power. However, it is explained in this chapter that the densification of the current network with additional small cells can only be energy efficient in high traffic. Furthermore, it is shown that non-hexagon deployed heterogeneous network might be optimal when the spatial traffic is not even.

Furthermore, relaying technique is investigated for rural scenario. The hybrid relaying scheme with capability of switching between DF and CF are achieving considerably energy savings if their locations are carefully chosen. Additionally, relay can also be used in hotspot to improve energy efficiency.

Some new solutions are proposed for future architecture evolution. One of them is the idea of separation of control and data plan as shown in Figure 6.17. In this split concept, the user terminals exchange data traffic with small cells and signalling traffic with large base stations which act like a umbrella covering multiple small cells. This concept provides additional degree of freedom and has potentials to improve the resource usage efficiency and energy efficiency because the small cells no longer need to transmit/receive signalling messages frequently and, therefore can be put into "deep sleep" mode to save energy.

Another proposed idea is the federated network where the network is virtualized in a federated manner through infrastructure sharing among operators, with the capability in the provisioning of ample opportunity for competition and differentiation among involved actors. The designed architecture and management mechanisms exploit increasing variation of traffic

Figure 6.17   Control/data plane splitting

distribution among operators and ensure transparent service provisioning to end users. These potential candidates for future architectures are currently drawing more and more attention from both academia and industry.

References

[1] A. Penttinen, Chapter 10 – Network Planning and Dimensioning, Lecture Notes: S-38.145 – Introduction to Teletraffic Theory, Helsinki University of Technology, Fall, 1999.

[2] AriaNetworks white paper, "Seven best practices for successful mobile network planning," [online]. Available: http://www.aria-networks.com.

[3] S. Xie, "Planning, design and building large-scale networks at campus," in Proc. ICCRD, pp. 495–498, 2011.

[4] https://www.ict-earth.eu/Earth project Summary leaflet.

[5] Press Release, EU Commissioner Calls on ICT Industry to Reduce Its Carbon Footprint by 20% as Early as 2015, MEMO/09/140, 2009.

[6] http://www.greentouch.org.

[7] C. Lamour, "Energy consumption of mobile networks," The Basestation eNewsletter, 2008.

[8] F. Richter, A. J. Fehske, and G. P. Fettweis, "Energy efficiency aspects of base station deployment strategies for cellular networks," in Proc. of VTC2009-Fall, Anchorage, Alaska, USA, pp. 1–5, 2009.

[9] M. A. Marsan, L. Chiaraviglio, D. Ciullo, and M. Meo, "Optimal energy savings in cellular access networks," in Proc. of ICC2009 Communications Workshops, Dresden, Germany, pp. 1–5, 2009.

[10] K. S. Neier-Hellstern, "The analysis of a queue arising in overflow models," IEEE Trans. Commun., vol. 37, no. 4, pp. 367–372, 1989.

[11] L. N. Singh and G. R. Dattatreya, "A novel approach to parameter estimation in Markov-modulated Poisson processes," in 2004 IEEE Emerging Technologies Conference (ETC), Richardson Texas, 2004.

[12] F. Richter, A. J. Fehske, and G. P. Fettweis, "Energy efficiency aspects of base station deployment strategies for cellular networks," in Proc. of VTC2009-Fall, Anchorage, Alaska, USA, pp. 1–5, 2009.

[13] K. S. Neier-Hellstern, "The analysis of a queue arising in overflow models," IEEE Trans. Commun., vol. 37, no. 4, pp. 367–372, 1989.

[14] EARTH project, "D2.2 – definition and parameterization of reference systems and scenarios," [online]. Available: https://www.ict-earth.eu/publications/publications.html.

[15] EARTH project, "D3.1 – most promising tracks of green network technologies," [online]. Available: https://www.ict-earth.eu/publications/publications.html.

[16] Y. Qi, M. Imran, and R. Tafazolli, "On the energy aware deployment strategy in cellular systems," in Proc. of PIMRC 2010, pp. 363–367, 2011.

[17] Technical Specification Group Radio Access Network, "Evolved universal terrestrial radio access (E0UTRA); LTE radio frequency (RF) system scenarios," 3rd Generation Partnership Project (3GPP), Tech. Rep. TS 36.942, 2008–2009.

[18] EARTH project, "D2.3 - Energy efficiency analysis of the reference systems, areas of improvements and target breakdown" [online]. Available: https://www.ict-earth.eu/publications/deliverables/deliverables.html.

[19] EARTH project, "D3.3 – Final report on green network technologies" [online]. Available: https://www.ict-earth .eu/publications/deliverables/deliverables.html.

[20] J. N. Laneman, D. N. C. Tse, and G. W. Wonell, "Capacity diversity in wireless networks: efficient protocols and outage behavior," IEEE Trans. Inform. Theory, vol. 50, no. 12, pp. 3062–3080, 2004.

[21] J. N. Laneman, Cooperative diversity in wireless networks: algorithms and architectures, PhD thesis, Massachusetts Institutes of Technology, 2002.

[22] 3GPP TR 36.814 v1.5.1 (2009-12), Further Advancements for E-UTRA, Physical Layer Aspects.

[23] M. Kats and S. Shamai, "Relay protocols for two colocated users," IEEE Trans. Inform. Theory, vol. 52, no. 6, pp. 2329–2344, Jun. 2006.

[24] Y. Qi, R. Hoshyar, M. A. Imran, and R. Tafazolli, "H^2-ARQ-relaying: spectrum and energy efficiency perspectives," IEEE JSAC, vol. 29, no. 8, pp. 1547–1558, 2011.

[25] Y. Qi, M. Ali Imran, S. Dario, and F. Roberto, "On the development opportunities for increasing energy efficiency in LTE-advanced with relay nodes," in Proc. WWRF, 2012.

[26] Y. Qi, M. Ali Imran, R. Demo Souza, and R. Tafazolli, "On the Optimization of Distributed Compression in Multi-Relay Cooperative Networks," accepted by IEEE Trans. on Vehicular Technology, 2015.

7

Green Radio

Taewon Hwang[1], Guowang Miao[2], Hyunsung Park[1], Younggap Kwon[1]
and Nageen Himayat[3]

[1] *Yonsei University, Department of Electrical and Electronic Engineering, Seoul, Korea*
[2] *KTH Royal Institute of Technology, Communications Department, Stockholm, Sweden*
[3] *Intel Corporation, Intel Labs, Santa Clara, USA*

7.1 Energy-Efficient Design for Single-User Communications

Shannon theory provides some insights into the fundamental limits on minimum energy per bit required for reliable transmission for a single-user link. Based on the Shannon capacity formula, the data rate of an ideal band-limited additive white Gaussian noise (AWGN) channel is given as follows:

$$R = W \log_2 \left(1 + \frac{Pg}{WN_o} \right), \quad \text{[bits/sec]}$$

where W is the channel bandwidth, P the transmission power, g the channel power gain, and N_o the noise spectral density. From a system-design perspective, g and N_o are usually given and W and P can be controlled by an appropriate transceiver design. To save energy, the transceiver should be designed to maximize the number of information bits reliably delivered per unit energy consumption, that is, the power should be chosen to maximize

$$u = \frac{W \log_2 \left(1 + \frac{Pg}{WN_o} \right) t}{Pt} = \frac{W \log_2 \left(1 + \frac{Pg}{WN_o} \right)}{P}, \quad \text{[bits/Joule]}$$

where t is the time used to send the bits, u is named the energy efficiency, with a unit bits/Joule. Note that choosing the power is the same as choosing the data rate and they are related through the power-rate function, for example, the Shannon capacity formula here. Therefore, another way formulating the problem is to write u as the function of the data rate R and the result

Green Communications: Principles, Concepts and Practice, First Edition.
Edited by Konstantinos Samdanis, Peter Rost, Andreas Maeder, Michela Meo and Christos Verikoukis.
© 2015 John Wiley & Sons, Ltd. Published 2015 by John Wiley & Sons, Ltd.

should be the same. It can be easily proved that u decreases in P and the lowest power should be used. Therefore, the highest energy efficiency is achieved when $P \to 0$, that is, $R \to 0$ and infinitely long time should be used for data transmission, and

$$u^* = \lim_{P \to 0} \frac{W \log_2 \left(1 + \frac{Pg}{WN_o} \right)}{P} = \frac{g}{N_0 \log 2}.$$

Similarly, if W is a variable, then u increases in W. The highest energy efficiency is achieved when $W \to \infty$, and

$$u^* = \lim_{W \to \infty} \frac{W \log_2 \left(1 + \frac{Pg}{WN_o} \right)}{P} = \frac{g}{N_0 \log 2}.$$

In both of the aforementioned cases, the spectral efficiency is zero, indicating a trade-off between spectral and energy efficiency.

The ideal analysis above ignores channel impairment and practical issues such as delay spread, frequency selectivity of the channel, phase noise, nonlinearity of power amplifiers, and other wideband RF circuits. Furthermore, in addition to transmission power, a device will also incur additional circuit power that is relatively independent of the transmission rate [1, 2]. Thus a fixed energy cost of transmission is incurred, which must also be accounted for when designing energy-efficient transmission systems. In the following sections, we consider energy-efficient design for single-user transmission, accounting for the impact of circuit power. We first consider the case of flat fading wireless channels before addressing the more complex case of frequency-selective channels.

7.1.1 Energy-Efficient Transmission in Flat Fading Channels

To facilitate the understanding, we first consider a special case that the channel is experiencing flat fading. This happens in narrowband communications. We examine the basic relationship between energy efficiency and channel gain, circuit power, and system bandwidth. In this section, we investigate the optimal transmission power level maximizing the bits/Joule metric, accounting for the circuit power, P_c, consumed during transmission. The total energy consumption considering the circuit power is given as follows:

$$E = (P + P_c)t,$$

and the system energy efficiency is given as follows:

$$u(P) = \frac{W \log_2 \left(1 + \frac{Pg}{WN_o} \right)}{P + P_c}.$$

It can be easily shown that u is strictly quasi-concave in P, as shown in Figure 7.1 [3]. For a strictly quasi-concave function, if a local maximum exists, it is also globally optimal. Besides, u is first strictly increasing and then strictly decreasing in P. Therefore, there is a

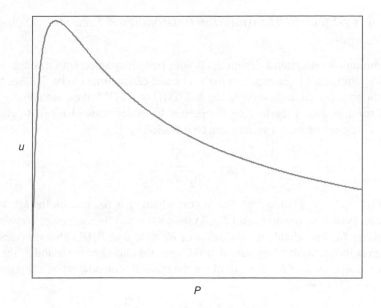

Figure 7.1   Quasi-concavity of the energy-efficiency function

unique optimal P^*, that is, a unique data rate R^*, that maximizes the energy efficiency and the first-order derivative of $u(P)$ at P^* is zero. P^* can be expressed as

$$P^* = \left(P^* + \frac{WN_0}{g} \right) \log \left(1 + \frac{P^* g}{WN_0} \right) - P_c.$$

In general, there is no closed-form expression of P^* and numeric methods such as the bisection method or the gradient assisted binary search (GABS) in [3] can be used to search for P^*. The basic idea of GABS is to first find a range that includes P^* and then use the bisection method to narrow down the range to the desired accuracy.

Further analyzing the energy-efficiency function, we can see that it increases with the bandwidth, W, indicating higher signal bandwidth can improve the transmitter energy efficiency. In a multiuser system, this means more system bandwidth should be allocated to users desiring energy efficiency [1].

It is shown in Ref. [3] that when the channel gain increases, higher transmission data rate should be used to improve energy efficiency. Furthermore, when the transmitter has a higher circuit power, higher transmission data rate should also be used. This is because with higher data rate, the duration the device has to be on can be decreased to reduce the circuit power consumption. When circuit power dominates power consumption, which is usually true with short-range communication, the highest data rate should be used to finish the transmission as soon as possible and then the device can be switched into a lower circuit power state, for example, sleep mode, as soon as possible. This technology has been commonly used in most medium access control (MAC) layer energy-efficient designs. When the circuit power is negligible, which is usually true with long-range communication such as satellite communications, the lowest data rate should be used, which coincides with the results in Refs. [4] and [5].

7.1.2   Energy-Efficient Transmission in Broadband Frequency-Selective Channels

Current communication systems design deals with frequency selectivity through subdividing the bandwidth into small segments, where the channel can assumed to be flat. So, for an ideal channel-orthogonalization technology such as MIMO or OFDM, the channel may be divided into K subchannels, each experiencing flat fading. Consider static channels to gain insights. The energy efficiency of such a system can be modeled by

$$u(R) = \frac{\sum r_k}{P_T(R) + P_c},$$

where $R = [r_1, r_2, \ldots, r_K]$ is the data rate vector where r_k is the rate on the kth subchannel, $\sum r_k$ the total system throughput, and $P_T(R)$ the total transmission power consumed by the power amplifier for the reliable transmission of R. Note that $P_T(R)$ characterizes all power consumption components that may vary depending on the data rate vector and P_c the remaining ones that are independent of R. An example of the transmission power consumption for coded QAM system is [3]

$$P_T(R) = \sum_{i=1}^{K} \left(e^{\frac{r_i}{W}} - 1 \right) \frac{N_0 W \mu}{g_i \gamma},$$

where μ is the SNR gap that defines the gap between the channel capacity and a practical coding and modulation scheme, and γ the power amplifier efficiency and depends on the design and implementation of the transmitter. It can be easily verified that $P_T(R)$ for almost all communication systems is strictly convex and monotonically increasing in R and in the following we make this assumption.

As shown in [3], $u(R)$ is strictly quasi-concave in R. In addition, it is either strictly decreasing or first strictly increasing and then strictly decreasing in any r_i, that is, the local maximum of $u(R)$ for each r_i exists at either 0 or a positive finite value. For strictly quasi-concave functions, if a local maximum exists, it is also globally optimal. Hence, a unique globally optimal transmission rate vector always exists and the necessary and sufficient condition of the optimal R can be found by setting the first-order derivative of $u(R)$ to be zero. When each subchannel achieves the Shannon capacity, the optimal power allocation is a dynamic energy-efficient water-filling approach. With this approach, while the absolute value of power allocation is determined by the maximum energy efficiency $u(R^*)$, which relies on both the circuit power and channel state, the relative differences of power allocations on different subchannels depend only on the channel gains on those subchannels.

The discussions above have not considered quality of service (QoS) assurance or resource constraints. If the global optimal transmission meets the QoS requirements and resource constraints, it can be used. Otherwise, the transmission should be adapted right on a subset of the boundary conditions and the optimality is guaranteed because of the strict quasi-concavity of the energy-efficiency function. Some examples of energy-efficient designs considering the constraints can be found in Refs. [3] and [6].

Similar to the energy-efficient transmission in flat fading channels, there are no closed-form expressions of the globally optimal transmission setting. Using appropriate approximation techniques, suboptimal closed-form link adaptation can be obtained. In Ref. [7], by using time-average energy efficiency, closed-form link adaptation is obtained with the knowledge of historical link energy efficiency in the past time slots and the performance is very close to the globally optimal one, depending on how fast the channel varies.

7.2 Energy-Efficient Design for Multiuser Communications

While it is important to investigate energy-efficient design for a single-link, wireless transmission is inherently multiuser in nature, with several mechanism defined for multiple access. Multiuser communications can simultaneously support a large number of users satisfying their QoS requirement, and therefore it significantly enhances the system performance as compared with single-user communications. In multiuser systems, the system resources must be divided among multiple users. For parallel transmission of symbols, multiuser MIMO allocates spatial degrees of freedom brought by the use of multiple antennas to multiple users, whereas OFDMA distributes subchannels obtained by dividing the entire bandwidth to multiple users. In cognitive radio, unlicensed secondary users can access the spectrum licensed to primary users by exploiting the idle times of the primary users. On the other hand, in cooperative systems multiple users can cooperate to enhance the transmissions of other users. In this section, we discuss energy-efficient designs of multiuser techniques such as multiuser MIMO, OFDMA, cognitive radio, and cooperative relay transmission.

7.2.1 Multiuser MIMO

For a single-user MIMO (SU-MIMO) system with M antennas at the base station (BS) and N antennas at the mobile user, the capacity gain is approximately $\min(M, N)$ times that of single-input single-output (SISO) systems [8]. In typical cellular systems, multiple antennas can be easily deployed at the BS, but the number of antennas at the mobile users is limited because of the size and cost constraints. Therefore, the capacity gain of SU-MIMO is usually limited by the number of antennas at the mobile user. An alternative to SU-MIMO is multiuser MIMO (MU-MIMO). In MU-MIMO systems, the BS serves multiple users simultaneously in the same frequency and time slots by spatially multiplexing the users' data using multiple antennas. The sum capacity of MU-MIMO grows linearly with $\min(M, nN)$, where n is the number of users served simultaneously, so an M-fold increase in the sum capacity can be obtained as long as nN is larger than M. When the number of users is larger than M, a scheduler can be employed to select up to M users. By opportunistically transmitting to the selected users having good channel conditions, multiuser diversity can be obtained [9]. Besides, by carefully allocating resources such as power, antennas, and subcarriers to the scheduled users, the performance of the MU-MIMO system can be improved. The user scheduling and the resource allocation can significantly improve the performance of MU-MIMO systems. In this subsection, we discuss user scheduling and resource allocation schemes to improve the energy efficiency of MU-MIMO systems.

We consider an energy-efficient user-scheduling problem in a downlink MU-MIMO system. With a zero-forcing precoder at the BS, the energy efficiency related to a user is given as:

$$u_i = \frac{\log_2 \left(1 + \frac{\gamma_i P_{T_i}}{\sigma^2} \right)}{P_{T_i} + P_C},$$

where γ_i is the effective channel power gain from the BS to the user, P_{T_i} is the transmit power that the BS consumes for the user, σ^2 is the noise power, and P_C is the circuit power of the BS. Due to the zero-forcing precoder, γ_i of a user depends on the channels of the other scheduled users as well as its own channel. The user scheduler that maximizes the energy efficiency of the BS is given as follows:

$$S'_n = \underset{S_n \in \Omega}{\text{argmax}} \sum_{i \in S_n} u_i,$$

where S_n is a set of scheduled users who are serviced simultaneously from the BS and Ω is the collection of all possible scheduled user sets. The number of possible scheduled user sets is $\binom{N_u}{k}$, where N_u is the number of all users and $k = |S_n| \leq M$ is the number of the scheduled users. Since the above scheduler does not consider the fairness, the cell-edge users, whose channel conditions are usually bad, are hardly selected.

Denote T_i to be the accumulated throughput for the ith user. The proportional-fair energy-efficient user scheduler

$$S'_n = \underset{S_n \in \Omega}{\text{argmax}} \sum_{i \in S_n} \frac{u_i}{T_i},$$

can balance the cell-average energy efficiency and the cell-edge energy efficiency by increasing the priorities of the users that have been served less (with lower T_i) [10].

Now, we discuss energy-efficient power allocation. For a given set of scheduled users, $r_i = \log_2 \left(1 + \frac{\gamma_i P_{T_i}}{\sigma^2}\right)$ is independent of the powers of the other scheduled users' due to the zero-forcing precoding at the BS and therefore, optimal P_{T_i} that maximizes $\sum_{i \in S_n} \frac{u_i}{T_i}$ is the one that maximizes $\frac{u_i}{T_i}$. Moreover, since r_i is strictly concave in P_{T_i} and $P_{T_i} + P_C$ is convex in P_{T_i}, u_i is strictly quasi-concave in P_{T_i} [11, 12]. Therefore, the optimal P_{T_i} is unique.

In an uplink MU-MIMO system where each user has multiple antennas, the energy efficiency of the users is written as

$$u = \frac{\sum_i \sum_k r_{ik}}{\sum_i \left(\sum_k P_{T_{ik}} + P_{a_i} + P_c\right)} \tag{7.1}$$

where r_{ik} and $P_{T_{ik}}$ are the data rate and transmit power for the kth stream of the ith user, respectively, P_{a_i} is the antenna-related circuit power consumption of the ith user, and P_c is the power consumption of circuit components that are independent of the antenna circuit operations. If the users use zero-forcing precoders and the BS uses a zero-forcing receiver, the data rate for the kth stream of the ith user is given as follows:

$$r_{ik} = \log_2 \left(1 + \frac{\gamma_{ik} P_{T_{ik}}}{\sigma^2}\right),$$

where γ_{ik} is the equivalent channel power for the kth stream of the ith user and σ^2 is noise power. Since r_{ik} is a strictly concave function and sum of strictly concave functions is a strictly concave function, the numerator of u is also a strictly concave function. Similarly, the denominator of u is a convex function. Therefore, u is a strictly quasi-concave function of $\{P_{T_{ik}}\}$ and there exists a unique optimal $\{P_{T_{ik}}\}$ that maximizes u.

While using a large number of active antennas is always beneficial for increasing the spectral efficiency, it can decrease the energy efficiency because it requires more antenna-related circuit power consumption. Therefore, an improved circuit management scheme that can turn off circuit operations of the user-antennas whose energy efficiency is too low can improve the energy efficiency [13].

In some cases, the problem of maximizing energy efficiency can be solved more efficiently by using the Dinkelbach method [12]. When the numerator and the denominator of an objective function are concave and convex, respectively, the objective function can be transformed to an equivalent objective function in subtractive form which is a concave function. For example, the problem of maximizing the energy efficiency in Eq. (7.1) can be solved by the Dinkelbach method as follows. First, set $q = 0$ and calculate as:

$$P'_{T_{ik}} = \operatorname*{argmax}_{P_{T_{ik}}} \sum_i \sum_k r_{ik} - q \sum_i \left(\sum_k P_{T_{ik}} + P_{a_i} + P_c \right).$$

Then, unless $\sum_i \sum_k r_{ik} - q \sum_i \left(\sum_k P'_{T_{ik}} + P_{a_i} + P_c \right) \approx 0$, update q by using the equation

$$q = \frac{\sum_i \sum_k r_{ik}}{\sum_i \left(\sum_k P'_{T_{ik}} + P_{a_i} + P_c \right)}.$$

This process repeats until $\sum_i \sum_k r_{ik} - q \sum_i \left(\sum_k P'_{T_{ik}} + P_{a_i} + P_c \right) \approx 0$.

In a MU-MIMO system, if the precoders and the receivers cannot fully eliminate the interference between users, the data rate of a user is affected by the powers of the other users as well as its own power. Moreover, the data rate of a user is not a concave function of transmit powers of all the users. In this case, the optimization problem is nonconvex and a brute force approach can be used to obtain a global optimal solution. By using some approximations, we can obtain a low-complexity suboptimal solution of the problem. For example, Ref. [14] considers a downlink MU-MIMO system where the BS adopts the MRT precoder. Since the object function is coupled and not concave, finding the optimal solution is difficult. For this reason, the energy efficiency is approximated to a lower bound which is a quasi-concave function. The problem of optimizing the quasi-concave energy efficiency function has been discussed above.

7.2.2 Orthogonal Frequency Division Multiple Access (OFDMA)

In OFDMA, the entire bandwidth is divided into a number of subchannels to transmit symbols for multiple users in a parallel fashion. By assigning each subchannel to the best user and adapting the rate and power according to its channel condition, the system throughput can be optimized. For these characteristics, OFDMA has been adopted by 4G standards such as 3GPP LTE and IEEE 802.16 WiMAX. In this subsection, we discuss adaptive subcarriers, power and rate allocation schemes for OFDMA to enhance the energy efficiency.

Consider an uplink OFDMA system in a flat fading channel. The energy efficiency of the ith user is given as follows:

$$u_i(c_i, P_{T_i}) = \frac{c_i \log_2 \left(1 + \frac{g_i P_{T_i}}{\sigma^2} \right)}{c_i P_{T_i} + P_c},$$

where c_i is the number of subcarriers allocated to the ith user, P_{T_i} is transmit power of the ith user on each subcarrier, g_i is the channel power gain of the ith user, σ^2 is the noise power, and P_c is the circuit power of the user.

For a given subcarrier allocation c_i to a user, the unique optimal power $P^*_{T_i}$ that maximizes u_i can be easily found because u_i is a quasi-concave function of P_{T_i}. Then, the optimal energy efficiency u^*_i and the corresponding rate r^*_i can be determined by using $P^*_{T_i}$. The optimal energy efficiency u^*_i increases with the channel power gain g_i. That is, if $g_i \leq g_j$ for $c_i = c_j$, then $u^*_i \leq u^*_j$ because

$$u^*_i = \frac{c_i \log_2 \left(1 + \dfrac{g_i P^*_{T_i}}{\sigma^2} \right)}{c_i P^*_{T_i} + P_c} \leq \frac{c_j \log_2 \left(1 + \dfrac{g_j P^*_{T_i}}{\sigma^2} \right)}{c_j P^*_{T_i} + P_c} \leq \frac{c_j \log_2 \left(1 + \dfrac{g_j P^*_{T_j}}{\sigma^2} \right)}{c_j P^*_{T_j} + P_c} = u^*_j.$$

Also, since the energy efficiency of the ith user can be rewritten as

$$u_i = \frac{\log_2 \left(1 + \dfrac{g_i P_{T_i}}{\sigma^2} \right)}{P_{T_i} + \dfrac{P_c}{c_i}},$$

increasing c_i is equivalent to decreasing the circuit power of the user. Therefore, energy efficiency increases with the number of allocated subcarriers. Since the number of total subcarrier is limited, $\sum_i c_i \leq K$, a proper subcarrier allocation is required to maximize the sum energy efficiency of users. Consider a toy OFDMA system with two users and $K = 10$ subcarriers. As c_1 increases, u_1 also increases while u_2 decreases because $c_2 = 10 - c_1$ decreases. As shown in Figure 7.2, there exits optimal subcarrier allocation that maximizes the sum energy efficiency of users.

For a given power allocation, the optimal subcarrier allocation c_i's that maximize sum energy efficiency of users $\sum_i u_i$ can be uniquely obtained because u_i is strictly concave in c_i and sum of strictly concave functions is a strictly concave function. An energy-efficient power and subcarrier allocation that can significantly improve the overall network energy efficiency is introduced in Ref. [1].

The above energy-efficient resource allocation for OFDMA can be extended to frequency-selective channels. In this case, dynamic power allocation across the subcarriers is required as well as power allocation across the users because even the subcarriers allocated to the same user experience different channel conditions. Each user's optimal power allocation across the subcarriers follows the water-filling algorithm, and the total power of a user should be determined to maximize its energy efficiency. Besides, the subcarrier allocation for the frequency-selective channel needs to decide which subcarriers are to be allocated to which users as well as the number of subcarriers to be allocated to each user. For these reasons, the subcarrier allocation for frequency-selective channels is much more complex in computation than that for flat fading channels. To reduce the complexity of subcarrier allocation, a suboptimal iterative subcarrier allocation that maximizes the minimum energy efficiency of a user rather than maximizing the system energy efficiency is proposed in Ref. [15].

If we use a nonexclusive subcarrier allocation, which allows allocating a subcarrier to more than one user, the performance of the OFDMA system can be further improved. However,

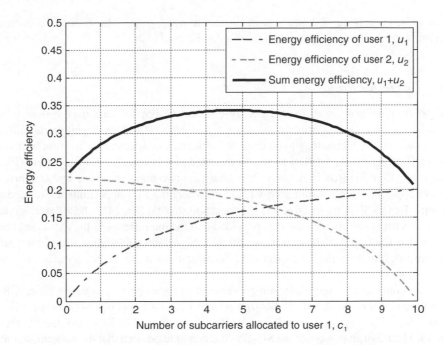

Figure 7.2   Sum energy efficiency of a two-user OFDMA system, $c_1 + c_2 = K = 10$

the nonexclusive subcarrier allocation makes the power and subcarrier allocation problem that maximizes sum energy efficiency of users to be coupled and nonconvex. Therefore, convex optimization techniques cannot be employed. In this case, a game-theoretic approach can be used. For example, Ref. [16] proposes a noncooperative game to obtain a low-complexity suboptimal energy-efficient power and subcarrier allocation. Here, the utility functions are chosen to satisfy the conditions of a potential game, which always admits Nash equilibrium [17]. The best response of each user can be found as follows. For a given subcarrier allocation, each user finds its optimal power that maximizes its utility. Each user can find the optimal subcarrier allocation that maximizes its utility function by the exhaustive search.

The above energy-efficient techniques for OFDMA systems assume perfect channel state information (CSI) and neglect the energy consumptions necessary for channel estimation. However, to be more precise, energy consumption in transmitting the pilot signals for channel estimation needs to be considered in the energy efficiency. Consider a downlink OFDMA system with M users and multiple subcarriers. Pilot symbols are periodically placed in the frequency domain and shared by the users. Denote α and β_i to be the pilot power and the data power of the ith user, respectively. The achievable rate of the ith user, C_i, is a function of pilot power α and its data power β_i [18]. The energy efficiency can be written as:

$$u = \frac{\sum_i C_i(\alpha, \beta_i)}{\alpha + \sum_i \beta_i + P_c},$$

where P_c is the circuit power at the BS. Cleary, the rate C_i increases as the pilot power α increase. Also, C_i is increasing in β_i. Our interest is to find α and β_i's that maximize u under the constraint of $\alpha + \sum_i \beta_i \leq P_{\max}$. It is shown in Ref. [18] that u is not jointly quasi-concave in α and β_i's, but u is quasi-concave with respect to $\alpha, \beta_1, \cdots \beta_M,$. In this case, a coordinate search that alternately finds optimal α and β_i's can be used [19].

7.2.3   Cognitive Radio

In the previous sections, we covered energy-efficient design for multiuser system wherein no specific spectrum access priority was assigned to users. However, cognitive radio (CR) is another potential key technology to increase the efficiency of spectrum utilization by allowing unlicensed secondary users (SUs) to access the spectrum licensed to primary users (PUs) as long as the QoS of the PUs is ensured [20]. There are two main spectrum access approaches in CR (1) opportunistic spectrum access [21, 22], where the SUs opportunistically access the spectrum when the PUs are inactive and (2) spectrum sharing [23, 24], where the SUs concurrently access the spectrum with the PUs provided that the interference to the PUs is kept below an acceptable level. In this subsection, we discuss the resource allocation techniques such as power control, beamforming, and spectrum allocation to increase the energy efficiency of the CR networks.

First, we consider an energy-efficient power control for a spectrum-sharing based CR network, where an SU network with K_s SU transmitters (SU-TXs) and one SU-receiver (SU-RX) coexists with a PU network consisting of K_p PU transmitters (PU-TXs) and one PU receiver (PU-RX). The transmit powers of the SU-TXs are constrained such that the maximum interference caused by all the SU-TXs to the PU-RX is below an acceptable level. The kth transmitter (among those $K_s + K_p$ transmitters) sends bits at a common bit rate R in packets consisting of L information bits and $M - L$ overhead bits. In this case, the energy efficiency of the kth transmitter can be written as:

$$u_k(\mathbf{p}) = R \frac{L}{M} \frac{(1 - e^{\text{SINR}(p)})^M}{p_k},$$

where $p_k \in [0, P_{k,\max}]$ is the transmit power of the kth transmitter, $\mathbf{p} = [p_1, \ldots, p_{K_p+K_s}]$ is transmit power vector, and $(1 - e^{\text{SINR}(p)})^M$ is the probability that a packet is correctly received. Note that $P_{k,\max}$ is the transmit power limit if the kth transmitter is a PU-TX and the precalculated maximum allowed transmit power to ensure the QoS of the PUs if the kth transmitter is an SU-TX [25].

If each transmitter selfishly chooses its transmit power to maximize its energy efficiency based on its local information, the power control problem can be formulated as a noncooperative game represented by $\mathcal{G} = \left\{ \mathcal{K}, \{S_k\}_{k=1}^{K_p+K_s}, \{u_k\}_{k=1}^{K_p+K_s} \right\}$, where $\mathcal{K} = \{1, \ 2, \ldots, K_p + K_s\}$ is the set of players (PU-TXs and SU-TXs), $S_k = [0, P_{k,\max}]$ is the set of the kth player's strategy (transmit power p_k), u_k is the kth player's utility (energy efficiency). The utility u_k can be shown to be strictly quasi-concave in p_k. If the utility function of each player is continuous in strategy vector $\mathbf{p}$ and quasi-concave in its strategy p_k, then the noncooperative game has at least one Nash equilibrium as long as the strategy set is compact and convex. Therefore, the above power control game has at least one Nash equilibrium. Moreover, the Nash equilibrium can be shown to be unique. Since u_k is strictly quasi-concave in p_k, the best response of the kth transmitter, or the p_k that maximizes u_k for a given set of other transmitters' powers is the minimum of (1) its power limit and (2) the power that satisfies $\frac{\partial u_k}{\partial p_k} = 0$. The best response function of this game can be shown to be a standard function [25]. A function $f(\mathbf{p})$ is standard if for all

$\mathbf{p} \geq \mathbf{0}$, it has (1) positivity: $f(\mathbf{p}) > 0$, (2) monotonicity: if $\mathbf{p} \geq \mathbf{p}'$, then $f(\mathbf{p}) > f(\mathbf{p}')$, and (3) scalability: For all $\alpha > 1$, $\alpha f(\mathbf{p}) > f(\alpha \mathbf{p})$, where the vector inequality $\mathbf{p} \geq \mathbf{p}'$ is an inequality in all components. If the best response of each player is a standard function, then the noncooperative game has a unique Nash equilibrium [26].

Next, we consider energy-efficient beamforming in a CR MU-MIMO system, where an SU-TX transmits messages to K SU-RXs sharing the spectrum with multiple pairs of PU-TX and PU-RX. The energy efficiency of the SU-TX is given by

$$u = \frac{\sum_{k=1}^{K} r_k}{\sum_{k=1}^{K} tr(\mathbf{A}_k \mathbf{A}_k^H) + P_c},$$

where $\mathbf{A}_k$ is the beamformer for the kth SU-RX, r_k is the rate of the kth SU-RX, and P_c is the circuit power at the SU-TX. The problem of finding the optimal $\mathbf{A}_k$'s that maximize u under the constraints on (1) maximum interference power to the PU-RXs, (2) maximum power of the SU-TX, and (3) the minimum throughput requirement is a nonconvex problem, and therefore hard to solve directly. We can construct an equivalent quasi-concave problem [27]. First, for a given power $p = \sum_{k=1}^{K} tr(\mathbf{A}_k \mathbf{A}_k^H)$ of the SU-TX, the optimal $\mathbf{A}_k$'s that maximize the system throughput $\sum_{k=1}^{K} r_k$ under the above constraints are found. Then, the energy efficiency u can be expressed as a function of p, that is,

$$u = \frac{R(p)}{p + P_c},$$

where $R(p)$ is the system throughput as a function of p. There is no closed-form expression of $R(p)$, but it can be shown that $R(p)$ is a concave function of p. Since a concave function divided by a convex function is quasi-concave, the energy efficiency u is a strictly quasi-concave function of p [11, 12]. In this case, the optimal p that maximizes u can be found numerically, for example, using the golden section method [27].

Now, we consider an energy-efficient spectrum and power allocation in a heterogeneous CR network that consists of a cognitive macro base station (MBS), multiple macro SUs (MSUs), and multiple femto base stations (FBSs). Each FBS is assumed to provide service to one femto SU (FSU). The cognitive MBS first purchases spectrum resources from multiple primary networks and then allocates the spectrum resources to MSUs and FBSs to maximize its revenue. The spectrum resource purchase and allocation problem can be formulated as a three-stage Stackelberg game. In Stage 1, the multiple primary networks perform a price competition game, where each primary network selfishly determines the spectrum selling price that maximizes its revenue. As shown in Ref. [28], the price competition game has a unique Nash equilibrium. In Stage 2, the cognitive MBS decides the bandwidth of the spectrum to purchase from each primary network, allocates the purchased spectrum to MSUs or FBSs, and performs energy-efficient power allocation for the MSUs to maximize its revenue. In Stage 3, each FBS performs power allocation that maximizes its energy efficiency. The energy efficiency of each FBS is shown to be strictly quasi-concave in its power [28]. Therefore, the optimal power of each FBS is the minimum of (1) its peak power and (2) a local maximizer of the energy efficiency. It is shown in Ref. [28] that the Stackelberg game has a unique Stackelberg equilibrium. The unique Stackelberg equilibrium can be numerically obtained by a gradient-based iterative algorithm in Ref. [28].

7.2.3.1 Cooperative Relay

In a cooperative relay system, relays assist the transmissions between sources and their corresponding destinations. With decode-and-forward (DF) protocol, each relay decodes the received signal, re-encodes the message bit, and then forwards the re-encoded signal. On the other hand, with amplify-and-forward (AF) protocol, each relay forwards the received signal with amplification only. Cooperative relaying can significantly enhance the network coverage and capacity because it reduces the path loss by shortening the transmission distances and enables cooperative diversity that mitigates the detrimental effect of fading in wireless networks [29]. This subsection considers enhancing the overall energy efficiency of relay systems. Natural questions are (1) what is optimal power control? (2) what is optimal number of relays to be used? and (3) what is optimal deployment of relays? In the following, we discuss energy-efficient designs in relay systems to answer these questions.

First, we consider a relay system where K relays using DF protocol are deployed in a serial fashion between a source and a destination. In the wideband regime, the data rate of the ith link is $r_i = \frac{ch_i P_i}{d_i^\alpha}$, where P_i is the transmit power of the ith link, d_i is the distance of the ith link, α is the path loss exponent, h_i is the small fading gain for the ith link, and c is a constant. The overall energy efficiency of the system is

$$u(\mathbf{P}, \mathbf{d}) = \frac{R(\mathbf{P}, \mathbf{d})}{\sum_{i=0}^{K} P_i + P_{c,\text{tot}}}, \tag{7.2}$$

where R is the overall data rate between the source and the destination, $\mathbf{P} = [P_0, \dots, P_K]$, $\mathbf{d} = [d_0, \dots, d_K]$, and $P_{c,\text{tot}}$ is the total circuit power dissipated by the source, the relays, and the destination. Since the nodes are serially concatenated, the overall rate R is dominated by the worst link, that is, $R = \min_{0 \le i \le K} r_i$. Therefore, each node set its power $P_i = \frac{R d_i^\alpha}{ch_i}$ to achieve $r_0 = \cdots = r_K = R$; otherwise, any node with higher power will waste its energy. For given link distances $\mathbf{d}$, the energy efficiency u increases with R, and maximizing u is equivalent to maximizing R [30]. Therefore, the worst link needs to transmit at its maximum allowable transmit power $P_{\max}$.

Now, we consider adjusting link distances $\mathbf{d}$ as well as link powers $\mathbf{P}$ to maximize the energy efficiency. For given link powers $\mathbf{P}$, the optimal link distances $\mathbf{d}^*(\mathbf{P}) = [d_0^*(\mathbf{P}), \dots, d_K^*(\mathbf{P})]$ need to satisfy $R = r_0 = \cdots = r_k$. Inserting the obtained $\mathbf{d}^*(\mathbf{P})$ into Eq. (7.2), the energy efficiency $u(\mathbf{P}, \mathbf{d}(\mathbf{P}))$ can be shown to be strictly quasi-concave in $\mathbf{P}$ [3]. Therefore, each node uses the minimum of (1) its peak power and (2) the power at which the first-order derivative of $u(\mathbf{P}, \mathbf{d}(\mathbf{P}))$ is zero.

To study the relationship between the energy efficiency and the number of deployed relays, we consider the case of $h = h_0 = \cdots = h_K$. In this case, optimal transmit powers satisfy $P_0 = P_1 = \cdots = P_K$ and as shown in Ref. [30], the energy efficiency is increasing with P_0. Therefore, if there is no transmit power limits, each node transmits with infinite P_0. Then, the energy efficiency increases with the number of relays. The reason is as follows. The total energy consumption linearly increases with the number of relays. However, the achievable rate is exponentially increasing with the number of relays because (1) the link distances are inversely proportional to the number of relays and (2) the achievable data rate is exponentially decreasing with the link distances. Therefore, the energy efficiency increases as the number of relays increases when there is no transmit power limits.

Next, we consider energy-efficient power control in a relay network where an AF relay helps the transmissions of the multiple source–destination pairs. The multiple sources transmit at the same time using the same frequency band, and there exist interferences among the multiple source–destination pairs. Finding the optimal source powers that maximize the sum of the energy efficiencies of the sources is a coupled problem, and the tools of the convex optimization cannot be applied. In this case, the game-theoretic approach can be used. Assuming there is no cooperation between the sources, the source-power control problem can be formulated as a noncooperative power control game for the sources. In a noncooperative game, Nash equilibrium is a state where no player can improve its utility by changing only its own strategy unilaterally. It is shown in Ref. [31] that the power control game always has at least one Nash equilibrium. Moreover, when there is no direct links between sources and destinations, the Nash equilibrium is shown to be unique.

Besides, relay selection is another way to improve the energy efficiency of a relay system in a fading environment when there are multiple relays. The energy efficiency can be improved by opportunistically selecting the best relay among the multiple relays deployed. Some examples of energy-efficient relay selection schemes can be found in Refs. [32] and [33].

7.3 Summary and Future Work

This chapter has reviewed energy-efficient transmission and resource allocation techniques for wireless communications. First, the energy efficiency of the point-to-point AWGN channel was investigated from an information-theoretic perspective ignoring the circuit power. In this case, the energy efficiency strictly decreases in the transmit power, and therefore, the highest energy efficiency is achieved when the transmit power is zero. If the circuit power is considered, however, the energy efficiency is strictly quasi-concave in the transmit power. Therefore, there exists a unique nonzero optimal transmit power. In frequency-selective channels, OFDM can be employed to divide the entire bandwidth into multiple parallel subchannels, each experiencing flat fading. For a given total transmit power, maximizing the energy efficiency is equivalent to maximizing the spectral efficiency, whose solution is well known to be the water-filling algorithm. The total transmit power or the water level further needs to be adjusted to maximize the energy efficiency.

Besides, we have discussed the energy-efficient designs for the multiuser techniques such as MU-MIMO, OFDMA, cognitive radio, and cooperative relay transmission. In MU-MIMO, user scheduling is an important issue because it can significantly enhance the system performance by exploiting the multiuser diversity. For example, the energy-efficient proportional-fair user scheduling can balance the cell-average energy efficiency and the cell-edge energy efficiency. In OFDMA, energy-efficient subcarrier allocation schemes were investigated. If a user is allocated more subcarriers, the rate increases, but the power consumption also increases. Since the energy efficiency related to a user is strictly concave in the number of allocated subcarriers c_i, the optimal c_i is unique. If the subcarriers are allowed to be allocated nonexclusively to more than one user, interuser interference occurs. A noncooperative game was considered where each user selfishly chooses its subcarriers and powers to maximize its energy efficiency. If the utilities of the users are designed properly, the existence of Nash equilibrium can be guaranteed. In CR systems, secondary users should meet the QoS of primary users as well as improving their energy efficiencies. In a CR system where there are multiple secondary users, interuser interference occurs. In this case, a game-theoretic approach can be used, where each secondary user selfishly chooses its transmit power to maximize its own energy efficiency while protecting the QoS of the primary users. In cooperative relay systems, choosing the number of deployed relays and their locations is important as well as allocating powers to improve the energy efficiency. For a serially concatenated relay system, the performance bottleneck is the link with the worst SNR. Therefore, the optimal relay deployment and power allocation should ensure that all the links achieve the same data rate. If there is no transmit power limits, the energy efficiency increases with the number of relays deployed.

So far, many energy-efficient techniques have been proposed. However, there are still some important issues that need to be investigated. Most of the existing energy-efficient techniques for cellular system have considered only the single-cell environment. Extending the energy-efficient designs to multicell environments is important because multicell techniques can efficiently mitigate intercell interference and improve the system performance. Also, most of the energy-efficient OFDMA techniques considered only single antenna systems. However, OFDMA techniques can be combined with MIMO techniques to enhance the spectral efficiency. Therefore, energy-efficient MIMO-OFDMA techniques need to be investigated. Finally, most of the existing works on the energy-efficient design assumed perfect channel estimation and ignored the energy consumption for the channel estimation. To evaluate the energy efficiency more precisely, however, the energy consumed for the channel estimation needs to be considered.

References

Single-user references

[1] G. W. Miao, N. Himayat, Y. Li, and D. Bormann, "Energy-efficient design in wireless OFDMA," in Proc. IEEE ICC 2008, pp. 3307–3312, 2008.

[2] A. Y. Wang, S. Cho, C. G. Sodini, and A. P. Chandrakasan, "Energy efficient modulation and MAC for asymmetric RF microsensor system," in Int. Symp. Low Power Electronics and Design, Huntington Beach, CA, pp. 106–111, 2001.

[3] G. W. Miao, N. Himayat, and Y. Li, "Energy-efficient link adaptation in frequency-selective channels," IEEE Trans. Commun., vol. 58, no. 2, pp. 545–554, 2010.

[4] F. Meshkati, H. V. Poor, S. C. Schwartz, and N. B. Mandayam, "An energy-efficient approach to power control and receiver design in wireless networks," IEEE Trans. Commun., vol. 5, no. 1, pp. 3306–3315, 2006.

[5] B. Prabhakar, E. U. Biyikoglu, and A. E. Gamal, "Energy-efficient transmission over a wireless link via lazy packet scheduling," in Proc. IEEE Infocom 2001, Vol. 1, pp. 386–394, 2001.

[6] C. Xiong, G. Y. Li, S. Zhang, Y. Chen, and S. Xu, "Energy- and spectral-efficiency tradeoff in downlink OFDMA networks," IEEE Trans. Wireless Commun., vol. 10, no. 11, pp. 3874–3886, 2011.

[7] G. Miao, N. Himayat, G.Y. Li, S. Talwar, "Low-complexity energy-efficient scheduling for uplink OFDMA," IEEE Trans. Commun., vol. 60, no. 1, pp. 112–120, 2012.

Multiuser MIMO references

[8] D. N. C. Tse and P. Viswanath, Fundamentals of Wireless Communication. Cambridge, UK: Cambridge University Press, 2005.

[9] P. Viswanath, D. N. C. Tse, and R. Laroia, "Opportunistic beamforming using dumb antennas," IEEE Trans. Inf. Theory, vol. 48, no. 6, pp. 1277–1294, 2002.

[10] L. Liu, G. Miao, and J. Zhang, "Energy-efficient scheduling for downlink multi-user MIMO," in Proc. IEEE ICC 2012.

[11] S. Schaible, "Minimization of ratios," J. Opt. Theory Appl., vol. 19, no. 2, pp. 347–352, 1976.

[12] W. Dinkelbach, "On nonlinear fractional programming," Management Sci., vol. 13, pp. 492–498, 1967. Available: http://www.jstor.org/stable/2627691.

[13] G. Miao, "Energy-efficient uplink multi-user MIMO," IEEE Trans. Wireless Commun., vol. 12, no. 5, pp. 2302–2313, 2013.

[14] D. W. K. Ng, "Energy-efficient resource allocation in OFDMA systems with large numbers of base station antennas," IEEE Trans. Wireless Commun., vol. 11, no. 9, pp. 3292–3304, 2012.

OFDMA references

[15] C. Xiong, G. Y. Li, S. Zhang, Y. Chen, and S. Xu, "Energy-efficient resource allocation in OFDMA networks," IEEE Trans. Commun., vol. 60, no. 12, pp. 3767–3778, 2012.

[16] S. Buzzi, G. Colavolpe, D. Saturnino, and A. Zappone, "Potential games for energy-efficient power control and subcarrier allocation in uplink multicell OFDMA systems," IEEE J. Sel. Areas Commun., vol. 6, no. 2, pp. 89–103, 2012.

[17] D. Monderer and L. Shapley, "Potential games," J. Games Econ. Behav., vol. 14, no. 0044, pp. 124–143, 1996.

[18] Z. Xu, G. Y. Li, C. Yang, S. Zhang, Y. Chen, and S. Xu, "Energy-efficient power allocation for pilots in training-based downlink OFDMA systems," IEEE Trans. Commun., vol. 60, no. 10, pp. 3047–3058, 2012.

[19] J. C. Bezdek, R. J. Hathaway, R. E. Howard, C. A. Wilson, and M. P. Windham, "Local convergence analysis of a grouped variable version of coordinate descent," J. Optim. Theory Appl., vol. 54, no. 3, pp. 471–477, 1987.

Cognitive radio references

[20] J. Mitola, "Cognitive radio: an integrated agent architecture for software defined radio," Ph.D. Dissertation, KTH Royal Inst. of Technol., Stockholm, Sweden, 2000.

[21] S. Haykin, "Cognitive radio: brain-empowered wireless communications," IEEE J. Sel. Areas Commun., vol. 23, no. 2, pp. 201–220, 2005.

[22] Y.-C. Liang, Y. Zeng, E. C. Y. Peh, and A. T. Hoang, "Sensing-throughput tradeoff for cognitive radio networks," IEEE Trans. Wireless Commun., vol. 7, no. 4, pp. 1326–1337, 2008.

[23] T. A. Weiss and F. K. Jondral, "Spectrum pooling: an innovative strategy for the enhancement of spectrum efficiency," IEEE Commun. Mag., vol. 42, no. 3, pp. S8–S14, 2004.

[24] S. Choi, H. Park, and T. Hwang, "Optimal beamforming and power allocation for sensing-based spectrum sharing in cognitive radio networks," IEEE Trans. Veh. Technol. vol. 63, no. 1, pp. 412–417, 2014.

[25] S. Buzzi and D. Saturnino, "A game-theoretic approach to energy-efficient power control and receiver design in cognitive CDMA wireless networks," IEEE J. Sel. Topics Signal Process., vol. 5, no. 1, pp. 137–150, 2011.

[26] R. D. Yates, "A framework for uplink power control in cellular radio systems," IEEE J. Sel. Areas Commun., vol. 13, no. 7, pp. 1341–1347, 1995.

[27] J. Mao, G. Xie, J. Gao, and Y. Liu, "Energy efficiency optimization for cognitive radio MIMO broadcast channels," IEEE Commun. Lett., vol. 17, no. 2, pp. 337–340, 2013.

[28] R. Xie, F. R. Yu, H. Ji, and Y. Li, "Energy-efficient resource allocation for heterogeneous cognitive radio networks with femtocells," IEEE Trans. Wireless Commun., vol. 11, no. 11, pp. 3910–3920, 2012.

Relay references

[29] J. Laneman, D. Tse, and G. Wornell, "Cooperative diversity in wireless networks: efficient protocols and outage behavior," IEEE Trans. Inf. Theory, vol. 50, no. 12, pp. 3062–3080, 2004.

[30] G. Miao and A. Vasberg, "Energy efficiency in the wideband regime," in Proc. IEEE WCNC 2013.

[31] A. Zappone, Z. Chong, E. A. Jorswieck, and J. Buzzi, "Energy-aware competitive power control in relay-assisted interference wireless networks," IEEE Trans. Wireless Commun., vol. 12, no. 4, pp. 1860–1871, 2013.

[32] G. Lim and L. J. Cimini, Jr.,, "Energy-efficient cooperative relaying in heterogeneous radio access networks," IEEE Wireless Commun. Lett., vol. 1, no. 5, pp. 476–479, 2012.

[33] O. Amin and L. Lampe, "Opportunistic energy efficient cooperative communication," IEEE Wireless Commun. Lett., vol. 1, no. 5, pp. 412–415, 2012.

8

Energy-Efficient Operation and Management for Mobile Networks

Zhisheng Niu and Sheng Zhou
Department of Electronic Engineering, Tsinghua University, Beijing, China

8.1 Principles

8.1.1 NM Should Be in a Holistic Manner

The continuously growing demands for ubiquitous and broadband access to the Internet brings the explosive development of information and communication technology (ICT) industry, which has become one of the major sources (responsible for 2–10%) of worldwide energy consumption and is expected to increase further in the future. In the meantime, we have witnessed a consistent increase in the number of mobile terminals, especially in the coming era of Internet of Things, which has triggered more complex and higher energy-consumed signal processing technologies. According to the portfolio analysis of the total energy consumption in a typical mobile network[1], it is reported that nearly 75% comes from the base station (BS) side and, inside a BS, nearly 70% of energy is consumed by baseband processing, power amplifiers, and air conditioners in order to keep the BS working (i.e., providing the coverage) even though there is no any traffic in the cell. Therefore, only through the reduction of transmitting power does not help too much for the total energy savings, and so are the incremental approaches such as slim base stations or smart cooling technologies. A more ambitious and system-wide solution is expected if some lightly loaded BSs can be turned into sleep mode or completely

---

[1] https://www.ict-earth.eu/default.html

Green Communications: Principles, Concepts and Practice, First Edition.
Edited by Konstantinos Samdanis, Peter Rost, Andreas Maeder, Michela Meo and Christos Verikoukis.

switched off so that the corresponding power amplifiers and air conditioners can also be shut down during that time.

In the contrary, the existing wireless networks are usually dimensioned for performance optimization without enough consideration of energy efficiency. Specifically, the so-called worst-case network planning philosophy has been widely adopted in order to provide quality-of-services (QoS) guarantee even during the period with peak traffic. As a result, during low traffic periods such as nights or holidays or in some sparse spots where the traffic load is temporarily getting very low due to the user mobility, many BSs are under-utilized but still, by being active, consume a great amount of power. Considering the fact that the nonworking time (including holidays and nighttime) is in fact more than half of the year, the wasted energy of the existing cellular networks is remarkable. This is even more severe for future mobile communication networks where the size of cells will be getting smaller and smaller (e.g., micro- or picocellular) in order to accommodate more high data rate users and increase the frequency reuse factor, which will further increase the dynamics of the traffic in a specific cell. Therefore, it will be very important to have the transmitting power (and, therefore, energy consumed) of network nodes adapt to the traffic variation, including completely switching off some BSs when the traffic load is lower than a threshold.

Furthermore, the dominant traffic in wireless networks has been shifting from mobile voice to mobile data and further to mobile video in the future [1]. This transition is in fact one of the key drivers of the evolution to new mobile broadband standards like 3G, WiMAX, LTE, and LTE-Advanced, resulting in the coexistence of macro-, micro-, pico-, and femtocells, that is, heterogeneous cellular networks. That is to say the scarce spectrum resources have to be segmented into many independent pieces and each cellular network has to provide full cover-age to their corresponding users by itself. Such a redundant deployment will definitely waste the spectrum as well as energy resources furthermore. In addition, with the development of mobile data and video, it is predicted that the traffic volume of mobile services in 2015 could be as much as 100–1000 times of the existing ones, and among which two-third (2/3) could be mobile video traffic. As the spectrum in wireless networks is limited in nature, this will lead to the widespread use of complex channel coding and modulation techniques for advanced interference mitigation. However, using such techniques typically implies accepting higher power consumption not only from the transceiver but also from the complete radio access network. As a result, the contribution of wireless networks to the global carbon footprint is forecast to double over the next ten (10) years.

To deal with this challenge, the traditional physical- and MAC-layer capacity-enhancement approaches are no more sufficient and efficient. A system- or network-level approach is needed, including rethinking about the existing cellular structure. Specifically, the network manage-ment in the future should be able to optimize the usage of the scarce spectrum segmented into heterogeneous networks (e.g., 2G, 3G, 4G, WiFi) in a global way and, meanwhile, shift the optimization goal from spectrum efficiency to energy efficiency, i.e., GREEN. Alternatively speaking, green communications should not simply mean lowering down the transmitting power or improving the energy efficiency of a transmission link, but a holistic approach from the whole network scope. By introducing more collaborations among the neighboring nodes in a wireless network as well as among the heterogeneous networks, the radio resources that are segmented into different nodes and networks can be more efficiently utilized.

8.1.2   NM Should Involve More Cognition and Collaboration

As shown earlier subsection, to overcome the scarcity of wireless spectrum and meet the ever-growing user demand, convergence of multiple wireless networks is a promising way to efficiently utilize wireless resources. Rather than the costly "clean slate" redesign of a one-fit-all wireless system, it is more practical to realize collaboration across these independently developed wireless systems. As a result, many academic entities and international standard organizations have been improving the spectrum efficiency of single-radio access networks with collaboration technology. In other words, the radio resources (even though it is licensed) should be used in a more open and collaborative way so that a harmonized radio ubiquitous system could be realized. As a result, the network managements in the heterogeneous radio systems should CHORUS together rather than sing independently. Otherwise, extra energy will be needed to combat with the noises and/or interferences.

For instance, the seminal work [2] has suggested to use other nodes to relay the signal of a dedicated transmission. The additional diversity can reduce transmission power and increase data rate. Through another route, multi-base station (Multi-BS) collaboration technology [3] is proposed by designing downlink signals collaboratively across BSs, where signals from other cells are used to assist transmission instead of acting as interference, and thus the spatial degrees of freedom (DoFs) are fully utilized. However, one must face excessive signaling and backhaul overhead to collect and exchange channel state information (CSI), user data, and synchronization information among multicells. Therefore, collaboration schemes with limited scale are proposed [4] as compromises.

Research efforts have also triggered the standardization of the collaboration techniques, such as CoMP (coordinated multipoint processing) and Relaying in 3GPP [5]. Nevertheless, their performance in large network is unknown, and small-scale collaboration schemes still suffer from boundary effects, that is, performance of users at the edge of collaboration regions will degrade. Scalable approaches applicable for large systems are valuable and yet to be discovered. Moreover, the benefits of seamless interworking of heterogeneous radio access networks are inherent because of the diverse advantages of heterogeneous networks. Network selection can be triggered by the user terminals through vertical handover, or admission control managed by the network side [6]. People also consider implementing new types of BSs with different coverage capabilities, that is, femtocell and picocell, to bring heterogeneity into single radio networks [7], and this improves coverage performance, interference suppression, and radio resource allocation flexibility. For heterogeneous networks, current studies mainly provide access flexibility, while ultimately it is desired to optimally choose the radio network for each packet so that the system resource is efficiently managed in a finer granularity. The challenges of promoting this kind of "smooth" networking lay in the optimal design in both the vertical (protocol) and the horizontal (heterogeneous cells) directions, and this requires a large amount of information cognition and exchange.

8.1.3   NM Should Be More Adaptive to Traffic Variations

As shown in Figure 8.1 [8], the traffic in a cellular network is typically unbalanced, changing not only in time domain but also in spatial domain. Generally speaking, holiday or weekend

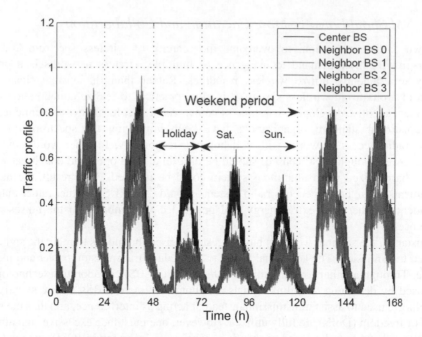

Figure 8.1   Traffic dynamics both in time domain and spatial domain

traffic is lower than weekdays and nighttime traffic is much lower than the daytime. During
the daytime, the so-called peak traffic period is only a small portion of the whole day. On the
other hand, the traffic in different regions may also be very different due to the user mobility
and bursty nature of data and video applications. For example, the business areas may be very
heavily loaded during the daytime but lightly loaded during the nighttimes, which leads to
unbalanced traffic load among neighboring BSs even during the daytime. Therefore, if the
capacity is planned based on the peak traffic load for each cell, there will always be some cells
under light load, while others are under heavy load. In this case, any static cell deployment
will not be optimal as traffic load fluctuates. Such kind of unbalanced traffic distribution can
be even more serious as the next-generation cellular networks move toward smaller cells such
as microcells, picocells, and femtocells.

Another trend of mobile traffic is that mobile data and video will dominate the whole networks [1]. On one hand, compared with voice traffic, data and video traffic are typically more
bursty and dynamic and, therefore, consume more spectrum and energy resources. On the other
hand, they can tolerate some delay in general and, furthermore, are more in point-to-multipoint
fashion (many people may be interested in the same content in a short time period) rather
than just point-to-point communication mode. As a result, it will not be energy-efficient to
provide mobile data and video services in a real-time and point-to-point way as for voice
traffic.

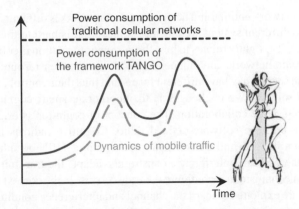

Figure 8.2 Conceptual figure of the framework TANGO

In short, the traffic dynamics can in fact provide some opportunities for energy savings. As shown in Figure 8.2, if we can trace the traffic variation and make the network resources (including transmitting power and other equipment's power) in a cell or the whole cellular network adapted to it, a great amount of energy could be saved. Also, if we can know a priori that many users are requesting the same data or video content from the server, a multicast session could be initiated accordingly so that the transmission energy could be saved by transmitting the content only in one time. This is to say the network panning and operation in green cellular networks should be more traffic-aware, i.e., TANGO (Traffic-aware Network planning and Green Operation). This just looks like a gentleman (traffic needs) and a lady (power and other radio resources) are dancing together in a harmonious way. In reality, more and more BSs have been equipping the self-organizing functionality or even with sleep mode. For instance, Verizon has started a Technical Purchasing Requirement (TPR) guideline by applying the Telecommunication Equipment Energy-Efficiency Rating (TEEER) methodology to all the network components since 2009. Alternatively speaking, if the network components cannot meet the TEEER criteria, they will not be purchased no matter how cheap they are and how excellent the performance is.

8.2 Architectures

8.2.1 Paradigm Shift to CHORUS

The most existing works for either inter or intrasystem collaboration are only applicable to limited scale, or merely provide access flexibility among heterogeneous networks. Generic solutions to fully exploit the features of different radio technologies in accordance with the channel and traffic variations are desired. However, realizing such scalable collaboration with fine resource granularity faces following challenges. First, collecting a large amount of system information across network entities and heterogeneous networks is difficult, not only due to the huge mass, but also different system architectures. Second, it is hard to find the collaboration opportunities, which also tightly relates to how the system information is collected. Last but not least, as different radio networks are independently designed, their resources, such as power, frequency, time, and space, may have different forms and thus their control interfaces also vary.

One solution to solve above challenges is to incorporate smart cognition methods, and jointly operate it with node collaboration. The concept of cognition is originally applied for cognitive radio [9], built on software-defined radio. Cognitive radio is mainly targeted at sensing the spectrum hole to find transmission opportunities [10] for unlicensed users, and then according to the CSI and interference constraints, adapts the transmit power with spectrum management mechanism to reconfigure the transceivers. In the context of cognitive radio, cognition methods are exploited to learn the channel and interference conditions, and thus various spectrum sensing technologies are proposed [10]. Moreover, the cognition behavior is also characterized from information theoretical perspectives [11] as side information that comprises knowledge of transceiver activities, channels, codebooks, and messages of other transceivers that share the spectrum.

Inspired by these applications, we *broaden* the cognition targets from spectrum and channels in cognitive radio to the whole heterogeneous networks, crossing physical-layer resources and upper-layer traffics. In our design, the core enabling method to realize scalable network collaboration that harmonizes different wireless resources is *cognitive synergy*. Based on the philosophy of smart cognition, as what is generally required in cognitive radio, *distributed* inspection of the system state across multiple layers and *self* organization/control of network entities can be realized with low overhead, which is the key to large scale networking. Moreover, cognitive synergy provides unified virtualization of the resources from heterogeneous networks, which is valuable for traffic offloading and collaborative communications among heterogeneous networks. We also make cognition interactions with collaboration so that

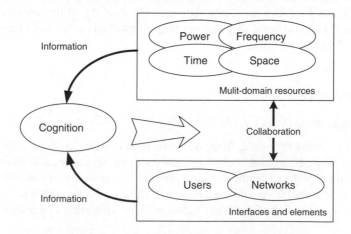

Figure 8.3   Our vision on the new paradigm of a ubiquitous radio network

node collaboration assists cognition, and is also reconfigured according to the cognition results.

Our basic vision on the new paradigm of a ubiquitous radio network is depicted in Figure 8.3, which is explicitly embodied with the framework design CHORUS. All the features shown in Figure 8.3 are reflected in the framework, and our focus is on how to achieve the optimal collaboration gain and environmental friendliness by cognitive synergy in ubiquitous radio network environments. To overcome the difficulties of collaborating transceivers and networks over large scale, cognitive synergy is exploited to inspect the status of multiple radios and layers. Based on the cognition result, harmonic collaboration control is applied to optimize the global performance. The CHORUS framework is designed to work with all available wireless access networks, such as LTE/LTE-A, 3G, WiMAX, WLAN, and so on. This enables CHORUS to make maximum coverage and synthetically enroll various permitted services.

8.2.1.1   Architecture of CHORUS

The architecture of CHORUS is presented in Figure 8.4. The core is a CHORUS server, which controls the procedure of CHORUS. The CHORUS server is a conceptual body in the network and can be either implemented in the gateway, such as radio network controller (RNC), or distributed in the BSs, or in a mixed way. Note that the way of implementation also indicates that the algorithms of CHORUS are realized in a centralized or distributed way. The CHORUS server has two engines. One is the **cognitive engine**, and it communicates with the entities in the wireless access network and collects information about the environment and users. It also analyzes this information, and stores the results in the other engine: **profile database**. The profile database stores the processed information, provided by the cognitive engine, about user terminals, access networks, wireless resources, and environment. In addition, the cognitive engine consists of the **data analyzer** and **cognition controller**. The data analyzer reprocesses the system information based on its raw version gathered from the network side, of which the results are utilized by the cognition controller for making network control decisions.

The network entities sense the environment, such as channel conditions, spectrum occupancy, and so on. They also interact with user terminals, collect their QoS requirements, and provide admission control and service regulations. The collected information will be first locally saved at the network entities. Then these information will be collected by the **cognitive engine** assisted by network node collaboration, which reduces the overhead of collecting the large amount of system information. On the other information route, the profile database delivers back the cognition results for collaboration control and resource reconfiguration to the network side. The contents delivered to the network also include how the network should sense the environment and get the user status more efficiently. The detailed operation flow of CHORUS, especially the cognition part, is presented as follows.

8.2.1.2   Work Flow of CHORUS

An exemplary working flow of CHORUS is shown in Figure 8.5. The core procedure of CHORUS operation is synergetic cognition, which guarantees scalable collaborative

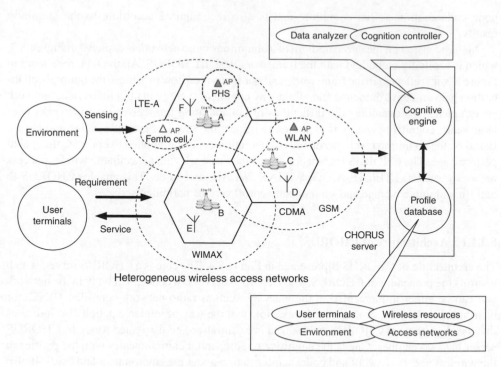

Figure 8.4   The architecture of CHORUS framework

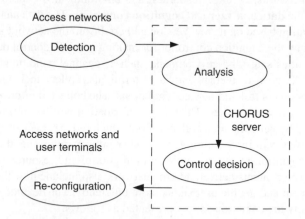

Figure 8.5   Example of CHORUS operation procedures

communications. There are four major steps, that is, **Detection**, **Analysis**, **Decision** and **Reconfiguration**.

- **Detection** is performed by access network entities to get necessary data and information from user terminals or environment, for corresponding cognition. same as the way CHORUS server is implemented, the detection can be locally accomplished by the network elements, or by a central controller with feedback from network elements. The single-link detection aims to adapt the transmission strategy according to channel conditions and spectrum usages, for instance, adaptive modulation and coding. Multiple-link detection is to gather information from interactive links in network, and the information consists of network attributes and parameters that affect the user experience. For heterogeneous networks, the resource occupancy across different networks is detected to provide resource virtualization of networks. Note that in the process of **Detection**, network entities also *collaborate* to provide their information for data analysis efficiently.
- After sufficient information is ready, data **Analysis** is executed by data analyzer based on the profile data saved in the profile database. The data analyzer uses certain techniques, such as data mining and learning algorithms, to find valuable information from the raw version collected from **Detection**. It also investigates the equivalent resource consumption of certain operations for heterogeneous networks. Network entities can also *collaborate* to estimate state information or predict the impact of possible network operations to assist data analysis. The data analyzer also in turn updates the profile database with the new cognition results.
- Afterward, the instructions of network **Control** and **Reconfiguration** decisions are made by cognition controller and sent to networks and user terminals. The on-demand collaborative control is harmonized with the system state. Over time, the cognition controller can refine which information of the system is most valuable for collaboration so that the **Detection** is optimized. This closed-loop for detection and cognition is vital for scalable information collection.
- Based on the decisions made by cognition controller, the network **Reconfiguration** guarantees the efficient collaboration form that fits mostly to the current status of wireless resources and user requirements.

8.2.1.3   Relationship between Cognition and Collaboration

In CHORUS, the unique feature is that cognition methods interact with collaboration. Network entities collaborate to investigate and collect system information, after which the cognition results will help reconfigure the collaboration scale, way of collaboration, andinformation required to perform collaboration, and so on. In other words, collaboration is tied with all the major steps of cognition procedure in CHORUS. This is also the reason why "cognition" appears jointly with "synergy" as the core idea of CHORUS. With node collaboration, CHORUS allows distributed resource control schemes, and the information flow of CHORUS can also have different styles. For example, as will be shown in the next section, *distributed negotiation* for dynamic collaboration is an effective way of transferring information and making joint control decisions. Inherently, due to the distributed feature, the overhead of information exchange is reduced, and it is important to scalable networking.

8.2.2   Paradigm Shift to TANGO

8.2.2.1   Adjusting the Working Mode of Base Stations

As shown in Figure 8.2, the traditional network planning and operation is mainly based upon the assumption that user requests may happen *anytime* and at *anyplace*. This is in fact also the dream that people are having about the mobile communications. As a result, the existing cellular networks were mostly designed to keep the transmitting power always-on in order to guarantee the cell coverage as well as provide the appropriate services if the request happens. This is clearly not energy-efficient because the user requests occur only *sometime* and *someplace* in practical situation. It is, therefore, reasonable to keep the cell coverage by a minimum number of BSs and then adjust the working mode (active or sleep) of the remaining BSs in accordance with the traffic variations. This is equivalent to adjusting the BS density of the cellular networks or, in other words, the network resources should only be provided on demand whenever there is such a need. As a result, the network architecture should be flexible enough to this adaptation. Apparently, this paradigm shift has a great potential for the energy saving in cellular networks. Of course, when some BSs are switched off or in the sleep mode, radio coverage and service provisioning are taken care of by the devices that remain active, i.e., BS cooperation is crucial.

BS sleeping is drawing more and more attention in recent years. Reference [12] gives a static BS sleep pattern according to a deterministic traffic variation pattern over time. However, neither the randomness nor the spatial variation of the traffic is considered. Reference [13] proposes a resource on-demand (RoD) strategy for high-density centralized WLANs, where a cluster-head AP takes care of the whole coverage in the cluster so that other APs in the cluster can be switched off when the traffic load is low. However, the channel model of WLANs is quite different from that of cellular networks where path-loss effect is dominating. Therefore, dynamic clustering algorithm by considering the BS collaboration is needed.

8.2.2.2   Adjusting the Cell Size

Cell size in cellular networks is in general fixed based on the estimated traffic load. However, as discussed earlier, the traffic load can have significant spatial and temporal fluctuations and, therefore, keeping the cell size fixed is not energy efficient. In other words, the cell size should be adjusted dynamically according to traffic conditions as well as the situation of neighboring BSs in a collaborative way. This is the concept of the so-called *cell zooming* proposed in Ref. [14], where two typical zooming patterns by increasing the transmitting power of those BSs that remain active and the corresponding cell planning results are shown. Unlike the power control on link layer, which does not actually change the cell size, cell zooming is a technique on network layer that changes the cell size by adjusting the transmit power of control signals. The detailed description of the cell zooming technique is shown in the following section.

8.2.2.3   Adjusting the Service Mechanism

As mentioned earlier, mobile data and video traffic will dominate the future networks. Unlike the voice traffic, which is typically delay-sensitive and symmetric in uplink and downlink, data and video traffic is in general loss-sensitive and asymmetry in uplink and downlink. But majority of the existing cellular networks were designed to accommodate voice traffic mainly, i.e., if the capacity of the networks is not enough, they try to increase the capacity anyway (and, therefore, consume more power) or simply reject the requests (and, therefore, deteriorate the QoS). In other words, the existing cellular networks are not so friendly to data and video traffic. As the IP-based data and video traffic can tolerate some delay in nature, they can be served in an opportunistic way in fact. Specifically, they can be served during the period when the channels are in good condition or the network is lightly loaded. For some data and video contents that many users are requesting in a short period, they can in fact be served by using multicast or broadcast [15] so that the same contents are not necessarily transmitted multiple times. Rather, they can be cached in between [16] so that the users who have the same request can get the service locally without going to the source node every time. In such a way by adjusting the service mechanism properly, the energy can be saved dramatically. Consequently, the cellular network architecture should be flexible enough to provide such differentiated services in an appropriate manner.

8.3   Implementation Examples

8.3.1   *CHORUS by Scalable Collaboration*

As previously mentioned, the idea of cognition in CHORUS is generalized by inspecting status and parameters of multiple scales. The cognition method is also jointly optimized with network collaboration, which enables efficient system resource utilization for information collection and exchange. This is beneficial but the most crucial barrier for network collaboration for both homogeneous and heterogeneous networks is the mass information exchange. In what follows, we firstly show the benefits of exploiting *cognitive synergy* in CHORUS through two case studies that realize **Scalable Collaboration** and **Ubiquitous Access**, correspondingly. Then, some implementation issues will be discussed.

8.3.1.1   **A Decentralized BS Dynamic Clustering Scheme**

Enabled by smart cognition, the network can harmonize the whole process of collaboration with two key capabilities: reducing the overhead to collect necessary information for collaborative communication and adapting the appropriate form and scale of collaboration. The first one is reflected in the **Detection** and **Analysis** steps, while the second is enabled through the **Decision** and **Reconfiguration** steps. We illustrate how these two capabilities are realized with our research on distributed dynamic BS clustering [17] and dynamic BS sleeping [18] for cellular systems. Specifically, we will show how to exploit cognitive synergy to realize scalable collaboration with high spectrum efficiency and reduced energy consumption.

Implementing collaborative communication is in fact a trade-off between performance gain and resource consumption for information sharing. Taking BS collaboration for example, by designing downlink signals cooperatively among BSs, signals from other cells are used to assist transmission instead of acting as interference, and thus BS collaboration can substantially increase the spectrum efficiency of the cellular network [17]. However, due to practical constraints like synchronization, and backhauling, only a limited number of BSs are allowed to collaborate [17], where the collaborating BSs formulate a BS *cluster*, and the whole network is

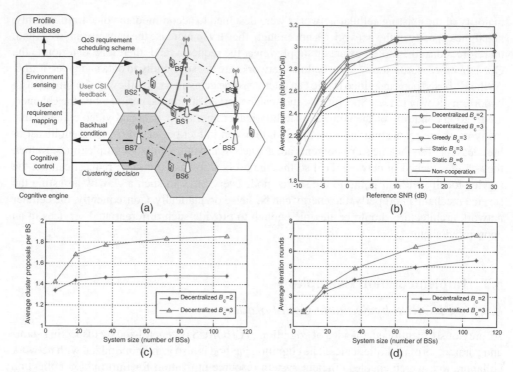

Figure 8.6 CHORUS application for BS collaboration, where B_c is the cluster size. (a) Architecture of dynamic BS clustering under CHORUS framework. (b) Average sum-rate with dynamic BS clustering. (c) Feedback overhead. (d) Calculation complexity. (simulation figures from Ref. [17])

divided into disjointing clusters. Even with fixed BS clustering, users at cluster edge still suffer from severe inter-cluster interference. Therefore, adaptively tuning the BS clustering structure is more preferable. In fact, dynamic clustering itself is challenging as cluster formation relies on scheduling decisions, precision and scale of user CSI feedback, user distribution, and so on, which in turn incur extra overhead. In our work, CHORUS reduces these overhead and provides suitable BS clustering formation structures.

The architecture of BS collaboration with the assistance of CHORUS is shown as Figure 8.6, where the BSs are separated into three difference clusters. The conceptual elements in Figure 8.6 have the following realizations. The cognitive engine is implemented in each BS and according to the CSI and user QoS requirements, for the **Detection** step it abstracts the preferable choices of clustering companions of each BS b into a preference function $\hat{R}(b, c)$, which represents the expected rate of the associated users if b joins cluster c. Then the value of a cluster c, denoted by $V(c)$, is defined as the sum of $\hat{R}(b, c)$ from each component BS. The unique feature of this piece of work is that the cognitive **Analysis** and **Control Decision** are merged in the form of distributed BS negotiation about their preference functions and cluster values. BS clustering can be accomplished with low complexity and in a *distributed* way. Detailed algorithm can be found in Ref. [17]. The sum-rate performance is shown in Figure 8.6. It is observed that BS collaboration enjoys significant throughput gain over single BS transmission, and dynamic clustering scheme outperforms static clustering substantially. Moreover, from Figure 8.6, one can see that both the feedback and calculation overhead scale very slowly with the network size: The feedback overhead is almost irrelevant to the network size when the network size is large enough, and the calculation overhead scales approximately logarithmically with the network size, while the complexity of conventional centralized greedy algorithm scales quadratically with the network size. Therefore, CHORUS shows its ability of maintaining the scalability of network collaboration. Furthermore, with cognition synergy, scalable CSI feedback optimization is investigated in Ref. [19], where we combine the dynamic clustering with the consideration of limited feedback bits. We design a feedback set adaptation scheme by the cognition of most valuable BS CSI feedback bits, and successfully achieve a good trade-off between collaboration gain and quantization precision, which further reduces collaboration overhead.

BS collaboration can also be used to save energy. Reducing the energy consumption has become one of the key features of future network design, and it has been shown that BSs cost most of the energy of the access network [8]. Therefore, switching on and off BSs *on demand* can be more efficient. In accordance with BS sleeping, cell sizes of active BSs should also be tuned to guarantee the network coverage, namely cell zooming [14]. Both of above highly rely on the precise cognition of the network temporal–spatial traffic conditions. BS sleeping requires BS collaboration so that users in sleeping cells are taken care of by the active BSs. In addition, taking signaling overhead, device lifetime and switching energy consumption into account, frequent BS mode switching should be avoided.

In CHORUS, we combine the traffic cognition and the BS sleep control. As shown in Figure 8.7, they are implemented in the cell zooming server. For cognition operation, here the neighboring BSs negotiate to exchange traffic information and working state, and then collaborate to determine the optimal BS working mode. The neighboring BS negotiation based **Detection** and **Analysis** steps greatly reduce the complexity and the cost for working mode switch: The full consideration of all network states has the exponential complexity with the network size, while with CHORUS, we are able to reduce it to linear information scaling [18]. Next, the **Control Decision** step is realized as a dynamic programming with the per-stage cost

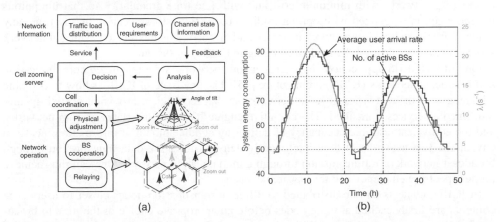

(a)   (b)

Figure 8.7 CHORUS application for smart dynamic BS energy saving. (a) Architecture of cell zooming under CHORUS framework. (b) Number of active BSs compared with average traffic intensity in time space with different traffic configurations

at stage i

$$g_i = \sum_{m=1}^{M} [|s_i^{(m)} - u_i^{(m)}|E_i + C_s u_i^{(m)} + h(\tilde{P}_i^{(m)})] \tag{8.1}$$

where $s_i^{(m)} \in \{0, 1\}$ is the BS working state, $u_i^{(m)} \in \{0, 1\}$ is the action of BS m, and $\tilde{P}_i^{(m)}$ is the blocking probability of a new call arrived in cell m. So, the per-stage cost is a combination of operation energy $s_{i+1}^{(m)} E_i$, switching cost $C_s u_i^{(m)}$, and blocking probability penalty $h(\tilde{P}_i^{(m)})$ of all the cells. The object is to find the optimal policy that minimizes the total cost of all stages

$$\{u_{i,\,opt}^{(m)}\}_{i=0,\ldots,N-1}^{m=1,\ldots M} = \arg \min_{\{u_i^{(m)}\}_{i=0,\ldots,N-1}^{m=1,\ldots M}} \sum_{i=0}^{N-1} g_i \tag{8.2}$$

The optimal decision of each BS is approximated as a function of local cognition results of the traffic in its own coverage as well as its first-tier neighboring BSs. An iterative algorithm is proposed to find the suboptimal decisions. The network energy consumption, represented by the number of active BSs, well matches the traffic intensity as shown in Figure 8.7.

8.3.1.2 A Ubiquitous Heterogeneous Radio Access Scheme

Scalable collaboration can happen not only intra-system, but more importantly in an inter-system way, of which ubiquitous access is one of the key functions. Ubiquitous is identified and articulated as a new computing paradigm, where the network is connected at any place, any time, and with any object. Current user equipments are capable of reconfiguration to access multiple networks with different protocols. While at the network side, the independent design hinders the ubiquitous access. The implementation of such network highly depends on the *virtualization* of the network resources and dynamic reconfiguration/reorganization of terminals and networks. When the collaboration happens between different networks, traffic can be split and conveyed over heterogeneous networks smoothly. In this section, we show how the **Detection**, **Analysis**, and **Decision** steps in CHORUS support the **Reconfiguration** of existing networks with minimum cost and with fine time granularity so that ubiquitous interconnection is realized with optimal radio, power, time-slot allocation at any time, any place. With the unified internetwork interface and protocol, it is flexible to design subnetwork satisfying new application demand and joining in CHORUS system.

One of CHORUS applications in heterogeneous network is the integrated communication and broadcast networks (ICBN) architecture [15]. The key issue of ICBN is how to combine advantages of the two types of radio networks so that an efficient provision of high-quality multimedia services is ensured. The cognitive engine identifies the states of both networks, and the equivalent resource occupancy of delivering certain amount of information bits from any of the two networks, it then intelligently delivers downlink data transmission through broadcast network or retransmission through communication network. One can thus achieve higher spectrum efficiency and thus system capacity.

In ICBN, contents can be distributed to different transmission nodes closer to users. The contents are firstly cached at these nodes before an appropriate choice of the time to broadcast/mulitcase contents to users within its coverage. For example, when relay stations (RSs) are introduced, caching with multicast can also help reduce the transmission cost, when the need for the same content of different users spreads over time [16]. In some content delivery

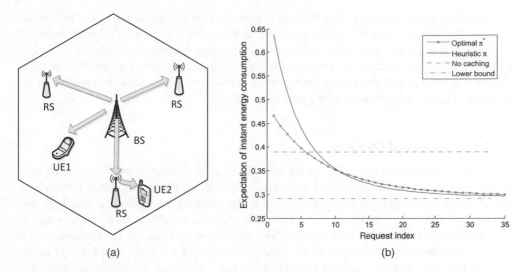

Figure 8.8 CHORUS application for smart caching assisted by relays. (a) Relay caching scheme in cellular network. (b) Cumulative average of transmission energy consumption, where we compare the optimal policy and a proposed heuristic with the baseline policy without relay caching, and the energy consumption lower bound to any policy is achieved when all the necessary contents are pre-stored in the relay

services such as video-on-demand service, multiple requests of the same content induce abundance of insignificant traffic if the delivery is naive unicast. However, as users may appear or request the content at different time, introducing relay caching can replace costy repeating re-broadcast with low power short distance transmission. A single cell in a cellular network with RSs is depicted as Figure 8.8. When a user requests a piece of content, the BS may transmit the content to the user directly or may broadcast the content to both the RSs and the UE. Contents can be cached into the buffer of RSs in order to serve the nearby users with future requests of the same content. When a piece of content that has already been cached in an RS is requested by a user in the RS's coverage, it can be transmitted from the RS to the user directly.

For this kind of dynamic caching, the key problem is to identify whether or not to cache the requested content in the RS, as the buffer size of RSs is limited. The probability of a piece of content being requested, that is, the popularity, is introduced to **Analyze** the content. The popularities of contents in the buffer are stored in the profile database, and the cognitive engine makes the **Control Decision** to determine whether to cache a newly requested content or not. As the system runs, the cognitive engine learns and **Reconfigures** the optimal caching policy through stochastic dynamic programming. The average energy consumptions of transmission with the optimal policy and the derived heuristic policies are shown in Figure 8.8. Energy consumption is saved by 15.3% after the caching period.

8.3.2 TANGO by Cell Zooming

In this section, the concept of cell zooming is introduced, which is to adaptively adjust the cell size according to the traffic load fluctuations. Cell zooming can not only solve the problem of traffic imbalance, but also reduce the energy consumption in cellular networks. Techniques such as physical adjustments, BS cooperation, and relaying can be used to implement cell zooming. Through the numerical examples, we show that the proposed cell zooming algorithms can leverage the trade-off between energy saving and blocking probability. The algorithms also save a large amount of energy when traffic load is light, which can achieve the purpose of green cellular network in a cost-efficient way.

8.3.2.1 Concept and Challenges

Cell size in cellular networks is in general fixed based on the estimated traffic load. However, as mentioned earlier, the traffic load can have significant spatial and temporal fluctuations, which bring both challenges and opportunities to the planning and operating of cellular networks. This subsection introduces a concept of *cell zooming*, which adaptively adjusts the cell size according to traffic load, user requirements, and channel conditions. The implementation issues of cell zooming are then presented. Finally, a use case of cell zooming for energy saving is investigated. Centralized and distributed cell zooming algorithms are developed, and simulation results show that the proposed algorithms can greatly reduce the energy consumption, which leads to green cellular networks.

An example of cell zooming is illustrated in Figure 8.9. It is a cellular network with five cells. One central cell is surrounded by four neighboring cells. BSs are located at the respective center of the cells, denoted by hollow squares; MUs are randomly distributed in the cells, denoted

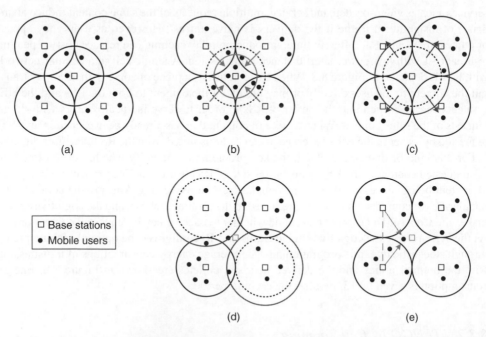

Figure 8.9 Cell zooming operations in cellular networks: (a) Cells with original size; (b) Central cell zooms in when load increases; (c) Central cell zooms out when load decreases; (d) Central cell sleeps and neighboring cells zoom out; (e) Central cell sleeps and neighboring cells transmit cooperatively

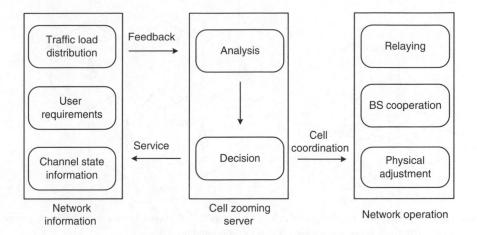

Figure 8.10   Framework of cell zooming

by solid dots. When some MUs move into the central cell and make it congested, the central cell can zoom in to reduce the cell size and, therefore, release from the congestion (Figure 8.9(b)). On the contrary, if some MUs move out of the central cell and cause the neighboring cells congested, the neighboring cells can zoom in and the central cell zooms out to avoid any possible coverage hole. If the neighboring cells are designed to have high capacity, and therefore not necessarily zoom in, the central cell can also choose to sleep to reduce the energy consumption. In this case, the neighboring cells can either zoom out to take care of the coverage as in Figure 8.9(d), or serve the left MUs by transmitting cooperatively as in Figure 8.9(e). This example shows that cell zooming has the potential to achieve green cellular networks.

Implementing cell zooming in cellular networks needs to introduce some new components and corresponding functionalities to current network architecture. The framework of cell zooming is illustrated in Figure 8.10. There is a cell zooming server (CS), which controls the procedure of cell zooming. The CS is a virtual entity in the network, which can be either implemented in the gateway or distributed in the BSs. The CS will first sense the network state information for cell zooming, such as traffic load, channel conditions, user requirements, and so on. The sensing process can be realized by specific control messages. After collecting the information, the CS will analyze whether there are opportunities for cell zooming and make decisions. If a cell needs to zoom in or zoom out, it will coordinate with its neighbor cells with the help of CS. Then these cells will either zoom in or zoom out by network operations such as physical adjustment, BS cooperation, and relaying.

Techniques
Many techniques can be used to implement cell zooming. A simple and straightforward way is to adjust the physical parameters such as the transmit power of BSs. Besides physical adjustment, other techniques can also be used for cell zooming, as illustrated in Figure 8.11. A detailed discussion of the techniques used for cell zooming is given as follows.

Physical Adjustment: Adjusting physical parameters of network deployment can help to implement cell zooming. Cells can zoom out by increasing the transmit power of BS, and

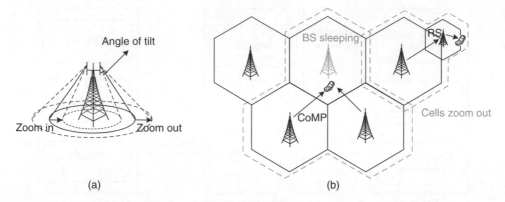

Figure 8.11 Techniques to implement cell zooming: (a). Cell zooms in or zooms out with physical adjustments; (b). Cells zoom out through BS cooperation and relaying

vice versa. Furthermore, antenna height and antenna tilt of BSs can also be adjusted for cells to zoom in or zoom out (Figure 8.11(a)). Such adjustments need the help of additional mechanical instruments.

BS Cooperation: BS cooperation means multiple BSs form a cluster, and cooperatively transmit to or receive from MUs, which is also named as CoMP (Coordinated Multipoint transmit/receive) in 3GPP LTE-A (Long-Term Evolution-Advanced) [5]. The newly formed cluster is a new cell from MUs' perspective, whose cell size is the sum of the original size of the BSs in cooperation. The size can be even larger, as BS cooperation can reduce inter-cell interference. In this case, cells zoom out to improve the coverage (Figure 8.11(b)).

Relaying: Relay stations (RSs) are deployed in cellular networks to improve the performance of cell-edge MUs, which is also an important technique in 3GPP LTE-A. The cell with RSs zooms out as shown in Figure 8.11(b). RSs can also be deployed near the boundary of two neighboring cells. In this case, RSs can relay the traffic from the cell under heavy load to the cell under light load. The former cell zooms in, and the latter cell zooms out.

BS Sleeping: When a BS is working in sleep mode, the air conditioner and other energy-consuming equipments can be switched off. BS sleeping can largely reduce the energy consumption of cellular network. In this case, the cell with BS working in sleep mode zooms in to 0, and its neighbor cells will zoom out to guarantee the coverage.

Benefits

Cell zooming can provide various benefits in cellular networks. Firstly, cell zooming can be used for load balancing by transferring traffic from cells under heavy load to cells under light load. Secondly, cell zooming can be used for energy saving. Contrary to the usage for load balancing, here cells zoom in to zero when the traffic load is light enough. Some BSs work in sleep mode, and the neighbor cells zoom out accordingly to guarantee the coverage. Therefore, cell zooming can both disperse load for load balancing and concentrate load for energy saving. In both cases, the resources are allocated to match the traffic distribution, however, the load transfer direction is opposite. It is a challenging problem to decide when to disperse load for load balancing and when to concentrate load for energy saving.

User experience can also improved by cell zooming, such as throughput, battery life, and so on. Techniques like BS cooperation and relaying can reduce the inter-cell interference, mitigate impact of shadowing and multipath fading, and reduce handover frequency. The techniques can also be jointly used. For example, in the scenario of isolated cell coverage, when cells zoom out by adjusting physical parameters such as antenna tilt, there will be more overlap among the cells. This provides opportunities for BS cooperation so that more MUs can achieve higher diversity gain, and coverage is also improved. As user requirements are better satisfied, there is no need for upgrading the network frequently, and this will reduce the operational cost of network operators.

Power control in cellular networks has been studied extensively in the literature (see the references in Ref. [20]). Power control can help to ensure efficient spatial reuse and minimize energy consumption. These functionalities is quite similar to that of cell zooming. However, cell zooming is different from power control in many ways. Power control focuses on the link-level performance and transmit power consumption, while cell zooming techniques focus on the network-level performance and energy consumption of the whole network. Power control does not actively change the cell size, while cell zooming actively changes the cell size by adjusting the transmit power of control signals.

Challenges

There exist many challenges to implement cell zooming. To make cell zooming efficient and flexible, traffic load fluctuations should be exactly traced and fed back to the CS. However, significant spatial and temporal fluctuations make it a challenging problem. One possible way to model the fluctuations is to divide it into long-term scale fluctuations and short-term scale fluctuations. The long-term scale fluctuations reflect the variation of traffic arrival rate, whose timescale is hours or days. The short-term scale fluctuation reflects the random arrival of users, whose timescale is seconds or minutes. It would be an interesting topic to find other models for the spatial and temporal traffic load fluctuations.

Compatibility is another challenging issue. Some of the techniques of cell zooming are not supported by current cellular networks, such as the additional mechanical equipments to adjust the antenna height and tilt, BS cooperation, and relaying techniques. Implementing cell zooming also needs to change current structure of network management. For example, feeding back the network information for cell zooming requires special control channels.

Cell zooming may cause other problems, such as inter-cell interference and coverage holes. When some neighboring cells zooms out together, there will be more inter-cell interference among them. If BS cooperation is infeasible, additional interference management schemes are needed to reduce the interference. Cell zooming may also produce coverage holes. When cells zoom in and zoom our, some areas in the network are possible have no coverage. In order to provide service to newly arrival MUs, the neighboring cells need to zoom in so as to cover these areas.

8.3.2.2 Centralized and Distributed Algorithms

In this subsection, a usage case of cell zooming for energy saving in cellular network is investigated. When traffic load is light, some cells can work in sleep mode to save energy, and other cells take care of the coverage. There have been many related studies about BS sleeping in

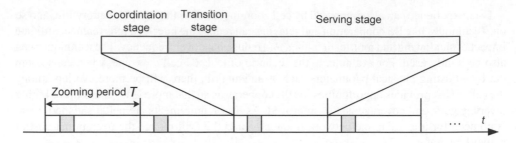

Figure 8.12   The process of cell zooming algorithms

cellular networks. In Ref. [12], a predefined BS sleeping scheme is presented according to a deterministic traffic variation pattern over time. Another similar work considers switching off some microcells at night hours while guaranteeing the blocking probability below a given target [21]. In these solutions, the sleeping pattern is fixed and the traffic intensity is assumed to be uniformly distributed over the whole network. In this article, we consider cellular networks with spatial and temporal traffic load fluctuations, and develop dynamic cell zooming algorithms for energy saving.

Consider a densely deployed cellular network in which the coverage of BSs overlaps and traffic load fluctuates over time and space. Assume there are M BSs, and all the BSs are assumed to have the same energy consumption. Each BS has two working modes: *active* mode with energy consumption P^a and *sleeping* mode with power consumption P^s, where P^a is usually much larger than P^s. MUs arrive at the network according to a Poisson process, and each MU will be associated with one BS upon its arrival. The sojourn time for each MU is exponentially distributed, and the rate requirement is fixed for each MU, denoted by r_i for MU i. The spectral efficiency is ω_{ij} when MU i is associated with BS j. Therefore, the bandwidth needed is given by $b_{ij} = r_i/\omega_{ij}$. We assume the spectral efficiency is independent of the associations among other BSs and MUs. The total bandwidth for BS j is B_j. When a new MU arrives, if there is not enough bandwidth to be allocated, the MU will be blocked. We are interested in two objectives, minimizing the energy consumption and minimizing the blocking probability. If there are more cells working in sleep mode, more energy will be saved, however, it also leads to larger blocking probability. Therefore, there is a trade-off between the two objectives.

As the mode transition of BSs will last for a period of time, during which the cells cannot provide service to MUs, thus frequent mode transition is infeasible in practice. In our cell zooming algorithms, time is divided into cell zooming periods, and the length of each period is T. Each period consists of three stages: coordination stage, transition stage, and serving stage, as shown in Figure 8.12. In the coordination stage, the CS collects necessary network state information for cell zooming, and makes decisions. Our proposed cell zooming algorithms will also work during this stage. In the transition stage, cells change their working modes, and complete the handoff process if needed. In the serving stage, cells fix their working mode, and provide service to current and newly arrival MUs in the network. We assume the length of coordination stage and transition stage are much shorter than serving stage, so the energy consumption depends on the work mode of cells in the serving stage.

Intuitively, in order to minimize the number of active BSs, traffic load should be concentrated to a few BSs so the left BSs under light load can be switched off. Following the intuition, two cell zooming algorithms are proposed. The first one is a centralized algorithm, in which all the channel conditions and user requirements in the network are collected by the CS, and resource allocation and cell zooming operations are performed in a centralized way. The second one is a distributed algorithm. Each MU will select the BS to be associated with by itself based on the information provided broadcasted by the BSs. Generally speaking, the centralized algorithm requires more signaling overhead, but can achieve better performance compared with the distributed one. The details of the two cell zooming algorithms are given as follows.

Centralized Algorithm

In the centralized cell zooming algorithm, MUs feed back channel conditions and rate requirements to the BSs during the coordination stage. The CS will collect all these information together with BSs' bandwidth limitation. After receiving updates from all the MUs and BSs, CS will generate a 0–1 matrix $X = [x_{ij}]$, where $x_{ij} = 1$ means MU i is associated with BS j, otherwise $x_{ij} = 0$. As each MU can only be served by one BS, the sum of each column in X is 1. The main idea of the algorithm is to switch off the BSs under light load as far as possible. As many MUs arrive during the serving stage, each active BS will reserve some bandwidth for the newly arrived MUs. Denote the proportion of bandwidth reserved in BS j as α_j, where $\alpha_j \in [0, 1]$. Initially, the idle bandwidth for BS j is given by

$$\tilde{B}_j = (1 - \alpha_j)B_j \qquad (8.3)$$

Denote the set of MUs associated with BS j as $\mathcal{M}_j$. The traffic load of BS j is given by

$$L_j = \sum_{i \in \mathcal{M}_j} \frac{b_{ij}}{B_j} \qquad (8.4)$$

The detailed procedure of the algorithm is described as follows.

- Step 1: Initialize all the L_j to be 0 and all the elements in matrix X to be 0.
- Step 2: For each MU i, find the set of BSs who can serve MU i without violating the bandwidth constraints, which means $L_j B_j + b_{ij} \leq \tilde{B}_j$. If the set is empty, MU i is blocked. Otherwise, associate MU i with a BS j that has the highest ω_{ij} in the set. Update L_j and X after each association.
- Step 3: Sort all the BSs by the ratio of $L_j B_j$ to $\tilde{B}_j$ by increasing order. All the BSs with the ratio 0 will zoom in to zero and work in sleep mode in the following serving period. For other BSs, find the BS j with the smallest ratio, and reassociation the MUs in $\mathcal{M}_j$ to other BSs in the network. If no MU is blocked, update X and go to Step 3. Otherwise, output X and end the procedure.

Distributed Algorithm

To reduce the information exchange and signaling overhead, we also propose a distributed cell zooming algorithm, in which each MU will select the BS by itself according to the measured channel conditions and BSs' traffic load. In the distributed algorithm, BSs also reserve bandwidth for newly arrival MUs as in centralized algorithm. In practice, traffic load information and bandwidth reservation parameters can be obtained by broadcasting control signals from BSs. Intuitively, each MU will select the BS with high load and high spectral efficiency. We define a preference function if MU i is to be associated with BS j as

$$U(\omega_{ij}, L_j, \alpha_j) = \begin{cases} \dfrac{\omega_{ij}(L_j B_j + b_{ij})}{\tilde{B}_j} & L_j B_j + b_{ij} \leq \tilde{B}_j \\ 0 & L_j B_j + b_{ij} > \tilde{B}_j, \end{cases} \tag{8.5}$$

which means MUs prefer those BSs with high load and high spectral efficiency, but the load cannot exceed a predefined threshold. The procedure of distributed cell zooming algorithm is described as follows.

- Step 1: Initialize all the L_j to be 0 and all the elements in matrix $\mathbf{X}$ to be 0.
- Step 2: For each MU i, find the set of BSs who can serve MU i without violating the bandwidth constraints, which means $L_j B_j + b_{ij} \leq \tilde{B}_j$. If the set is empty, MU i is blocked. Otherwise, associate MU i with a BS j that has the highest $U(\omega_{ij}, L_j, \alpha_j)$ in the set. Update L_j and $\mathbf{X}$ after each association.
- Step 3: Repeat Step 2 until there is no update of $\mathbf{X}$, then output $\mathbf{X}$ and end the procedure.

In the distributed algorithm, no coordination among BSs is needed; therefore, much signaling overhead is reduced. The distributed algorithm works in an iterative way. The convergence of the distributed algorithm is guaranteed if any two MUs take no action simultaneously. This is because the BS selection set of each MU is finite. After the algorithm converges, the BSs with no association will work in sleep mode during the serving stage.

8.3.2.3 Performance Evaluation

The proposed dynamic cell zooming algorithms are evaluated in a scenario with time-varying traffic distribution. The simulation layout is 10 by 10 hexagon cells wrapped up to avoid boundary effect (Figure 8.13). The cell radius is set to 200 m, and assume that each BS can extend its coverage to at most 400 m. We only consider path loss for the channels between BSs and MUs, according to ITU microcell test environment [5]. Power consumption is 400 W for BSs in active mode and 10 W for BSs in sleep mode. The bandwidth of each BS is 5 MHz. MUs arrive in the network according to a Poisson process, and the average sojourn time of each MUs is 1 minute. To evaluate the algorithms in cellular networks with spatial traffic load fluctuations, three hotspots with relatively higher load than other areas are generated, as shown in Figure 8.13. A new MU arrives in each hotspot with probability 5% respectively, and their locations follow normal distribution with mean at central point of each hotspot and standard deviation R. The others are uniformly placed in the whole area. The rate requirement of each MU is 122 kbps. The cell zooming period T is set to be 1 hour, and all the simulation results are averaged over 100 cell zooming periods.

In the simulation, we set the reservation parameters the same for all BSs with $\alpha_i = \alpha$, then tune the value of α and calculate the average energy consumption. When α increases, more bandwidth is reserved and the cell zooming algorithm becomes more conservative. This will result in more BSs working in the active mode, and less blocking probability can be achieved. Therefore, by tuning α, we can leverage the trade-off between energy consumption and quality of service. The simulation results in Figure 8.14 verify our analysis. For a given arrival rate, there is a trade-off curve of energy consumption versus outage probability for each algorithm.

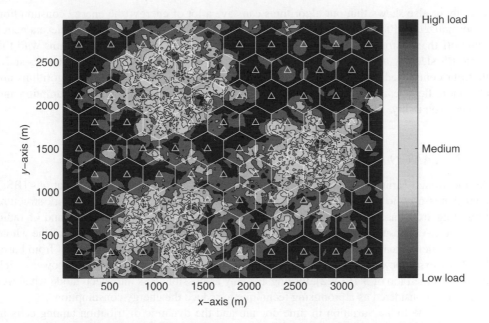

Figure 8.13 Traffic distribution in the tested cellular network layout

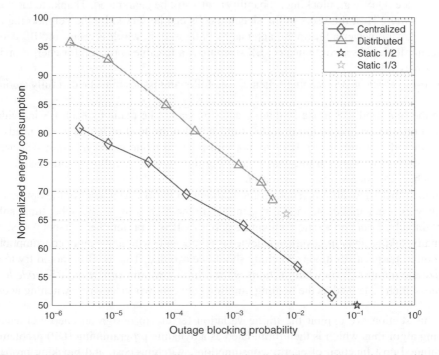

Figure 8.14 Energy outage trade-off of centralized and distributed cell zooming algorithms

The figure also shows that our algorithms can save a lot of energy (the energy consumption is normalized to 100 if all the BSs are active). The centralized algorithm can achieve a better trade-off than distributed algorithm. We also compare the cell zooming algorithms with the static BS sleeping algorithm, which switches off 1/2 or 1/3 of all the BSs. The results show that our centralized algorithm performs better than the static algorithm. Our algorithms are also more flexible as they can freely leverage the trade-off between energy consumption and outage probability.

8.3.3   TANGO by Adaptive BS Sleeping

As mentioned earlier, in cellular networks, the energy consumption of base stations (BSs) contributes 60% to 80% of the whole network [12], and will increase as network structure migrating from macrocell to micro- and picocells to meet the increasing demand of radio resources. As a result, the energy consumption of BSs becomes a major portion of the whole network energy consumption. As the energy consumption of a BS mainly comes from baseband processor, power amplifier, air-conditioner, and so on, rather than transmit power which only takes the ratio of 3.1% [22], turning as many as possible BSs into sleep mode whenever possible is considered as a promising technique to reduce the energy consumption.

In fact, due to the variation in time domain and the dynamic distribution among cells in space domain [23], there are opportunities for some BSs to turn to sleep mode when the traffic load in their coverage is low. However, when BSs turn to sleep mode, radio coverage and quality of service (QoS, e.g., blocking probability) must still be guaranteed. Thanks to the concept of *cell zooming* [14], the users in the sleeping cells can be served by the neighboring active BSs by transmit power adjustment, antenna reconfiguration, wireless relay, and BS cooperation technologies. As a consequence, BS sleeping is a feasible approach for energy saving in cellular networks.

To design efficient BS sleeping schemes, the following issues must be carefully studied.

- On the one hand, BS mode switching decision cannot be made by each BS individually. Not only the load condition of a BS itself, but also the load of its neighbors needs to be considered. For instance, a BS may not turn to sleep while its neighboring BSs are over loaded, even if its own traffic load is low. For this reason, each BS should make its mode switching decision via BS cooperation.
- On the other hand, although cooperation among all the BSs can achieve the optimal sleep policy, it is not applicable in real system due to the high complexity. Suboptimal solution obtained by *local* cooperation among neighboring BSs is preferable.
- Finally, taking signaling overhead, device lifetime, and switching energy consumption into account, frequent BS mode switching should be avoided. That is, BSs should try to minimize the number of switching actions or, in other words, maximize the BS *mode holding time*, which is defined as the holding duration between two successive switching actions.

In this section, we exploit the traffic variation feature to design an energy-efficient BS sleeping algorithm, which is then formulated as a dynamic programming (DP) problem with a combined cost function of energy consumption, switching cost, and blocking probability penalty. To reduce the dimension of state space and that of action space, *per-cell Q-factor* based on the cooperation among neighboring BSs is introduced, and a low-complexity algorithm is proposed to find the suboptimal policy. In addition, to match the system with BS sleeping behavior, user association and handover algorithms are redesigned. Simulations demonstrate that, with the proposed algorithm, the active BS pattern well meets the time variation and the nonuniform spatial distribution of system traffic. Besides, the trade-off between the energy saving from BS sleeping and the cost of switching is well balanced by the proposed scheme.

8.3.3.1 System Model

Consider a downlink cellular network consisting of M BSs with universal frequency reuse. Let $\mathcal{M} = \{1, \ldots, M\}$ denote the set of BSs. The *maximum coverage* of BS m is the area where BS m can provide the required data rate, and the *cell* m is defined as the area that is nearest to BS m compared with other BSs. As depicted in Figure 8.15, the cell radius is R_c and the BS maximum coverage radius is R_b, which indicates that each BS is able to cover its neighbor cells. In the traditional cellular networks where all BSs are active, each cell is taken care of by its own BS. When some BSs turn to sleep, the actual BS coverage extends from their own cells to the neighbors with sleeping BSs. This is reasonable in urban scenarios, where BSs are densely deployed. The neighbors of BS m are denoted as $m(1), \ldots, m(B)$, where B is the number of neighbors ($B = 6$ in hexagonal cellular system). Denote $\mathcal{B}_m = \{m, m(1), \ldots, m(B)\}$ as the set of BSs, which can provide service to the users in cell m.

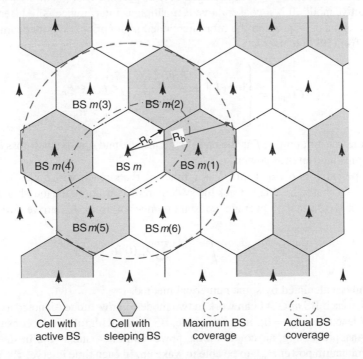

Cell with active BS   Cell with sleeping BS   Maximum BS coverage   Actual BS coverage

Figure 8.15 Cellular network architecture. The cell radius is R_c and the maximum coverage radius is R_b, which indicates the overlapped network structure. When some BSs turn to sleep, the active BSs extend their actual coverage from their own cells to the neighbors with sleeping BSs

The traffic arrives in cell m at time t as a Poisson process with intensity $\lambda_m(t)$. The traffic is assumed uniformly distributed in each cell, but asymmetric among different cells. Assume that the system has the statistic traffic information, that is, the average arrival rate $\lambda(t) = \{\lambda_m(t)\}_{m=1}^{M}$, which is a periodic function with period T (e.g., 24 hours). Each user has a minimum rate requirement r_0. All users arrive randomly and then remain stationary until the transmission is finished. The transmission duration of each user follows exponential distribution with mean $1/\mu$.

Assume that each active BS m has limited radio resource, that is, the maximum bandwidth $W_m^{\max}$. Notice that the bandwidth here is the generalization of wireless resources that the BS can allocate, that is, subcarriers or time-slots, and so on. If user k is associated to BS m, the corresponding bandwidth demand is

$$w_{mk}(t) = \frac{r_0}{C_{mk}(t)} \tag{8.6}$$

where $C_{mk}(t)$ is the spectrum efficiency (instantaneous peak rate per unit bandwidth). It varies over time due to shadowing, multipath fading, and so on. Nevertheless, as the long timescale performance is considered here, we ignore the fast fading effects, that is, we assume that the spectrum efficiency is constant during the transmission, which is determined by the large-scale path loss. The expressions are then simplified to C_{mk} and w_{mk} without time index. The inter-cell interference is assumed already being taken care of by certain reuse or interference management schemes. Interference power is averaged over all possible user positions assuming all the BSs are in active mode. It is a conservative estimation as interference is reduced when some BSs go to sleep mode. Based on the aforementioned assumptions, C_{mk} depends only on the distance l_{mk} from BS m to user k

$$C(l_{mk}) = \begin{cases} \log_2\left(1 + \dfrac{P_t \beta l_{mk}^{-\alpha}}{N_0}\right), & 0 < l_{mk} \le R_b, \\ 0, & l_{mk} > R_b, \end{cases} \tag{8.7}$$

where P_t is the transmit power; β is the path-loss constant, and α is the path-loss exponent; N_0 is the noise-plus-interference power.

The time period T is divided into $N+1$ time intervals (each with index i) as shown in Figure 8.16. Note that to be able to track the system traffic variation, the length of time interval $\tau^{(i)}$ varies with $\lambda(t)$ so that on average, constant number of users, K_τ, arrive during $\tau^{(i)}$. Then we have

$$K_\tau = \int_{t^{(i)}}^{t^{(i)}+\tau^{(i)}} \sum_{m=1}^{M} \lambda_m(t)\mathrm{d}t, \tag{8.8}$$

where $\tau^{(i)}$ can be calculated by some numerical method.

Assume that each BS $m \in \mathcal{M}$ can work in two modes: *active* mode (denoted as $s_m^{(i)} = 1$) and *sleep* mode (denoted as $s_m^{(i)} = 0$). In active mode, BS works with full power consumption $P_{\max}$ including signal processing, air-conditioning, power amplifier, and so on. In sleep mode, BS works with minimum power $P_{\min}$ to be able to wake up. In each time interval $\tau^{(i)}, i = 1, \ldots, N$, the system works in the fixed state $s^{(i)} = \{s_m^{(i)}\}_{m=1}^{M}$. The state space is

$$S = S_1 \times S_2 \times \ldots \times S_M, \tag{8.9}$$

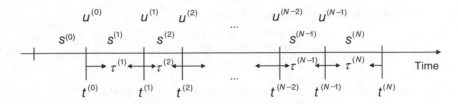

Figure 8.16 System operation over time. The network keeps a constant state $s^{(i)}$ in each time interval $\tau^{(i)}$, and operates action $\boldsymbol{u}^{(i)}$ at each time spot $t^{(i)}$

where $S_m = \{0, 1\}, m = 1, \ldots, M$ is the set of working modes of BS m.

At each time spot $t^{(i)}, i = 0, \ldots, N$, the BSs take the action $\boldsymbol{u}^{(i)} = \{u_m^{(i)}\}_{m=1}^M$. The action space is

$$\mathcal{U} = U_1 \times U_2 \times \ldots \times U_M, \tag{8.10}$$

where $U_m = \{0, 1\}, m = 1, \ldots, M$ is the set of actions of BS m. Denote $u_m^{(i)} = 1$ as the action that BS m switches its working mode and $u_m^{(i)} = 0$ otherwise, as that BS m maintains its working mode.

The overview of the system operation is as follows. The BSs work in the constant state $s^{(i)}$ during the time interval $\tau^{(i)}, i = 1, \ldots, N$, and the users are served by the currently active BSs. At each time spot $t^{(i)}, i = 0, 1, \ldots, N - 1$, the BSs decide whether to switch their working modes or not according to the action $\boldsymbol{u}^{(i)}$. If a BS switches from active mode to sleep mode, the associated users are shifted to the active neighbors. After the BS mode switching process is finished, the system goes into the next time interval $\tau^{(i+1)}$ with updated state $s^{(i+1)}$.

Generally speaking, the object of BS sleeping algorithm is to determine the action $\boldsymbol{u}^{(i)}, i = 0, \ldots, N - 1$ to minimize the system energy consumption given the initial state s^0 and statistic traffic information $\lambda(t)$, as well as to maintain a predefined blocking probability and avoid from frequent mode switching. Specifically, we formulate it as a dynamic programming problem, which is detailed in the next section.

8.3.3.2 Problem Formulation

A standard dynamic programming problem contains the following elements: state, action, state transition, and per-stage cost [24]. Note that the system state is actually $(s^{(i)}, \lambda^{(i)})$, where $\lambda^{(i)} = \{\lambda_m^{(i)}\}_{m=1}^M, \lambda_m^{(i)} = 1/\tau^{(i+1)} \int_{t^{(i)}}^{t^{(i+1)}} \lambda_m(t)dt$. We denote $s^{(i)}$ as the system state for notation convenience, as $\lambda^{(i)}$ is system-determined parameter and do not change with any action.

Given the current system state $s^{(i)}$ and the control action $\boldsymbol{u}^{(i)}$, the state transition is determined by

$$s^{(i+1)} = f(s^{(i)}, \boldsymbol{u}^{(i)}) = \{|s_m^{(i)} - u_m^{(i)}|\}_{m=1}^M. \tag{8.11}$$

The per-stage cost function $g(s^{(i)}, \boldsymbol{u}^{(i)})$ composed of three parts. The first part is the energy consumption of BS operation, which is calculated as

$$g_e^{(i)}(s^{(i)}, \boldsymbol{u}^{(i)}) =$$

$$\sum_{m=1}^M [s_m^{(i+1)} P_{\max} + (1 - s_m^{(i+1)}) P_{\min}] \tau^{(i+1)} \tag{8.12}$$

The second part is the BS mode switching cost

$$g_s^{(i)}(\boldsymbol{s}^{(i)}, \boldsymbol{u}^{(i)}) = \sum_{m=1}^{M} E_s u_m^{(i)} \tag{8.13}$$

where E_s is the cost of switching between active mode and sleep mode.

Finally, according to the optimization problem, blocking probability penalty should be integrated into the cost function. To relate the BS sleeping action with blocking probability, we define *system blocking probability* and *area blocking probability* next.

Definition 8.1 *The* system blocking probability *at state* $\boldsymbol{s}$ *is the probability that a newly arrived user* k' *is blocked, that is, none of the active BSs can provide the required bandwidth to this user:*

$$P_{\text{sys}}(\boldsymbol{s}) = \Pr\left\{ \bigcap_{\substack{m:\, s_m=1, \\ w_{mk'}<\infty}} \left(w_{mk'} + \sum_{k} x_{mk} w_{mk} > W_m^{\max} \right) \right\} \tag{8.14}$$

where binary variable $x_{mk} = 1$ *if user* k *is associated to BS* m, *and equals 0 otherwise. The summation is over all the existing users in the system.*

Definition 8.2 *The* area blocking probability *of area* A *is the conditional probability that a user* k' *arrived in* A *is blocked:*

$$P_a(A, \boldsymbol{s}) =$$

$$\Pr\left\{ \bigcap_{\substack{m:\, s_m=1, \\ w_{mk'}<\infty}} \left(w_{mj} + \sum_{k} x_{mk} w_{mk} > W_m^{\max} \right) | k' \in A \right\}. \tag{8.15}$$

As the network is divided into multiple cells, the relationship between system blocking probability and area blocking probability is given by the law of total probability

$$P_{\text{sys}}(\boldsymbol{s}) = \sum_{m=1}^{M} \Pr(k' \in A_m) P_a(A_m, \boldsymbol{s}) \tag{8.16}$$

where A_m is the area of cell m. As cell m can only be covered by the BSs in B_m, the area blocking probability of A_m can be approximated as

$$P_a(A_m, \boldsymbol{s}) \approx P_a(A_m, \tilde{\boldsymbol{s}}_m) \tag{8.17}$$

where $\tilde{\boldsymbol{s}}_m = \{s_m, s_{m(1)}, \dots, s_{m(B)}\}$ is the *local state* of cell m. It can be easily verified that a sufficient condition for $P_{\text{sys}}(\boldsymbol{s}) \le P_{\text{thr}}$ is

$$P_a(A_m, \tilde{\boldsymbol{s}}_m) \le P_{\text{thr}}, \quad \forall m \tag{8.18}$$

where P_{thr} is a given threshold.

The approximated area blocking probability $P_a(A_m, \tilde{s}_m)$ is derived in 8.3.3 and is summarized next:

$$P_a(A_m, \tilde{s}_m) = \prod_{n \in B_m, s_n=1} \left(\frac{\lambda_n'}{\lambda_n' + \mu} \right)^{K_n^{\max}} \tag{8.19}$$

where $\lambda_n' = \lambda_n + \sum_{j \in B_n \cap B_m, s_j=0} \lambda_j / I_j M_j^{\mathrm{on}}, R_j = R_c \sqrt{(B/I_j M_j^{\mathrm{on}} + 1)}, M_j^{\mathrm{on}} = \sum_{j' \in B_j \cap B_m} s_{j'}$, and $I_j = 1$ if $j = m$ and $I_j = 2$ if $j = m(b), b = 1, \ldots, B$, and $K_n^{\max}$ is expressed as

$$K_n^{\max} = \lceil x \rceil \tag{8.20}$$

The operator $\lceil \cdot \rceil$ rounds the real number to the nearest integer no smaller than it, where x stands for

$$x = \frac{R_c^2 W_n^{\max}}{2r_0} \left(1 + \sum_{j \in B_n \cap B_m, s_j=0} \frac{\lambda_j I_n}{I_j M_j^{\mathrm{on}} \lambda_n} \right)$$

$$\left(\int_0^{R_c} \frac{l \, dl}{C(l)} + \sum_{j \in B_n \cap B_m, s_j=0} \frac{\lambda_j I_n}{B \lambda_n} \int_{R_c}^{R_j} \frac{l \, dl}{C(l)} \right)^{-1} \tag{8.21}$$

Then the blocking probability penalty is calculated as a sum of area blocking probability penalty of all cells

$$g_b^{(i)}(s^{(i)}, u^{(i)}) = \sum_{m=1}^{M} h(P_a(A_m, f(\tilde{s}_m^{(i)}, \tilde{u}_m^{(i)})), P_{\mathrm{thr}}) \tag{8.22}$$

where $\tilde{u}_m = \{u_m, u_{m(1)}, \ldots, u_{m(B)}\}$ is *local action* of cell m, and the penalty function h is defined as

$$h(P_a(A_m, \tilde{s}_m), P_{\mathrm{thr}}) = \begin{cases} E_b P_a(A_m, \tilde{s}_m), & \text{if } P_a(A_m, \tilde{s}_m) > P_{\mathrm{thr}} \\ 0, & \text{else} \end{cases} \tag{8.23}$$

where E_b is a very large number.

In summary, the per-stage cost is given by

$$g^{(i)}(s^{(i)}, u^{(i)}) = g_e^{(i)}(s^{(i)}, u^{(i)}) + g_s^{(i)}(s^{(i)}, u^{(i)}) +$$

$$g_b^{(i)}(s^{(i)}, u^{(i)}), i = 0, 1, \ldots, N-1 \tag{8.24}$$

$$g^N(s^N) = 0 \tag{8.25}$$

In this paper, given the traffic variation function $\lambda(t)$ and the initial state $s^{(0)}$, we seek to minimize the total cost of all stages

$$\min_{\{u^{(0)}, \ldots, u^{(N-1)}\}} \sum_{i=0}^{N-1} g^{(i)}(s^{(i)}, u^{(i)}), \tag{8.26}$$

and find an optimal control policy $\boldsymbol{v} = \{v^{(0)}, v^{(1)}, \dots, v^{(N-1)}\}$ that satisfies

$$\boldsymbol{v} = \arg \min_{\{u^{(0)}, \dots, u^{(N-1)}\}} \sum_{i=0}^{N-1} g^{(i)}(s^{(i)}, u^{(i)}) \tag{8.27}$$

In the next section, we first present the standard DP algorithm. Then by reducing the state size using per-cell Q-factor, a low-complexity algorithm is proposed. Also, BS sleeping related user association and handover is discussed, as well as implementation issues.

8.3.3.3 Dynamic Programming Algorithm

A. General Solution

The cost minimization problem (8.26) can be solved by the standard DP algorithm [24] taking the form

$$J^N(s^{(N)}) = 0 \tag{8.28}$$

$$J^{(i)}(s^{(i)}) = \min_{u^{(i)} \in \mathcal{U}} [g(s^{(i)}, u^{(i)}) + J^{(i+1)}(f(s^{(i)} u^{(i)}))] \tag{8.29}$$

where $i = 0, 1, \dots, N - 1$. Proceeding backward induction of Eq. (8.29) from $N - 1$ to 0, the optimal cost is equal to $J^0(s^0)$ for the given $s^{(0)}$. Furthermore, if $\boldsymbol{v}^{(i)} = u^{(i)}(s^{(i)})$ minimizes the right side of Eq. (8.29) for each $s^{(i)}$ and i, the policy $\boldsymbol{v} = \{v^{(0)}, v^{(1)}, \dots, v^{(N-1)}\}$ is optimal.

Note that the cardinalities of $\mathcal{S}$ and $\mathcal{U}$ are both 2^M, which increase exponentially with the number of BSs M. Due to the *curse of dimensionality* (as termed in Ref. [24]), the computational requirement to obtain the optimal control policy is overwhelming if the network size is large. As a consequence, it is very difficult to implement the standard DP algorithm in practical systems. In the following, we introduce per-cell Q-factor estimation to reduce the size of state space and propose a low-complexity decision algorithm to simplify the process of action.

B. Q-factor and Space Reduction

Define the Q-factor as follows:

$$Q^{(i)}(s^{(i)}, u^{(i)}) = g^{(i)}(s^{(i)}, u^{(i)}) + J^{(i+1)}(f(s^{(i)}, u^{(i)})) \tag{8.30}$$

where $i = 0, 1, \dots, N - 1$. According to (8.29) and (8.30), we have

$$J^{(i)}(s^{(i)}) = \min_{u^{(i)} \in \mathcal{U}} Q^{(i)}(s^{(i)}, u^{(i)}) \tag{8.31}$$

$$Q^{(i)}(s^{(i)}, u^{(i)}) = g^{(i)}(s^{(i)}, u^{(i)}) +$$
$$\min_{u^{(i+1)} \in \mathcal{U}} Q^{(i+1)}(f(s^{(i)}, u^{(i)}), u^{(i+1)}) \tag{8.32}$$

The Q-factor $Q^{(i)}(s^{(i)}, u^{(i)})$ represents the cost of applying the action $u^{(i)}$ at the current state $s^{(i)}$ and applying the action $\arg \min_{u^{(i+1)} \in \mathcal{U}} Q^{(i+1)}(f(s^{(i)}, u^{(i)}), u^{(i+1)})$ at the next state $s^{(i+1)} = f(s^{(i)}, u^{(i)})$. To reduce the size of state space, we approximate the Q-factor as a sum of per-cell Q-factors, that is,

$$Q^{(i)}(s^{(i)}, u^{(i)}) \approx \sum_{m=1}^{M} Q_m^{(i)}(\tilde{s}_m^{(i)}, \tilde{u}_m^{(i)}) \tag{8.33}$$

where the per-cell Q-factor is

$$Q_m^{(i)}(\tilde{s}_m^{(i)}, \tilde{u}_m^{(i)}) = g_m^{(i)}(\tilde{s}_m^{(i)}, \tilde{u}_m^{(i)}) +$$
$$\min_{\tilde{u}_m^{(i+1)}} Q_m^{(i+1)}(f(\tilde{s}_m^{(i)}, \tilde{u}_m^{(i)}), \tilde{u}_m^{(i+1)}) \qquad (8.34)$$

where the per-cell per-stage cost is

$$g_m^{(i)}(\tilde{s}_m^{(i)}, \tilde{u}_m^{(i)}) = \frac{1}{B+1} \sum_{n \in B_m} \{[s_n^{(i+1)} P_{\max} +$$
$$(1 - s_n^{(i+1)}) P_{\min}] \tau^{(i+1)} + E_s u_n^{(i)} \} +$$
$$h(P_a(A_m, f(\tilde{s}_m^{(i)}, \tilde{u}_m^{(i)})), P_{\mathrm{thr}}) \qquad (8.35)$$

which includes the energy consumption and the switching cost of the BSs in B_m, and the area blocking probability of cell m. The per-cell Q-factor $Q_m^{(i)}(\tilde{s}_m^{(i)}, \tilde{u}_m^{(i)})$ only needs the information of the BSs in B_m, which indicates the limited cooperation among neighboring BSs. It can be recursively calculated for each BS $m \in \mathcal{M}$. The suboptimal control policy is then given by

$$\nu^{(i)}(s^{(i)}) = \arg \min_{u^{(i)} \in \mathcal{U}} \sum_{m=1}^{M} Q_m^{(i)}(\tilde{s}_m^{(i)}, \tilde{u}_m^{(i)}). \qquad (8.36)$$

Algorithm 1 Q-factor Recursion

1: Calculate $P_a(A_m, \tilde{s}_m^{(i)})$ for all $i = 1, 2, \ldots, N, \tilde{s}_m^{(i)} \in \{0, 1\}^{B+1}$.
2: Calculate $Q_m^{(N-1)}(\tilde{s}_m^{(N-1)}, \tilde{u}_m^{(N-1)}) = g_m^{(N-1)}(\tilde{s}_m^{(N-1)}, \tilde{u}_m^{(N-1)})$ for all $\tilde{s}_m^{(N-1)}, \tilde{u}_m^{(N-1)} \in \{0, 1\}^{B+1}$.
3: **for** $i = N - 2$ to 0 **do**
4:   Calculate $Q_m^{(i)}(\tilde{s}_m^{(i)}, \tilde{u}_m^{(i)}) = g_m^{(i)}(\tilde{s}_m^{(i)}, \tilde{u}_m^{(i)}) + \min_{\tilde{u}_m^{(i+1)}} Q_m^{(i+1)}(f(\tilde{s}_m^{(i)}, \tilde{u}_m^{(i)}), \tilde{u}_m^{(i+1)})$ for all

$\tilde{s}_m^{(i)}, \tilde{u}_m^{(i)} \in \{0, 1\}^{B+1}$.
5: **end for**

Remark 1 (State Space Reduction): The size of state space in each stage is substantially reduced from 2^M (exponential growth with respect to the number of BSs M) to $M2^{B+1}$ (linear growth w.r.t. M).

Although the size of state space is reduced by introducing per-cell Q-factor, the minimization progress (8.36), which requires exhaustive search over the action space $\mathcal{U}$, is still of high complexity. Based on the per-cell Q-factor, we propose an iterative decision-making algorithm. In each iteration, the BSs find the optimal local actions and update the global action one by one. Until the sum of per-cell Q-factors does not decrease, the iteration terminates. It is summarized in **Algorithm 2**.

Remark 2 (Action Space Reduction): In the greedy search step 6, the size of decision space is 2^{B+1}. Obviously, the iteration (from step 3 to step 10) converges in a finite number of iterations. Simulations show that the number of iterations is no more than 4 mostly. As a result, the action search complexity in each stage is reduced from $O(2^M)$ to $O(M2^{B+1})$.

Algorithm 2 Action Iteration

1: **for** $i = 0$ to $N - 1$ **do**
2:     Set $u^* = 0, Q_{min} = \infty, \hat{Q} = E_b$.
3:     **while** $\hat{Q} < Q_{min}$ **do**
4:         Set $Q_{min} = \hat{Q}$.
5:         **for** $m = 1$ to M **do**
6:             Find the optimal solution of the following problem

$$\tilde{v}_m^* = \arg\min_{\tilde{v}_m} \sum_{n=1}^{m} Q_n^{(i)}(\tilde{s}_n^{(i)}, \tilde{u}_n), \; \tilde{v}_m \in \{0, 1\}^{B+1}$$

            where u is determined as: $u_n = u_n^*$, if $n \notin B_m$; $u_n = |u_n^* - v_n|$, if $n \in B_m$.
7:             Update u^* as: $u_n^* = u_n^*$, if $n \notin B_m$; $u_n^* = |u_n^* - v_n^*|$, if $n \in B_m$.
8:         **end for**
9:         Update $\hat{Q} = \sum_{n=1}^{M} Q_n^{(i)}(\tilde{s}_n^{(i)}, \tilde{u}_n^*)$.
10:    **end while**
11:    Set $u^{(i)} = u^*, s^{(i+1)} = f(s^{(i)}, u^{(i)})$.
12: **end for**

C. Systematic Design

Recall that during the time interval $\tau^{(i)}, i = 1, \ldots, N$, the users are served by the currently active BSs. At each time spot $t^{(i)}, i = 0, 1, \ldots, N - 1$, if a BS switches from active mode to sleep mode, the associated users are shifted to the active neighbors. As the accessible BSs provide different signal strength and are of various load conditions, user association and handover should be carefully designed to optimize resource allocation.

User Association: Load balancing scheme is implemented for the user association to reduce system blocking probability. In the literature, load balancing has been extensively studied and some efficient scheduling methods have been proposed. We make use of the *load-aware cell-site selection scheme* [25]. If the user k arrives in cell m, its candidate serving BS set is $C_k = \{n \in B_m | s_n = 1\}$. Notice that the remaining bandwidth of the BS $n \in C_k$ maybe not enough to accept user k, that is, $W_n^{max} - W_n < w_{nk}$, where W_n is the allocated bandwidth of BS n. In this case, the user should try the BSs in C_k one by one. The proposed user association algorithm is stated in **Algorithm 3**.

Note that $| \cdot |$ is the cardinality of a set. This scheme can be operated in a distributed manner. User selects the serving BS with high channel quality and low traffic load as well. Consequently, the asymmetric load across the whole system is implicitly balanced.

User Handover: At each time spot $t^{(i)}, i = 0, 1, \ldots, N - 1$, the users which associate with the BSs turning from active mode to sleep mode, should change their association to neighboring active BSs. To minimize the number of droppings, the handover algorithm also takes the idea of load balancing. Initialize $\mathcal{H}^{(i)} = \{k | x_{mk} = 1, s_m^{(i)} = 1, s_m^{(i+1)} = 0, m \in \mathcal{M}\}$. The handover algorithm is presented in **Algorithm 4**.

After the algorithm terminates, the remaining users in $\mathcal{H}^{(i)}$ are dropped. Due to the nature of load balancing, the number of dropped users is minimized.

Algorithm 3 User Association

1: Set up a list $L_k = \{n_1, n_2, \dots | n_j \in C_k\}$ with $c_{n_1} \geq c_{n_2} \geq \dots$, where $c_n = C(l_{nk})W_n^{\max}/W_n$.
2: Set flag = 0.
3: **for** $j = 1$ to $|C_k|$ **do**
4:   **if** $W_{n_j}^{\max} - W_{n_j} \geq w_{n_jk}$ **then**
5:     $x_{n_jk} = 1, W_{n_j} = W_{n_j}^{\max} - w_{n_jk}$, flag = 1; break.
6:   **end if**
7: **end for**
8: **if** flag $\neq$ 1 **then**
9:   No BS is selected, user k is blocked.
10: **end if**

Algorithm 4 User Handover

1: Set up a BS-user pair list $L^{(i)} = \{(m_1, k_1), (m_2, k_2), \dots | s_{m_j}^{(i+1)} = 1, k_j \in \mathcal{H}^{(i)}, C(l_{m_jk_j}) > 0\}$
   with $c_{(m_1,k_1)} \geq c_{(m_2,k_2)} \geq \dots$, where $c_{(m,k)} = C(l_{mk})W_m^{\max}/W_m$.
2: Set $j = 0$.
3: **while** $\mathcal{H}^{(i)} \neq \emptyset$ and $j < |L^{(i)}|$ **do**
4:   Set $j = j + 1$.
5:   **if** $W_{m_j}^{\max} - W_{m_j} \geq w_{m_jk_j}$ and $k_j \in \mathcal{H}^{(i)}$ **then**
6:     $x_{m_jk_j} = 1, W_{m_j} = W_{m_j}^{\max} - w_{m_jk_j}, \mathcal{H}^{(i)} = \mathcal{H}^{(i)}\backslash\{k_j\}$.
7:   **end if**
8: **end while**

D. Implementation Issue

As the proposed algorithm offers an off-line solution, different policies are implemented for different traffic variation pattern. A typical example is that policies for workday and weekend should be distinguished.

In real networks, the statistic features of traffic distribution and variation may change. For instance, the increase of the total number of subscribers enhances the average traffic intensity; a newly opened business center will become a new hotspot in the daytime, which changes the traffic distribution in space domain. To be able to track the long-term variation of traffic, the system should establish a dataset and record the number of calls in each cell. Depending on the statistic information obtained from the dataset, the system can operate the proposed algorithm to update the BS sleep pattern whenever necessary.

8.3.3.4 Simulation Study

The simulation layout is 10 by 10 hexagon cells with wrap-up to avoid boundary effect, which is shown in Figure 8.17. The cell radius is $R_c = 200$ m and the BS's maximum coverage is $R_b = 520$ m. We set the power consumption $P_{\max} = 1 \times 10^3$ W and $P_{\min} = 0$. The maximum bandwidth is $W_m^{\max} = 5$ MHz, which is identical for all BSs. User rate requirement is

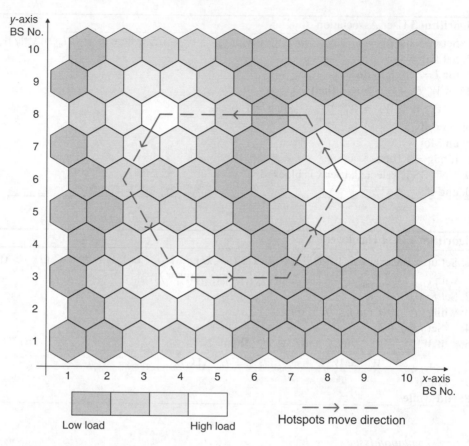

Figure 8.17 Simulation layout and traffic distribution. Three hotspots are formed and move along the dashed line anticlockwise every 24 hours a cycle. The highest load is $\lambda_h(t)$, and the others are $\alpha_l \lambda_h(t)$, $l = 1, 2, 3, 0 \le \alpha_3 \le \alpha_2 \le \alpha_1 \le 1$, respectively

$r_0 = 122$ kbps. Transmission duration parameter is $\mu = 1/180 s^{-1}$. The link parameters are set according to ITU microcell test environment [26]. The transmit power is $P_t = 41$ dBm. The noise-plus-interference power N_0 is calculated by setting the reference SNR at distance 200 m to be 0 dB. Path-loss model is $PL^{\mathrm{dB}}(l_{ik}) = 33.05 + 36.7\log_{10}(l_{ik})$. The number of user arrivals in each time interval is $K_\tau = 1 \times 10^4$. The blocking probability penalty $E_b = 1 \times 10^8$ J and the threshold $P_{\mathrm{thr}} = 1\%$. To simulate the asymmetric traffic distribution, three *hotspots* are formed in the network. The traffic distribution is configured as follows:

- Average arrival rate (or traffic intensity) of the whole network $\lambda_w(t) = \sum_{m=1}^{M} \lambda_m(t)$ varies according to the red-dashed line in Figure 8.18. The period of arrival rate is $T = 24$ hours.
- Three hotspots are generated and move along the dashed line shown in Figure 8.13 anti-clockwise every 24 hours a cycle.
- Set the arrival rates of the hotspot center cells as $\lambda_m(t) = \lambda_h(t)$. Then the arrival rates of the *l*st-tier of hotspot center cells are $\lambda_m(t) = \alpha_l \lambda_h(t)$, $l = 1, 2$, and the others are $\lambda_m(t) = \alpha_3 \lambda_h(t)$, where $0 \le \alpha_3 \le \alpha_2 \le \alpha_1 \le 1$.

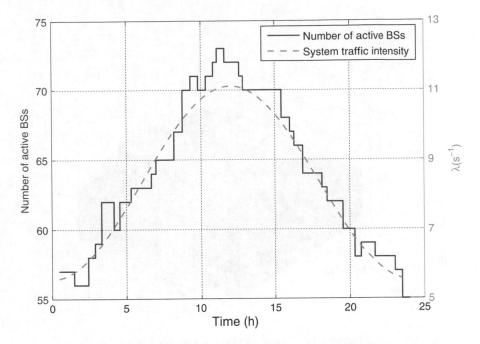

Figure 8.18   Number of active BSs compared with average traffic intensity in time space

The simulation is performed as follows. We first calculate the sleeping policy with respect to the statistic traffic information $\lambda(t)$ and the given initial state $s^{(0)}$. Then the random user arrival is generated in accordance with $\lambda(t)$ to test the performance of the policy obtained by the proposed algorithm. The initial state $s^{(0)}$ is set by activating half of the BSs in the network uniformly and then opening two more BSs in each hotspot.

A. BS Sleeping Pattern
In this part of simulation, parameter settings are: $E_s = 2.5 \times 10^5$ J/switch, $\alpha_1 = 0.88, \alpha_2 = 0.63, \alpha_3 = 0.50$.

BS mode switching behavior along with the average traffic intensity is presented in Figure 8.18. It is shown that our proposed DP algorithm tracks the variation of the average traffic intensity well in time domain. To illustrate the spatial consistency between the traffic distribution and the number of active BSs, we take the stage $i = 20$ as an example, and calculate $s'_m = s_m^{(i)} + \sum\limits_{m(j),j=1,\ldots,B} s_{m(j)}^{(i)}/2$ to imply the number of active BSs around each cell. Comparing Figure 8.19 with Figure 8.20, we can see that more BSs are active in the highly loaded area, while less BSs are active in the area with low load. Still, in the low load area, there generally are some active BSs in order to guarantee the network coverage. As a result, the active BS distribution well meets the spatial distribution of traffic intensity. In addition, the blocking probability in each time interval is maintained below the target (1%) almost all the time (see Figure 8.21). On average, more than 36% energy consumption is reduced. The average blocking probability over 24 hours is 0.3%, which shows that the area blocking probability estimation is conservative. More elaborate area blocking probability analysis can be performed to further improve the energy saving performance.

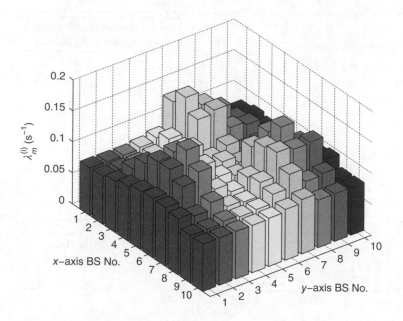

Figure 8.19 Traffic distribution in spatial domain

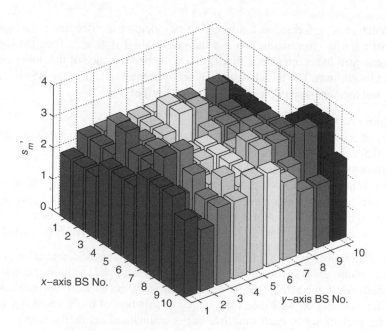

Figure 8.20 BS state distribution in spatial domain

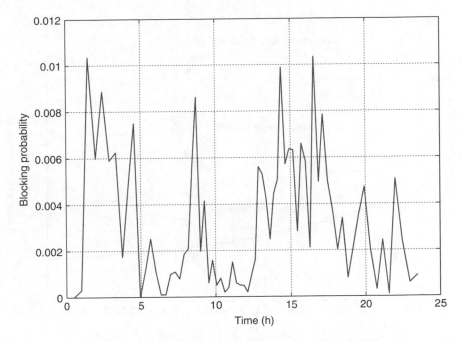

Figure 8.21 System blocking probability variation versus time

B. Comparing with Uniform BS Sleeping

We compare the proposed DP algorithm with the *uniform BS sleeping* approach proposed in Ref. [12], where active BSs are uniformly located in the network. We modify the uniform algorithm from binary pattern to multiple pattern. The number of active BSs are determined according to

$$N_{on}(t) = 55 + 5 \times \left\lfloor \frac{5(\lambda_w(t) - \lambda_{min})}{\lambda_{max} - \lambda_{min}} \right\rfloor \tag{8.37}$$

which is a function of average traffic intensity $\lambda_w(t)$, $\lambda_{min} < \lambda_w(t) < \lambda_{max}$. As a result, $N_{on}(t) \in \{55, 60, \cdots, 75\}$. Here $\lfloor \cdot \rfloor$ rounds the real number to the nearest integer no larger than it. The switch cost of the DP algorithm is fixed $E_s = 2.5 \times 10^5$ J/switch.

In Figure 8.22, the average number of active BSs and the average blocking probability are compared. We decrease the value of α_1, α_2, and α_3 to enhance the degree of asymmetric traffic distribution. By matching BS sleeping pattern with traffic intensity distribution from both time domain and space domain, the proposed DP algorithm outperforms the uniform one. Specifically, as the traffic distribution becomes more and more asymmetric, the performance gap becomes larger. That is, the gap of the average number of active BSs increases from 1.27 to 3.65, and the average blocking probability ratio of the uniform algorithm to the DP algorithm increases from 3.63 to 41.3.

The figure also shows the following result, which is a little bit surprising: both the number of active BSs and the blocking probability of the proposed DP algorithm decrease as the parameters α_1, α_2, and α_3 decrease. As a matter of fact, the blocking events mainly come from two aspects: (i) blocking caused by overloading in the hotspot cells; (ii) blocking caused

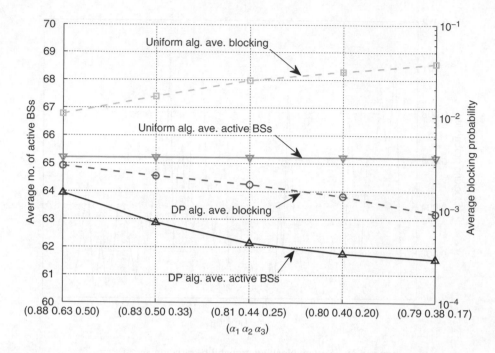

Figure 8.22   Comparison of proposed DP BS sleeping algorithm and uniform BS sleeping algorithm

by the high bandwidth requirement of the users in sleeping cells, which are denoted as *coverage edge* users. In current simulation settings, the system is in low traffic scenario. As a consequence, the blocking events caused by coverage edge users outweigh those caused by overloaded hotspot. As the traffic distribution becomes more and more asymmetric, more and more users are taken care of in the hotspots, and the number of coverage edge users becomes less. Hence, the blocking probability becomes lower and more BSs can sleep. Note that if the system traffic is extremely high on the contrary, the blocking events happened in hotspots outweigh those caused by coverage edge users, the blocking probability might increase.

C. Switching Cost
Extensive simulations are run to test the influence of the switching cost E_s on the performance. We set $\alpha_1 = 0.88, \alpha_2 = 0.63, \alpha_3 = 0.50$ for the simulations.

With different switching cost $E_s = 0, 2.5 \times 10^5, \ldots, 1 \times 10^6$ (J/switch), the average numbers of active BSs in the period of 24 hours are 62.9, 63.9, 65.2, 66.0, and 67.2 respectively. The blocking probability and the dropping probability are depicted in Figure 8.23. The proposed load balancing based algorithms for user association and handover are compared with the strongest signal based ones, where the selection criterion is simply $C(l_{mk})$ and the selection process is similar with **Algorithms 3** and **4**. The result shows that by effectively utilizing the wireless resources, load balancing is helpful for reducing the number of blocking and dropping events. It also illustrates that with the increase of E_s, the energy consumption increases, while both the blocking probability and the dropping probability decrease. It can be concluded that

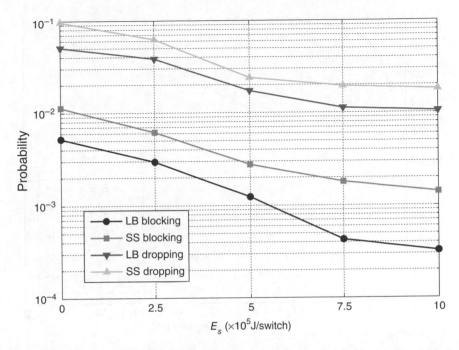

Figure 8.23 System blocking probability and dropping probability versus switching cost E_s. LB: proposed load balancing based BS selection and user handover algorithm; SS: strongest signal based algorithm

Table 8.1 Handover performance with different switching cost

E_s (J/switch)	0	2.5×10^5	5×10^5	7.5×10^5	1×10^6
No. of switching/stage	32.6	10.0	4.06	1.85	1.63
No. of handover/stage	385.7	108.5	41.74	17.73	16.03
No. of dropping/stage	19.6	4.16	0.72	0.20	0.17

there is a trade-off between the energy saved from turning BSs into sleep mode and the energy cost of BSs' mode switching. Because high switching cost prevents the BS switching from active to sleep, blocking probability is reduced.

Figure 8.24 shows the cumulative distribution function (CDF) of BS mode holding time. Without the switching penalty, more than 70% of BSs' measured mode holding time is less than 1 h. Such frequent mode switching maybe not acceptable for BS equipments in the real system. It not only consumes additional energy, but also brings large amount of handover, which causes exploding signaling overhead and user QoS degradation (shown in Table 8.1). This result explains the necessity of integrating switching cost into the total cost. As the value of E_s increases, the BS mode holding time increases accordingly, which shows that our algorithm well balances the trade-off between energy saving from sleep and cost from switching.

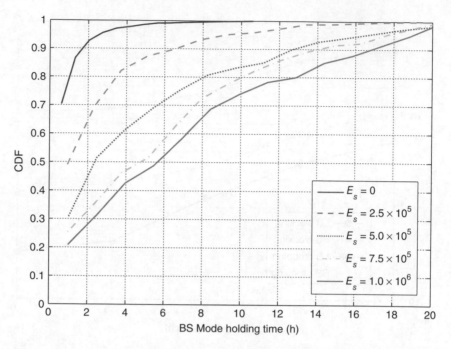

Figure 8.24 Cumulative distribution function of mode holding time with different switching cost E_s (J/switch)

8.4 Derivation of Area Blocking Probability

In this section, we ignore the stage index i for simplicity. According to the fact that a newly arrived user in cell m is blocked if all the active BSs in $\mathcal{B}_m$ are out of bandwidth, we have

$$P_a(A_m, \tilde{s}_m) = \Pr\left(\bigcap_{n \in \mathcal{B}_m, s_n = 1} W_n \geq W_n^{\max} \right)$$

$$= \prod_{n \in \mathcal{B}_m, s_n = 1} \Pr(W_n \geq W_n^{\max}) \tag{8.38}$$

where the second equality holds because the bandwidth utilizations of BSs are independent of each other.

If $s_n = 1, n \in \mathcal{B}_m$, the users in cell n can be served by BS n. The density of users in cell n is $K_n/\pi R_c^2$, where K_n is the number of users in cell n. Under the assumption that the user distribution in each cell is uniform, the bandwidth utilization of BS n for the users in cell n is calculated as

$$W_{nn} = \int_0^{R_c} w_{nk} \frac{K_n}{\pi R_c^2} 2\pi l \, dl$$

$$= K_n \frac{2r_0}{R_c^2} \int_0^{R_c} \frac{l \, dl}{C(l)} \tag{8.39}$$

If $s_j = 0, j \in B_m$, due to the nature of load balancing technique, the area of cell j is evenly assigned to its neighbor active BSs. Assume that the shifted traffic are uniformly distributed in the $1/B$ annulus sector with larger radius R_j and smaller radius R_c. The bandwidth requirement of the shifted traffic is

$$W_{jn} = \int_{R_c}^{R_j} w_{nk} \frac{K_j}{\pi R_c^2} \frac{2\pi}{B} l dl$$

$$= K_j \frac{2r_0}{BR_c^2} \int_{R_c}^{R_j} \frac{l dl}{C(l)} \tag{8.40}$$

where R_j is set to evenly assign the cell area to its neighbor active BSs, i.e.,

$$R_j = R_c \sqrt{\frac{B}{I_j M_j^{\text{on}}} + 1} \tag{8.41}$$

where $M_j^{\text{on}} = \sum_{j' \in B_j \cap B_m} s_{j'}$ is the number of active neighbor BSs of sleeping BS j, $I_j = 1$ if $j = m$ and $I_j = 2$ if $j = m(b), b = 1, \dots, B$. Note that local information sets $\tilde{s}_m$ and $\tilde{\lambda}_m$ do not contain the full local information of BS $m(b), b = 1, \dots, B$. Therefore, we introduce the parameter I_j assuming that the local information of BS $m(b)$ is symmetric, that is, the state and the arrival rate of BS $j' \in B_{m(b)} \setminus B_m$ are the same as those of BS $j' \in B_{m(b)} \cap B_m$.

In summary, the bandwidth utilization of active BS n is

$$W_n = W_{nn} + \sum_{j \in B_n \cap B_m, s_j = 0} W_{jn} I_n$$

$$= K'_n \gamma_n \tag{8.42}$$

where $K'_n = K_n + \sum_{j \in B_n \cap B_m, s_j = 0} I_n K_j / (I_j M_j^{\text{on}})$ is the total number of user served by BS n, and

$$\gamma_n = \frac{\frac{2r_0}{R_c^2} \left(\int_0^{R_c} \frac{l dl}{C(l)} + \sum_{j \in B_n \cap B_m, s_j = 0} \frac{\lambda_j I_n}{B \lambda_n} \int_{R_c}^{R_j} \frac{l dl}{C(l)} \right)}{1 + \sum_{j \in B_n \cap B_m, s_j = 0} \frac{\lambda_j I_n}{I_j M_j^{\text{on}} \lambda_n}}, \tag{8.43}$$

where we make use of the fact that $K_n / \lambda_n = K_j / \lambda_j$.

At the same time, the traffic load in cell n is evenly shifted to its neighbor active BSs. Similarly, we assume that half of the traffic of BS $m(b), b = 1, \dots, B$ is shifted to accessible active BSs in B_m. As a result, the traffic load of BS $n(s_n = 1)$ is

$$\lambda'_n = \lambda_n + \sum_{j \in B_n \cap B_m, s_j = 0} \frac{\lambda_j}{I_j M_j^{\text{on}}} \tag{8.44}$$

As the radio resource is shared by active users, the number of users K'_n associated with BS n evolves like the number of customers in a processor-sharing queue with Poisson arrivals and i.i.d. service times [27]. The key property of the processor-sharing queue is that the stationary distribution of the number of customers is insensitive to the distribution of service times. Hence, the stationary distribution of the number of active users is given by

$\Pr(K'_n = k) = (\rho_n)^k(1 - \rho_n)$ with mean $E[K'_n] = \rho_n/(1 - \rho_n)$, where ρ_n is the average traffic load of BS n. Applying Little's law [28], we get $E[K'_n] = \lambda'_n/\mu$, which results in $\rho_n = \lambda'_n/(\lambda'_n + \mu)$. Finally, the area blocking probability is expressed as

$$P_a(A_m, \tilde{s}_m) = \prod_{n \in B_m, s_n = 1} \Pr(K'_n \geq W_n^{\max}/\gamma_n),$$

$$= \prod_{n \in B_m, s_n = 1} \rho_n^{\lceil W_n^{\max}/\gamma_n \rceil} \tag{8.45}$$

Summarizing the equations derived earlier, we obtain the expression of the approximated area blocking probability.

References

[1] Cisco, "Global mobile data traffic forecast update 2010–2015."

[2] A. Sendonaris, E. Erkip, and B. Aazhang, "User cooperation diversity, part i: system description," IEEE Trans. Commun., vol. 51, no. 11, pp. 1927–1938, 2003.

[3] S. Shamai and B. M. Zaidel, "Enhancing the cellular downlink capacity via co-processing at the transmitter end," in *Proceedings of VTC 2001 Spring*, 2001.

[4] H. Huang, M. Trivellato, A. Hottinen, M. Shafi, P. J. Smith, and R. Valenzuela, "Increasing downlink cellular throughput with limited network mimo coordination," IEEE Trans. Wireless Commun., vol. 8, no. 6, pp. 2983–2989, 2009.

[5] 3GPP, "Further advancements for E-UTRA: physical layer aspects," Technical Specification on Group Radio Access Network (Release 9), TR 36.814 V 1.2.1, Third Generation Partnership Project (3GPP), 2012.

[6] W. Song and W. Zhuang, "Multi-service load sharing for resource management in the cellular/WLAN integrated network," IEEE Trans. Wireless Commun., vol. 8, no. 2, pp. 725–735, 2009.

[7] A. Yeh, S. Talwar, G. Wu, N. Himayat, and K. Johnsson, "Capacity and coverage enhancement in heterogeneous networks," IEEE Wireless Commun. Mag., vol. 18, no. 3, pp. 132–139, 2011.

[8] E. Oh, B. Krishnamachari, X. Liu, and Z. Niu, "Towards dynamic energy-efficient operation of cellular network infrastructure," IEEE Commun. Mag., vol. 49, no. 6, pp. 56–61, 2011.

[9] S. Haykin, "Cognitive radio: brain-empowered wireless communications," IEEE J. Sel. Areas Commun., vol. 23, no. 2, pp. 201–220, 2005.

[10] A. Ghasemi and E. S. Sousa, "Spectrum sensing in cognitive radio networks: requirements, challenges and design trade-offs," IEEE Commun. Mag., vol. 46, no. 4, pp. 32–39, 2008.

[11] A. Goldsmith, S. A. Jafar, I. Maric, and S. Srinivasa, "Breaking spectrum gridlock with cognitive radios: an information theoretic perspective," Proc. IEEE, vol. 97, no. 5, 2009.

[12] M. A. Marsan, L. Chiaraviglio, D. Ciullo, and M. Meo, "Optimal energy savings in cellular access networks," in *Proceedings of the IEEE ICC'09 Workshop, GreenComm.*, June 2009.

[13] A. P. Jardosh, K. Papagiannaki, E. M. Belding, K. C. Almeroth, G. Iannaccone, and B. Vinnakota, "Green WLANs: on-demand WLAN infrastructures," Mob. Networks Appl., vol. 14, no. 6, pp. 798–814, 2009.

[14] Z. Niu, Y. Wu, J. Gong, and Z. Yang, "Cell zooming for cost-efficient green cellular networks," IEEE Commun. Mag., vol. 48, no. 11, pp. 74–79, 2010.

[15] Z. Niu, L. Long, J. Song, and C. Pan, "A new paradigm for mobile multimedia broadcasting based on integrated communication and broadcast networks," IEEE Commun. Mag., vol. 46, no. 7, pp. 126–132, 2008.

[16] X. Wang, Y. Bao, X. Liu, and Z. Niu, "On the design of relay caching in cellular networks for energy efficiency," in *IEEE INFOCOM'11 workshop*, April 2011.

[17] S. Zhou, J. Gong, Z. Niu, Y. Jia, and P. Yang, "A decentralized framework for dynamic downlink base station cooperation," in *Proceedings of IEEE Globecom'09*, December 2009.

[18] J. Gong, S. Zhou, and Z. Niu, "A dynamic programming approach for base station sleeping in cellular networks," IEICE Trans. Commun., vol. 95B, no. 2, 2012.

[19] S. Zhou, J. Gong, and Z. Niu, "Distributed adaptation of quantized feedback for downlink network mimo systems," IEEE Trans. Wireless Commun., vol. 10, no. 1, pp. 61–67, 2011.

[20] M. Chiang, P. Hande, T. Lan, and C. W. Tan, "Power control in wireless cellular networks," *Foundations and Trends®* in Networking, vol. 2, no. 4, pp. 381–533. Now Publishers, 2008.

[21] L. Chiaraviglio, D. Ciullo, M. Meo, and M. A. Marsan, "Energy-aware UMTS access networks," in *Proceedings of the 11th International Symposium on Wireless Personal Multimedia Communications*, September 2008.

[22] H. Karl, "An overview of energy-efficiency techniques for mobile communication systems." Report of AG Mobikom WG7, September 2003. [Online]. Available: http://www.tkn.tu-berlin.de/\break fileadmin/fg112/Papers/TechReport_03_017.pdf.

[23] D. Willkomm, S. Machiraju, J. Bolot, and A. Wolisz, "Primary user behavior in cellular networks and implications for dynamic spectrum access," IEEE Commun. Mag., vol. 47, no. 3, pp. 88–95, 2009.

[24] D. P. Bertsekas, Dynamic Programming and Optimal Control, 3rd ed., Massachusetts: Athena Scientific, 2005.

[25] A. Sang, X. Wang, M. Madihian, and R. Gitlin, "A load-aware handoff and cell-site selection scheme in multi-cell packet data systems," in *Proceedings of Globecom'04*, pp. 3931–3936, December 2004.

[26] Ericsson, "Radio characteristics of the ITU test environments and deployment scenarios," [Online]. Available: http://ftp.3gpp.org/tsg_ran/WG1_RL1/TSGR1_56b/Docs/R1-091320.zip.

[27] T. Bonald and A. Proutiere, "Wireless downlink data channel: user performance and cell dimensioning," in *Proceedings of MobiCom'03*, pp. 339–352, September 2003.

[28] L. Kleinrock, Queueing Systems, John Wiley & Sons, 1976.

9

Green Home and Enterprise Networks

Łukasz Budzisz and Adam Wolisz

Telecommunication Networks Group, Technische Universität Berlin, Berlin, Germany

9.1 Home and Enterprise Networks Today

9.1.1 Similarities

In home and small office environments, Internet connectivity is most typically provided with a broadband access, accounting currently for more than 650 million subscribers worldwide [1]. Among different broadband technologies, DSL is the most widespread, with 58% (or about 380 million) share of the market. Other broadband technologies are cable (19%) and fibre, with both fibre to the home (FTTH) and other forms of fibre (FTTx) accounting for 22% of all broadband subscribers.[1] The remaining subscribers are shared among wireless broadband technologies, mainly satellite using K_a band (in Europe and the Americas) and other wireless radio technologies [1]. According to Ref. [1], the current outlook of the market sees the dominant position of the DSL technology firm, despite some customers moving to fibre that is the fastest growing segment of the market, with nearly 10% growth in the first quarter of 2013. Satellite broadband access also experiences a small rise (1.5% quarterly), providing Internet connectivity in the areas where traditional and more cost-effective fixed broadband access technologies are not available.

It is thus safe to state that most of the home environments on the user side consist of a gateway to the broadband access network, called here *home gateway* (H-GW), that is, modem, wireless LAN (WLAN) access point (AP) and router (usually WLAN AP and router are one

---

[1] With some regional variations, e.g., about 50% of broadband subscribers in Americas use cable broadband access due to the paid TV subscriptions, whereas Asia has the highest share of the FTTx market accounting for nearly 30% of all subscribers [1].

Green Communications: Principles, Concepts and Practice, First Edition.
Edited by Konstantinos Samdanis, Peter Rost, Andreas Maeder, Michela Meo and Christos Verikoukis.
© 2015 John Wiley & Sons, Ltd. Published 2015 by John Wiley & Sons, Ltd.

physical device, whereas modem may be physically separated), and on the ISP side they consist of an ISP modem occupying a port in one of the line cards (e.g., for DSL technology, a typical DSL line card serves 12–72 lines) that are within the control of an access multiplexer, for example, DSL access multiplexer (DSLAM) for the DSL access [2]. Further on the user side, the H-GW is interconnected with the plethora of devices belonging to household inhabitants (in most cases not more than 30 devices in a single home [3]) that can be found useful at home, for example, PCs, laptops, smart TVs, gaming consoles, audio equipment, DVD players, printers, and so on, with more and more new devices capable of being connected to the Internet coming to the market. All these devices are not necessarily connected directly, possibly only via Internet 'at large'. Very often, due to the coverage problems there is a need to extend the connectivity to the H-GW by means of repeaters, additional APs, and so on. What makes home networking environments extremely interesting is the multitude of available low-cost transmission technologies that are applicable in these scenarios [3]. In case of wired communication, Ethernet (IEEE 802.3) [4] is by far the most commonly used technology. Alternatively, power lines can also be used to transmit the data thanks to the power line communication (PLC) protocol [5]. Nevertheless, wireless communication technologies are gaining the edge, because of the convenience and flexibility they offer. According to a study from 2005 [6], already by then 52% of the US households with a computer network were using wireless technology, with this trend accelerating in the recent years. The most dominant wireless technologies being used in home environments belong to the IEEE 802.11 family (with majority of available products supporting one or multiple of IEEE 802.11a/b/g/n/ac standards) [7]. Another important wireless technology used in home networking is Bluetooth [8], especially in the context of applications that do not require excessively much bandwidth, or simply, as a replacement for cable connections. Recently, body area networks (BANs) are expanding quickly, with many new applications related to monitoring the human body functions and the surrounding environment [9]. Home automation [10] is another field gaining a huge growth of popularity, especially recently in the context of 'Internet of Things', and thus requiring a large number of sensor and actuators to communicate in order to fully automate and control home environments. Finally, a recent introduction of broadband femtocells provides good, alternative candidates to be applied as H-GWs [11]. Femtocells are customer-deployed, low-power base stations (BSs) using one or more commercial cellular standards and operating in the licensed part of the spectrum, in contrast to all the above-mentioned wireless standards that operate in the non-licensed industrial scientific medical (ISM) band. In home deployments, small and inexpensive femtocells can typically provide connectivity for up to four users [12]. According to the most recent market outlook presented by Small Cell Forum [12] the number of deployed consumer femtocells overtook the total number of macrocells in February 2013.

Typical home network scenario is summarized in Figure 9.1. Two parts of the network can be clearly distinguished. First, an Ethernet-based backbone that spreads between the ISP access network, H-GW and possible connectivity extensions (APs). The second part of the network, in most cases wireless, interconnects the individual devices with the H-GW, or if available, with APs extending the connectivity. Connection between additional APs and H-GW can be wireless, too.

With the broad adoption of laptops, smartphones and tablets in the recent years, enterprise environments have evolved from the purely Ethernet-based architectures to the networks that offer much more flexibility still being highly secure and reliable. New requirements mean supporting mobile users, flexible hosting of (wireless) guest users, offering the same set of services

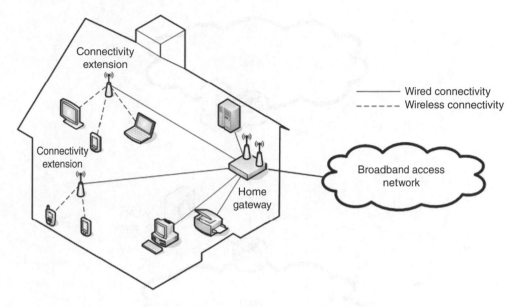

Figure 9.1   A typical home network scenario

at main sites as well as remote offices, just to name a few prominent examples. Nowadays, the structure of enterprise networks ensembles more that of home networks with Ethernet-based backbone and a wireless last hop. Typical deployments feature thousands of APs (e.g., [13] mentions densities up to 4300 WLAN AP per km^2), thus being called *dense WLANs*, in order to provide enough capacity in large buildings usually spanning multiple floors and/or buildings, or even larger complexes. In order to efficiently manage such networks, a centralized management scheme is deployed. The WLAN APs are just providing a simple point of the attachment for the users (so-called *thin APs*), with all (or most of) the management functions being moved to central controller(s) and being powered from switches via IEEE 802.3af power over Ethernet (PoE) ports. Furthermore, due to security policies posed by the companies that limit the scope of the officially allowed access technologies and applications, resulting network architectures are even more complex, including enterprise servers, and so on. Finally, thanks to the recent advances and broad adoption of infrastructures to support cloud computing and software as a service (SaaS), enterprise clouds are also very common in corporate network landscape [14]. The typical structure of an enterprise network is shown in Figure 9.2.

As for the last hop in enterprise networks, WLAN is by far the predominant wireless technology at the moment, with more than 50% of the organizations in the United States deploying WLANs [15]. The market of the enterprise WLAN is also growing very fast, currently experiencing a 32% annual growth rate. Among other wireless access technologies cellular networks are also becoming quite common. Furthermore, femtocells present a high potential and are quickly gaining this sector of the market, with plans to offer small cells as a service (SCaaS) by various operators [12]. Femtocells typically deployed in the corporate scenarios can serve up to 16 users. According to Small Cell Forum [12], in 2013 there were six different deployments for enterprise environments by European operators and another eight for enterprise/consumer sector in the United States and New Zealand.

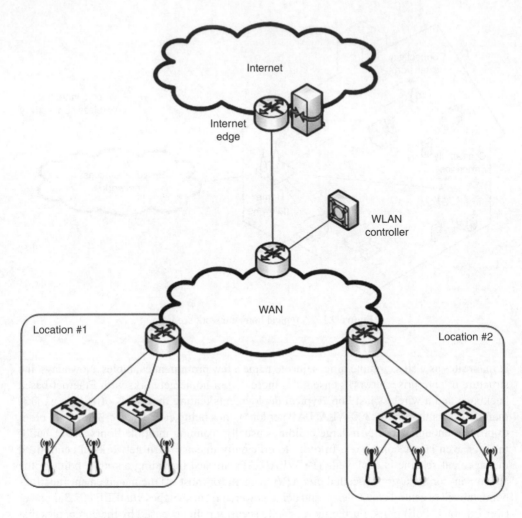

Figure 9.2   Typical enterprise network scenario (simplified version)

9.1.2 *Differences*

Obviously, the two discussed environments exhibit some differences as well. The most important ones will be discussed shortly.

The most evident differences are to be found within the network management approaches. In home networks, network management is split between the ISP and the individual end-user (network owner), with the latter having usually limited training and limited interest in getting deeply involved. This frequently results in poor network management – for example, frequent networking and security-relevant misconfigurations. On the contrary, it also poses strict requirements on the network devices to be easily configurable without requiring too complex setups, so-called *zero configuration*, and self-manageable, that is, being able to detect and adapt to changing network conditions [3]. Moreover, owners (administrators) of different networks have no incentive to co-operate in order to improve network deployment among different apartments, meaning no common interference management scheme is applied at all, leading to a lot of conflicts and inefficiencies.

In contrast to home scenarios and similarly to cellular networks, enterprise networks are now usually deployed under centralized management, with the usage of specialized tools (network design support tools, network management software, and so on for example, [16, 17]). Migration from autonomous to centralized architectures in corporate networks [18] is actually one of the signs reflecting the importance of an energy-efficient operation that is much easier to implement with a centralized scheme, despite the shortcomings that may be potentially introduced, for example, single point of failure, processing latency. Scope of management represented by various possible approaches provided by different vendors may differ significantly, including various proprietary solutions for load and interference management. Typically, two most common approaches to organize network management in enterprise WLANs may be distinguished, either (1) only time-critical functions, for example, exchange of management frames, are executed by the AP, whereas all the control and data traffic is routed to/from a central controller or, more commonly, (2) the central controller does not take complete control over all AP traffic, having, however, precise knowledge about the state of the network (AP notifies the controller using a separate protocol), that is, used channel, number of associated users, and so on. Hence, central controller has a global view of the network and can decide about the actual network configuration according to the traffic handled by the APs.

Another important point is that the corporate networks are designed to meet the performance objective (in terms of throughput and/or delay) and scaled to meet the peak of the user demand. Enterprise networks are thus heavily over-provisioned, with one or more WLAN controller(s), switches forming the Ethernet-based backbone and, more importantly, a dense WLAN with densities ranging up to 4300 APs per km^2 [13] contributing to the total energy bill. To make the things even worse the typical link utilization in the enterprise networks ranges from 1 to 5% [19–21], which is actually even lower than the utilization rates of home environments reported to be about 9% on average [2, 22–24].

Finally, deployment of femtocells has different aims in home and enterprise scenarios. In home environments, the main objective is to avoid leakages in the coverage of a single femtocell into the public space, whereas in enterprise deployments femtocells have to work together to jointly provide the increased capacity in a large building, group of buildings that are typical for enterprise or campus environments [25].

9.1.3 Perspectives

Growing popularity of wireless (mobile) devices heavily contributes to the increasing traffic demand, which in recent years has shown nearly exponential growth [26, 27]. Mobile videos (multimedia streaming) and web browsing are the main applications in both home and enterprise environments that contribute the most to this increase, with the first category foreseen to comprise nearly 70% of all mobile traffic by 2017 [26]. Another important driver is the requirement of ubiquitous connectivity, frequently coupled also with strict requirements regarding the quality of the Internet connection in terms of throughput and/or delay, which is especially important in corporate scenarios, for example, employees accessing the same set of corporate resources (services) from a corporate laptop, smartphone or any other device (virtual desktops). In order to fulfil all the aforementioned requirements there is a need to dramatically expand the existing network infrastructure which in enterprise scenarios would generate huge additional costs not only for installing additional equipment but also for its maintenance. As mentioned in Section 9.1.2, because of the low level of link utilization in both network types (1–5% in enterprise and about 9% in home networks), there would be a drastic increase in the densities of the APs that are required to serve such traffic. This, in turn, will raise severe concerns regarding energy consumption in such networks, which will further be discussed in the next sections of this chapter.

Nevertheless, several incentives to address this increasing traffic demand, differently from just adding additional infrastructure, can already be observed nowadays. In home networks it is the aggregation of the bandwidth from neighbouring WLAN APs. The idea itself is not new and despite serious security concerns several solutions have already been proposed in the literature, with the most prominent examples being: broadband hitch-hiking (BH^2) [2], coordinated upload bandwidth sharing (CUBS) [24], FastVAP [28], THEMIS [29], PERM [30] and COMBINE [31], to name a few. CUBS addresses increasing the upload bandwidth in order to improve the performance of peer-to-peer applications, whereas COMBINE provides a solution for collaborative downloading. Fast VAP and THEMIS demonstrate the feasibility of bandwidth aggregation with just a single virtualized WLAN card, and PERM discusses how to schedule flows among different available APs. For a more exhaustive survey of WLAN link aggregations see Ref. [2]. BH^2 as being energy-saving oriented solution will be discussed in Section 9.3.1. Nevertheless, practical, large-scale deployments of such wireless solutions remain to be seen. One example of an already deployed wired system is the 'apartment LAN' or 'A-LAN' deployed in South Korea [1]. A-LAN is defined as using a shared fibre or DSL connection to the apartment block with Ethernet-based distribution within the apartment block.

As stated already, enterprise environments will have to face not only the increasing traffic demand driven by rich multimedia content, for example, teleconferences including voice, high-quality video and data, but also the challenge of ubiquitous, device-agnostic connectivity. This will lead to a change in the concept of enterprise network, as it was previously described in Section 9.1.1. Network boundaries will be pushed beyond corporate firewalls (where they are nowadays) to form a so-called *extended enterprise network* that includes suppliers, vendors, partners, and so on, all working outside of the corporate premises and using any type of available access technology on any possible device (it is forecasted that up to 2020 every human could be associated with 70–100 IP devices), as pointed out in Ref. [32]. In enterprise environments it is network virtualization among many other Internet technology enablers that will play an important role in this paradigm shift. Another issue is the transformation from the network-centric solutions to more flexible architectures that could support ubiquitous connectivity (more human-centric approach) without compromising network security.

Finally, few words must be said about the future outlook of femtocells in the context of home and enterprise environments. According to Ref. [12], the femtocell market is expected to experience a significant growth over the next few years, reaching about 70 million of femtocell access points (FAP) in the market by 2017, growing nearly 40-fold from 2011 to 2016. Several enterprise deployments have already been made. As for 2013, there were 14 different enterprise deployments and more than 30 public customer deployments by different mobile operators worldwide, as reported in Ref. [12].

9.2 Home and Enterprise Networks in the Context of Green Wireless Networking

This section brings the 'green context' to the presented picture of home and enterprise networks. To this end, first the important metrics used in the context of green networking in wireless scenarios are introduced, and using these metrics an initial evaluation of the potential energy savings in both scenarios is performed.

9.2.1 Metrics for Green Communication

The forecasted increase in wireless traffic demand, mentioned in Section 9.1.3, has already raised severe concerns regarding energy consumption in both home and enterprise environments. To make this picture even worse, it has to be stated that until very recently, energy efficiency has not been part of the wireless network design considerations [33, 34]. Therefore, to fully understand the problem of energy consumption in wireless networks, it is first necessary to define the most important trade-offs related to the energy-efficient networking. The following discussion extends the framework presented in Ref. [27], addressing particularly home and enterprise environments and helps to define green metrics that are important in this context. From the perspective of system design challenges, there are four trade-offs that need to be discussed:

- Deployment efficiency versus energy efficiency
- Spectrum efficiency versus energy efficiency
- Bandwidth power trade-off
- Delay power trade-off

Deployment efficiency is usually measured in terms of achieved system throughput related to the total cost spent by the network operator on the deployment of the network, that is, both capital (CapEx) and operational (OpEx) expenditures. CapEx typically include infrastructure-related costs, whereas OpEx include operation and maintenance costs, among them energy bills. On the other hand, energy efficiency is defined as the system throughput per unit of energy consumption, or alternatively as energy spent per one user. According to Ref. [27], these two metrics very often lead to the opposite design criteria for network planning, for example, in order to reduce the number of deployed APs one would extend the area covered by one AP, with this in turn leading to a sub-optimal performance of the users placed at the edges of the coverage area of an AP. Typical enterprise networks are designed today to meet the peak of the traffic demand, thus being highly over-provisioned and not energy efficient. Nevertheless, the trade-off between deployment and energy efficiency is now becoming of outmost importance, with multiple reasons behind that. For enterprise networks, all the network deployment costs are paid by the company that is interested in having reliable, secure network capable to support rich multimedia applications. With traffic demand forecasted to further increase, there is a need to look for more sustainable solutions, improving the relation in the deployment and energy efficiency. Examples of such solutions will be further discussed in Section 9.3. The current picture will also change significantly with the wide deployment of femtocells in the enterprise scenarios. Having a heterogeneous network structure with one layer of the cells just providing the coverage, and the other layer providing required capacity would increase both cost efficiency and energy efficiency. In contrast, for home environments energy consumption of a WLAN AP or H-GW constitutes a very small part of the total energy bill that is paid by a residential user, thus giving the user a very little incentive to look for any improvements. Furthermore, to make things even worse, energy efficiency of the devices employed typically in home environments is very low, being two or three orders of magnitude lower than that of core network devices [2].

Spectrum efficiency is usually defined as throughput of the system per unit of the used spectrum, being widely accepted criterion in wireless network optimization [27]. In case of home and enterprise environments the trade-off between spectral and energy efficiency was not considered an issue as long as purely WLAN-based networks were considered. However, with the incorporation of additional wireless access technologies into the corporate scenarios, and especially with the introduction of femtocells to both enterprise and home environments this trade-off becomes a very important one, due to additional inter-cell interferences that are introduced. To this end, various solutions for mitigating inter-cell interferences in femtocell deployments will be discussed in Section 9.3.

Bandwidth and power used for the transmission are the two limited resources in wireless communication. Therefore, the relation between these two factors is important in the context of wireless resource management. Again, examples of such algorithms will be shown in Section 9.3. Whereas bandwidth power trade-off describes mostly solutions that are related to the physical layer, delay (or service latency) power trade-off refers to the application layer performance. In enterprise networks, it is one of the most important design goals to deliver a timely bounded service, often without much concerns about the power consumption. Therefore, in first place, there is not much space in which the delay can be compromised. For home environments, in contrast, there is a certain grade of flexibility in the function of the applications that are most commonly in use, for example, video and web browsing applications permitting some trade-off, whereas online gaming are less elastic.

9.2.2 Green Potential

As pointed out in Section 9.2.1, so far wireless networks have not been designed with the energy efficiency in mind. In order to quantify possible gains that can be achieved in both home and enterprise environments, let us first break down the energy consumption of different parts of each network, stating where lays the biggest potential to save the energy, under assumption of perfect management of each network.

As shown in Figure 9.2 (in Section 9.1), a typical deployment of an enterprise network features one or more WLAN network controller, few switches forming the Ethernet backbone and a huge number of WLAN APs, for example, 125 WLAN APs in a single-building corporate network and up to 5000 WLAN APs in a campus network [15], with a clear trend to further increase these numbers due to the increasing demand for high-quality multimedia traffic. According to the studies that survey power consumption of WLAN network elements, the usual power consumption figures are: up to 466 W for commercially available WLAN controllers (typical WLAN controller being able to manage up to 512 APs and 8192 users) [15], about 350 W for switches (a switch with 24–72 PoE ports) [15], and up to 10 W for a thin WLAN AP [15, 35]. Yet a single WLAN AP does not consume much energy, but due to their abundance, it is safe to state that, similarly to the cellular networks, the radio accessible part of the enterprise network consumes the most energy and thus the APs contribute the most to the total energy bill. For example, a total yearly energy consumption of an enterprise network consisting of 125 APs could be estimated with the data provided above to be about 21 MWh, out of which consumption of WLAN APs would yield nearly 11 MWh.

In the case of home networks, it is much more difficult to make such a precise balance of the total energy costs, due to the high variability in quantity and type of the devices used. Nevertheless, there are several studies aiming at evaluation of the energy consumption in home environments [36–38]. Spinney et al. [36] analyze the most recent trends of the energy consumption in households, taking into account also social and cultural perspective, stating that there was a huge paradigm shift in our perception of the role of communication in our social relations and habits at home with the introduction and mass deployment of wireless communication and devices such as laptops and smartphones. Interestingly, we tend to spend much more time, than in the era of desktops, using our wireless portable devices to work or for leisure, with emerging practices such as multitasking and always-on-ness. To that end, a very interesting comparison of the energy consumption of a desktop and laptop has been provided. On the basis of participant usage patterns and the energy consumption data available in Ref. [37], it was stated that a two-hour daily on activity incurs about 65% more of the consumed power for laptop (estimated as 1.7 kW per week) than for a desktop computer. The difference is mainly explained by the different usage patterns: desktop computer is completely switched off after completing the on activity, whereas laptop stays in the standby mode for a considerable amount of time, and so are the remaining elements of the network. One more important observation is being made; actually laptops are often not competing with desktops in the home environments, rather being just an additional layer of devices used for communication. Therefore, the total energy consumption in home environments has been recently increasing even more significantly.

Another very important aspect in the context of the green potential of home and enterprise environments is the link utilization. For both networks, the average link utilization has been reported to be very low, in order of 1–5% for the enterprise networks [19–21], and about 9% for home DSL links [2, 22–24]. Yet enterprise networks are highly over-provisioned in order to meet the peak of user demand, which occurs very rarely and may be related to only a small part of a corporate or campus network, for example, Jardosh et al. [39] quantify that around 10–80% of all WLAN APs in a corporate network and 20–65% for the campus deployment are idle during the entire month of the observation time; in addition, they also report more than 70% of all APs being idle for the time span of at least an hour in the corporate scenario and more than 50% in the campus scenario, respectively. To make things even worse, energy consumption of the network devices in both home and enterprise networks is *not proportional to the traffic load*, that is, large fraction of the energy is burned independently of the traffic load, and the lowest level of energy consumption per transmitted bit is achieved at the full load. Several experimental results illustrate how bad the situation actually is. Nguyen and Black [37] report the results of their measurements of home ADSL modems (two ADSL modems and one ADSL/WLAN modem) under different levels of traffic load stating that the difference between the idle state (powered on and not sending any data) and fully loaded state is actually minimal. Hlavacs et al. [38] report the same phenomena for residential (8 port Fast Ethernet switch) and corporate (24 port and 48 port Gigabit Ethernet switch) switches, consuming marginally more power when being loaded than in the idle state. Consumption of WLAN APs has also been analyzed in numerous research studies, for example, Refs. [35, 40], including broad range of possible devices supporting different versions of IEEE 802.11 standard, for example, with multiple input multiple output (MIMO) antennas (IEEE 802.11n), dual-band (2.4 and 5 GHz) operation, and so on, with a common conclusion that at least 70% of the energy was spent just to power the AP (idle state) [40]. Readers may further refer Refs. [35, 40] for more detailed surveys of WLAN AP power consumption of different vendors.

This degree of lack of load proportionality, of course, leads to very high levels of energy consumption per transmitted bit, being two or three orders of magnitude higher than that of core network devices [2]. Therefore, taking into account the huge number of the deployments of dense WLAN environments worldwide, as well as trends indicating further growth in both academia and industry deployments, there is a considerable potential for saving the energy in such networks.

9.3 Possible Savings in the Current Home and Enterprise Network Landscape

This section focuses on the reality of energy savings that are possible to achieve in the current state of home and enterprise networks. To that end, first a short overview of what can be done is given, including possible hardware improvements and network-level solutions. Next, an overview of the most important challenges and limitations preventing the implementation of proposed solutions is provided. And, finally, a detailed survey of the most prominent network-level solutions applicable in dense WLAN scenarios is presented.

9.3.1 Quick Survey of What Can be Done

Lack of load proportionality of network devices currently deployed in home and enterprise scenarios has been one of the main reasons why in the advent of forecasted traffic explosion serious concerns about energy consumption have been raised and spurred a considerable amount of research. All proposed solutions can be divided into two general groups: (1) hardware-level solutions and (2) network-level solutions. In addition, there is a third group of solutions aiming at improving energy efficiency at the user end devices that is important in the context of home networks and will also be discussed shortly.

Hardware-level solutions aim at improving directly the energy efficiency of a network. This means increasing the efficiency of single components, for example, power amplifiers, chipsets, (IEEE 802.11) radios, of network devices in order to decrease power consumed during the transmission or idle mode of operation. A detailed study for WLAN APs [40] has revealed that for the best devices under tests, still at least 70% of the energy was spent on base power consumption (load-independent consumption when the AP is idle). It is therefore reasonable to ask the following question: how much can be actually saved by further improving single components – where are the limits of possible energy savings with hardware-level solutions? In an effort to answer this question, we analyzed collections of power consumption data of WLAN APs made by various manufacturers that are presented in Refs. [35, 40]. The main conclusion that can be drawn is that with the current state-of-the-art devices, the overall power consumption is mainly attributed to the base power consumption, with radio transmission circuits contributing very little, sometimes close to none. Similar conclusions regarding load proportionality are drawn for switches that are currently deployed in enterprise networks, which tend to consume about 90% of the power in the idle mode [14]. Marsan et al. [33]

provide a further theoretical discussion of the above-mentioned question, considering possible energy savings with hypothetical devices, not possible to be constructed with the present state-of-the-art technologies and knowledge that have 90% of load proportionality. Of course, that brings savings when compared to the best devices available today, but even then the amount of energy that is wasted during the periods of low traffic load and in the idle mode is still huge, if we consider the size of a typical enterprise network. Therefore, hardware-level solutions cannot be seen as a viable option in case of dense WLANs, where most of WLAN APs stays idle for most of the time, and there is a need for more efficient low power operation or sleep modes (network-level solutions).

In the context of home networks, an important development is the definition of the low-power Bluetooth profile, within Bluetooth 4.0 specification [8]. However, this and other ongoing improvements to hardware are not able to keep the pace of the rising demand for energy consumption of end-user devices and thus advocate for software-level solutions, including all layers of protocol stack [41].

Network-level solutions, in contrast, try to dynamically adjust the number of active network devices, for example, Ethernet switches, WLAN APs or H-GWs, to the changing traffic load conditions. This means putting unnecessary devices to sleep or even completely switching them off in periods of low traffic load; therefore, network-level solutions are also referred to as resources-on-demand (RoD). To that end, there are several issues that must be addressed, namely, (1) in order to increase the energy efficiency of a network, low-load situations must be correctly recognized and potentially unnecessary devices must be selected, (2) sufficient number of network devices remaining in the operation must be assured in order to provide active users with the requested service, that is, sufficient bandwidth capacity, as well as to avoid coverage holes, that is, sufficient coverage, (3) any increase in the traffic demand must be timely detected and, if required, additional resources must be, also timely, provided. It is therefore quite typical for this type of approaches to define two layers of network devices: these that are absolutely necessary to provide basic operation of the network, that is, *basic coverage*, and additional layer of redundant devices that provide additional network capacity required in situations of heavy traffic load. Decision about switching network devices on and off may be taken either locally, independently of the other devices, or centrally in a coordinated manner, depending on the network architecture.

There are numerous network-level solutions that have been presented in the literature. In case of enterprise networks, the most prominent solutions for the wired backbone are based on energy-efficient Ethernet (EEE) [42]. EEE adaptively adjust Ethernet link data rate according to the current traffic conditions, and as different rates correspond to different amounts of consumed power more energy-efficient operation is achieved, for example, Refs. [21, 43]. Another example also based on EEE is green virtual LANs (VLANs), solution proposed to optimize energy efficiency of VLANs[2] that are nowadays commonly deployed in the enterprise scenarios [44]. Pointing out the most important deficiencies of the currently deployed VLANs, for example, placement policies, authors suggest an improved VLAN architecture, designed with energy efficiency as one of the optimization criteria. Energy inefficiency of the Ethernet switches is also tackled from the perspective of offloading users (which generate little or no traffic) to the wireless part of the enterprise network while putting remotely such unused switches to sleep, in a solution called smart wireless aggregation (SWA) or energy-efficient

---

[2] VLAN groups logically users belonging to the same category (e.g., employees in a division of a company) that are connected to LAN switches or routers despite their different physical location.

wireless (EEW) [45]. EEE and EEW can also be deployed jointly, as proposed in Ref. [21]. Finally, there is a huge group of network-level solutions focusing on the energy-efficient management of the wireless part of the enterprise networks, provided that this is the part of the network that consumes the most energy and has the biggest power saving potential. The most popular WLAN AP management techniques will be surveyed separately in Section 9.3.3. In addition, the most important solutions for future enterprise scenarios that include femtocells will be presented in Section 9.4.

In the context of home networks, the most important network-level solutions focus on the traffic aggregation in the heavily underutilized DSL links, similar to the idea of SWA. Examples such as broadband hitch-hiking (BH^2) [2] (solution already introduced in Section 9.1.3) or energy-efficient protocol for gateway-centric federated residential access [46] propose to reuse the overlapping coverage areas of the neighbouring WLAN APs in order to offload unused WLAN APs. A substantial number of WLAN APs (65–90% according to [2]) can be offloaded and thus switched off, bringing considerable power savings for a federated deployments of WLANs in future home environments. Other solutions for traffic aggregation in the DSL access that were mentioned in Section 9.1.3, do not focus on the energy optimization, however.

In addition, several recent works, for example, [47, 48], proposed approaches to reduce energy consumption in WLANs, having as a main goal increase in the battery lifetime of the end-user devices. These end-user approaches can be easily integrated with network-level energy-aware solutions mentioned here, and may have more importance in home scenarios. Nevertheless, these solutions are out of the scope of the analysis presented in this chapter.

9.3.2   Challenges and Limitations

In contrast to cellular networks, both enterprise and home networks have less potential for saving the energy (a single WLAN AP or H-GW is consuming relatively little energy in comparison to a single (macro) BS in cellular networks; see numbers provided in Section 9.2.2) and thus it is more difficult to employ power saving strategies there. Practically, only dense WLAN network in enterprise environments represents a scenario where employing network-level schemes to manage the energy is feasible [33]. In contrast, home networks due to their scattered nature (high level of distribution of the energy cost, with a very small cost per single user) and distribution of management among many different parties are not well suited for the adoption of energy-efficient approaches known from cellular or enterprise environments (with the only exception being federation of WLAN APs; nevertheless, large-scale deployments of federated WLANs are yet to be seen). Of course, one may argue that if the energy saving scheme proposed for distributed deployment is adopted on the mass scale in all households the cumulative energy that may be saved is significant, for example, Ref. [15]; however, such an assumption seems upfront very unrealistic.

But even for dense WLAN environments there are several important limitations that must be taken into account. The main limitations arise from the nature of centralized management scheme that is deployed, especially in the variant where the AP handles all non-time critical functions, namely, creation of a single point of failure, as well as increased processing latency [18] that may not be acceptable for some enterprise deployments. These limitations impose therefore new requirements on manufacturers to provide more flexible and reliable management schemes, facilitating a fast and accurate network re-configuration and being the most important challenges related to dense WLAN deployment nowadays.

Another important limitation is related to traffic aggregation solutions. Whereas traffic aggregation is feasible in enterprise scenarios [21, 43–45], aggregation of traffic in home scenarios may pose severe challenges due to the failure to induce sleep [2]. A DSL line can be put to sleep only if there is no traffic on the line at all, as there are no alternatives for Internet connectivity. Also, it takes about a minute to boot and synchronize the modem with DSLAM, and therefore traffic inactivity must be sufficiently long. However, it is very well-known that due to some periodic traffic updates to keep the network presence of various user applications there is a continuous light traffic flowing almost constantly and thus preventing sleep on idle.

9.3.3 Survey of On/Off Switching Mechanisms for Enterprise (Dense WLANs)

Solutions proposed in the literature to dynamically reduce power consumption of a dense WLAN in periods of very low demand can be divided into two fundamental groups: (1) homogeneous (relying entirely on IEEE802.11 technology) and (2) heterogeneous solutions (assuming support of other, additional network access technology, e.g., IEEE802.15.4, Bluetooth).

Following the general outlines of a design of RoD approaches presented in Section 9.3.1, there are two key aspects that must be addressed for homogeneous solutions, namely, (1) the assessment of the area covered by a given set of WLAN APs and (2) the assessment of the demand. Therefore, there are two issues that heavily influence the design process of such solutions: coverage models and AP placement pattern (also called in the literature the *area coverage problem*). Coverage models will not be discussed here; for a good survey of different approaches refer Ref. [49]. There are two obvious design choices for the area coverage problem: deterministic and random placement of WLAN APs [49]. Random AP placement, although more realistic, leads to a non-homogeneously covered area, which has been proved to be a NP-hard problem. Possible solutions include either tile-based approximation [50] or computationally expensive (and not easily scalable) optimization algorithms to approximate the coverage [51]. Deterministic placement of APs, with all its shortcomings, seems thus a viable alternative, for example, Refs. [52, 53].

One of the first examples of homogeneous solutions has been proposed in Ref. [39]. WLAN APs are grouped into clusters on the basis of the Euclidean distance. Within a cluster, only one AP, called *cluster head*, is powered on during the period of low load. The demand is estimated on the basis of the number of associated users, and once the threshold is exceeded, additional APs within the cluster are powered on to provide the capacity. Achieved power saving depends on the APs density, yielding 20–50% power saving for less dense scenarios and 50–80% for dense deployments. However, the criterion for forming the cluster is not very well tailored for wireless environments. Due to fading and interference, even adjacent APs may have significantly different performance, and consequently coverage holes may appear between the clusters. This problem has been addressed in an improved version of this solution, presented in Ref. [15], where detection of neighbouring APs and forming of clusters is based on RSSI criterion. Also, estimation of the user demand has been improved with the number of associated users being substituted with a channel busy fraction metric, that is, the fraction of time the channel is busy due to the transmission and inter-frame spacing, to better reflect the real traffic picture. It is reported that up to 53% and 16% power saving can be achieved for the low and high traffic conditions, respectively (although it is hard to compare these results to that achieved by the previously described solution due to different evaluation scenarios). Another cluster-based strategy, proposed in Ref. [54], is based on the usage patterns, derived from long-term measurements. These usage patterns are fed into two continuous-time Markov chain (CTMC) models of a cluster to calculate the required number of powered-on APs. This approach has been reported to save about 40% of the consumed power. Further, two other approaches use either integer linear programming (ILP) optimization [55] or heuristic (reducing the computational complexity of the former) [56] to adjust the density of the powered-on APs within a cluster. First of the proposed approaches provides a 63% power saving, whereas the latter further increases this gain by 10%.

An alternative approach has been recently presented in Refs. [52, 53]. Authors claim that the density of the WLAN APs can be very drastically reduced, on the basis of the following observation: to detect the user presence (with a given probability) it is actually sufficient that just one out of several Probe Request frames transmitted with the lowest bit rate is received within a desired delay. Once the user presence is discovered, additional APs can be switched on to provide the additional capacity required by the user. One of the main characteristics of the proposed algorithm is that the status of each AP can be determined in a distributed fashion. Initial evaluation demonstrates astonishing power saving: up to 98% of inactive APs can be switched off.

Heterogeneous approaches, in contrast, take advantage of multiple radio interfaces that are currently commonly available in the end-user devices. The underlying design assumptions in all solutions of this type are that (1) the alternative network provides the coverage in the entire area in question and (2) the position of the end-user devices is roughly known. Fulfilling these two assumptions guarantees that in the event of an increasing demand WLAN AP(s) can be timely powered on or user connectivity can be provided entirely with the alternative access network. Among several possible technologies to choose from, cellular is the one most often used in this context. One of the first solutions was proposed by Lee et al. [57], who suggested substituting WLAN beacon frames with cellular paging procedure in order to wake up on demand the additional WLAN APs that are in close proximity of the user device. With such an approach, WLAN APs can be very aggressively put to sleep (even all APs can be turned off!), guaranteeing considerable power savings. In the same way Bluetooth technology can be used to provide the basic coverage for low-rate transmissions, with WLAN interface being switched on only on detection of an excessive demand [58]. Using low-power Bluetooth interface traffic from nodes that do not require excessive bandwidth (forming a cluster) is aggregated at one node called *cluster head* that further pushes the traffic to the AP. Once one of the nodes requires more bandwidth, it switches on the more power consuming WLAN interface to communicate with the AP, whereas Bluetooth clusters are reorganized accordingly to reflect the change in the network topology. Finally, one of the most popular approaches presented so far in the literature proposes to additionally equip WLAN APs with IEEE 802.15.4 narrow-band radios that are used as 2.4 GHz band spectrum sensors [59]. Thanks to these sensors WLAN APs are able to detect radio activity (user sending WLAN beacon frames, attempting to connect to an AP) consuming considerably less power than the WLAN radio would (a 91% power saving is claimed). WLAN radios are activated on potential discovery of a user connection; however, the IEEE 802.15.4 sensors are not WLAN-technology-selective and might unnecessarily (a false positive) wake up WLAN APs if any other radiation is present in a channel.

To summarize, solutions that are based purely on IEEE802.11 technology, although sometimes coming close, for example, Refs. [52, 53], cannot reduce the density of the APs as aggressively as the ones based on the heterogeneous technologies. In the latter case, however, usage of multiple radio interfaces is required, which may make the deployment more complicated. Homogeneous strategies have little complexity in terms of deployment, control and management, but their main drawback is that they are not transparent to the users, meaning that the users may experience a slight performance degradation (e.g., delay) during switching on/off phases. A more detailed survey of on/off switching algorithms in the dense WLAN environments can be found in Refs. [53, 60] and interested readers should refer there for further details.

9.4   Possible Savings in Future Home and Enterprise Network

As argued in Section 9.1, the biggest challenge, as well as the biggest opportunity, for home and enterprise environments is related to the deployment of the femtocells on a massive scale (according to a study [61], 30% of corporate and 45% of household users experience poor indoor coverage). Here, we shortly analyze possible implications related to energy consumption, focusing in particular on the problem of interference management in dense deployments of femtocells.

9.4.1   Interference Management Techniques

Perspective of massive deployments of femtocells has increased concerns about their energy consumption. Therefore, quite considerable research efforts have been spent, with most of the work focusing on the interference management problem [62]. Before further discussion, let us first classify the interferences in the environment where femtocells are deployed (femtocells constitute another tier that is coexisting with the traditional access networks, e.g., cellular macro cells). To that end, two main groups can be distinguished [63]:

- *Co-tier interferences*, caused by network elements belonging to the same tier of a network, that is, interferences introduced by femtocells to the other femtocells, usually cells that are deployed in immediate proximity or sufficiently close to each other. Furthermore, there is a need to distinguish between the interferences in uplink direction (introduced by an end-user device) and downlink direction (introduced by a femtocell AP).
- *Cross-tier interferences*, which occur between network elements belonging to different tiers of a network, that is, interferences between femto and macro cells. Also, with further distinction between uplink (end device in a femtocell is a source of interferences to a macrocell, or end device using macrocell to a femtocell) and downlink direction (caused by a femtocell AP to a neighbouring macrocell or from the macrocell BS).

Interference management techniques are strictly tailored to the radio technology being used, for example, CDMA, OFDMA, as well as to the femtocell access mode (open access, closed subscriber group, or hybrid). Efficiency of the interference management schemes is usually highly dependent on the particular femtocell scenario. Whereas less dense femtocells deployments favour very simple schemes, the real challenge remains in enabling the dense femtocell deployments, given the ad hoc nature of the femtocells and high level of cross-tier interferences. The most important schemes for interference management proposed so far in the literature can be divided into three groups: (1) interference cancellation schemes, (2) interference avoidance schemes and (3) distributed interference management schemes [64].

Interference cancellation schemes deal with removing the influence of the interferences from the received signal. As such, they require knowledge of the characteristics of the interfering signal, as well as antenna arrays at the receiver end to cancel these interferences. Therefore, these solutions are targeted to be deployed at the macrocell BSs and femtocell APs, dealing with the uplink interferences.

Interference avoidance schemes include all techniques that may be deployed at the femtocell AP in order to help it self-optimize and self-configure, according to the current network conditions. To that end, a femtocell AP must be able to (i) sense the users in its vicinity, (ii) communicate with neighbouring femtocell APs and (iii) receive feedback from the users about the present conditions [65, 66]. One of the most prominent examples of interference avoidance techniques includes power control aided coverage optimization that deals with finding a compromise between the deployment efficiency and energy efficiency (recall the first of the metrics discussed in Section 9.2.1), taking into account indoor mobility patterns to construct adaptive coverage algorithms, for example, Ref. [66].

Finally, distributed interference management techniques are a subset of interference avoidance schemes, in which a femtocell AP disposes of limited information about the femtocell network, for example, in dense femtocell deployments it may not be feasible to distribute this information via backhaul. The optimization decisions must thus be taken locally, on the basis of the information that is sensed by a femtocell AP in question and its immediate neighbours. Distributor schemes are far more complex than the centralized ones and in order to be efficient may require additional knowledge, for example, about macrocell users that occupy the same spectrum.

Furthermore, more detailed surveys of the most important interference management techniques belonging to each of the three groups mentioned above can be found in Ref. [64, 65].

9.5   Conclusions and Future Outlook

Home and enterprise networks having similar structure and different management systems have different saving potentials. Practically speaking, the only current scenarios in which *significant* energy savings are possible to achieve are dense corporate WLANs. There, the dynamic WLAN AP on/off switching approaches may be applied, resulting in considerable reduction of the energy footprint while keeping the desired level of the QoS. Some further savings are possible in the Ethernet backbone of the corporate networks, where solutions based on EEE can also be deployed. For home scenarios, due to their scattered nature (high level of distribution of the energy cost, with a very small cost per single user) and distribution of management among many different parties, the perspectives for the adoption of energy-efficient approaches on a wide scale are rather low. Federated residential access scenarios (DSL aggregation) have yet to prove to be deployable on wider scale.

In future context, a successful deployment of femtocells in home and corporate scenarios may considerably change the current perspective. The denser deployment of the femtocells forecasted in the near future will require more efficient interference management schemes, especially for femtocells that are applied in a distributed manner. Moreover, more efficient use of (wireless) network resources should also be taken into perspective in the future picture of home and enterprise scenarios, and context-aware networking has the necessary potential to make it reality. Especially in home and enterprise networks, where building smart environments (more and more tailored to the user needs) would soon become reality, this issue is highly relevant and should be discussed in the near future.

References

[1] Point Topic Whitepaper, "World Broadband Statistics Q1 2013", June 2013. Available online: http://point-topic .com/wp-content/uploads/2013/02/Point-Topic-Global-Broadband-Statistics-Q1-2013.pdf. Last accessed: 01.12.2013.

[2] E. Goma, M. Canini, A. Lopez Toledo, N. Laoutaris, D. Kostić, P. Rodriguez, R. Stanojević and P. Yagüe Valentin, "Insomnia in the access: or how to curb access network related energy consumption", SIGCOMM Comput. Commun. Rev., vol. 41, no. 4, pp. 338–349, 2011.

[3] H. El Abdellaouy, D. Bernard, A. Pelov and L. Toutain, "Green home network requirements", In Proc. of 2012 IEEE International Energy Conference and Exhibition (ENERGYCON), pp. 896–902, 2012.

[4] D. Boggs and R. Metcalfe, "Ethernet: distributed packet switching for local computer networks", ACM Commun., vol. 19, no.7, pp. 395–404, 1976.

[5] A. Majumder and J. Caffery Jr., "Power line communications", IEEE Potentials, vol. 23, no. 4, pp. 4–8, 2004.

[6] D. Becker, "Wi-Fi takes over homes", January 2005. Available online: http://news.cnet.com/Wi-Fi-takes-over-in -homes/2100-1010_3-5544025.html. Last accessed: 01.12.2013.

[7] IEEE 802.11-2012, Wireless LAN Medium Access Control (MAC) and Physical Layer (PHY) Specifications, IEEE standards, 2012.

[8] Bluetooth Special Interest Group, "Bluetooth Core Specification version 4.0", 2010, Available on line: https://www.bluetooth.org/en-us/specification/adopted-specifications. Last accessed: 02.12.2013.

[9] S. Ullah, H. Higgins, B. Braem, B. Latre, C. Blondia, I. Moerman, S. Saleem, Z. Rahman and K. Kwak, "A comprehensive survey of wireless body area networks: on PHY, MAC, and network layers solutions", J. Med. Syst., vol. 36, no. 3, pp. 1065–1094, 2012.

[10] J. Gerhart, Home Automation and Wiring, McGraw-Hill Professional, 1999. ISBN 0070246742.

[11] J. Andrews, H. Claussen, M. Dohler, S. Rangan and M. Reed, "Femtocells: past, present, and future", IEEE J. Sel. Areas Commun., vol. 30, no. 3, pp. 497–508, 2012.

[12] Informa Telecoms & Media, "Femtocell Market Status", Femtoforum whitepaper, February 2013. Available online: http://www.smallcellforum.org/smallcellforum_resources/pdfsend01.php?file=050-SCF_2013Q1-market-status%20report.pdf. Last accessed: 01.12.2013.

[13] J. Cox, "Designing 'iPad WLANs' poses new, renewed challenges", Network World, March 2012. Available online: http://www.networkworld.com/news/2012/032612-ipad-wlans-257585.html. Last accessed: 01.12.2013.

[14] P. Mahadevan, S. Banerjee, P. Sharma, A. Shah, and P. Ranganathan, "On energy efficiency for enterprise and data center networks", IEEE Commun. Mag., vol.49, no. 8, pp. 94–100, 2011.

[15] A. P. Jardosh, K. Papagiannaki, E. M. Belding, K. C. Almeroth, G. Iannaccone, and B. Vinnakota, "Green WLANs: on-demand WLAN infrastructures", Mob. Netw. Appl., vol. 14, no. 6, pp. 798–814, 2009.

[16] Cisco Validated Design, "Campus Wireless LAN, Technology Design Guide", August 2013, Available online: http://www.cisco.com/en/US/docs/solutions/CVD/Aug2013/CVD-CampusWirelessLANDesignGuide-AUG13 .pdf. Last accessed: 01.12.2013.

[17] Cisco Whitepaper, "Save Energy in the campus or distributed office", 2013. Available online: http://www.cisco .com/en/US/prod/collateral/switches/ps10904/ps10195/ps13408/white-paper-c11-729777_ps10195_Products _White_Paper.html. Last accessed: 01.12.2013.

[18] T. Sridhar, "Wireless LAN switches - functions and deployment", Internet Protocol J., vol. 9, no. 3, pp. 2–15, 2006.

[19] K. Christensen, C. Gunaratne and B. Nordman, "The next frontier for communications networks: power management," Comput. Commun., vol.27, no. 18, pp. 1758–1770, 2004.

[20] A. Odlyzko, "Data networks are lightly utilized, and will stay that way", Rev. Netw. Econ., vol. 2, no. 3, pp. 210–237, 2003.

[21] P. L. Nguyen, T. Morohashi, H. Imaizumi and H. Morikawa, "A performance evaluation of energy efficient schemes for green office networks", In Proc. of IEEE Green Technologies Conference 2010, 2010.

[22] M. Dischinger, A. Haeberlen, K. Gummadi and S. Saroiu, "Characterizing residential broadband networks", In Proc. of 7th ACM SIGCOMM Conference on Internet Measurement (IMC'07), pp. 43–56, 2007.

[23] G. Maier, A. Feldmann, V. Paxson and M. Allman, "On dominant characteristics of residential broadband internet traffic", In Proc. of 9th ACM SIGCOMM Conference on Internet Measurement Conference (IMC'09), pp. 90–102, 2009.

[24] E. Tan, L. Guo, S. Chen and X. Zhang, "CUBS: coordinated upload bandwidth sharing in residential networks", In Proc. of 17th IEEE International Conference on Network Protocols (ICNP 2009), pp. 193–202, 2009.

[25] F. Mhiri, K. Sethom, R. Bouallegue and G. Pujolle, "AdaC: adaptive coverage coordination scheme in femtocell networks", in Proc. of 4th Joint IFIP Wireless and Mobile Networking Conference (WMNC), 2011, October 2011.

[26] Cisco Report, "Cisco Visual Networking Index: Global Mobile Data Traffic Forecast Update 2012-17", February 2013. Available online: http://www.cisco.com/en/US/solutions/collateral/ns341/ns525/ns537/ns705/ns827/white_paper_c11-520862.html. Last accessed: 01.12.2013.

[27] C. Yan, Z. Shunqing; X. Shugong and G. Li, "Fundamental trade-offs on green wireless networks", IEEE Commun. Mag., vol. 49, no. 6, pp. 30–37, 2011.

[28] S. Kandula, K. C.-J. Lin, T. Badirkhanli, and D. Katabi, "FatVAP: aggregating AP backhaul capacity to maximize throughput", 5th USENIX Symposium on Networked Systems Design and Implementation (NSDI '08), pp. 89–104, 2008.

[29] D. Giustiniano, E. Goma, A. Lopez Toledo, I. Dangerfield, J. Morillo, and P. Rodriguez "Fair WLAN backhaul aggregation", in Proc. of 16th Annual International Conference on Mobile Computing and Networking (MobiCom'10), pp. 269–280, 2010.

[30] N. Thompson, G. He, and H. Luo, "Flow scheduling for end-host multihoming", in Proc. of 25th IEEE International Conference on Computer Communications (INFOCOM'06), 2006.

[31] G. Ananthanarayanan, V. Padmanabhan, L. Ravindranath, and C. Thekkath, "COMBINE: leveraging the power of wireless peers through collaborative downloading", In Proc. of 5th International Conference on Mobile Systems, Applications and Services (MobiSys'07), pp. 286–298, 2007.

[32] P. Nedeltchev, "The new opportunities of enterprise networking", Cisco IT Trends in IT Article, Extended Enterprise Network, 2012. Available online: http://www.cisco.com/en/US/solutions/collateral/ns340/ns1176/borderless-networks/extended_enterprise_network.html. Last accessed: 01.12.2013.

[33] M. Ajmone Marsan and M. Meo, "Green wireless networking: three questions", In Proc. of 10th IFIP Annual Mediterranean Ad Hoc Networking Workshop (Med-Hoc-Net), pp. 41–44, 2011.

[34] T. Chen, Y. Yang. H. Zhang, H. Kim and K. Horneman, "Network energy saving technologies for green wireless access networks", IEEE Wireless Commun., vol. 18, no. 5, pp. 30–38, 2011.

[35] S. Chiaravalloti, F. Idzikowski, and Ł. Budzisz, "Power consumption of WLAN network elements," TKN Group, TU Berlin, Tech. Rep. TKN-11-002, 2011.

[36] J. Spinney, N. Green, K. Burningham, G. Cooper, and D. Uzzell, "Are we sitting comfortably? Domestic imaginaries, laptop practices, and energy use", Environ. Plann. A, vol. 44, no.11, pp. 2629–2645, 2012.

[37] T. Nguyen and A. Black. Preliminary study on power consumption of typical home network devices. Technical Report 07011A, Centre for Advanced Internet Architectures (CAIA), 2007. Available online: http://caia.swin.edu.au/reports/071011A/CAIA-TR-071011A.pdf. Last accessed: 01.12.2013.

[38] H. Hlavacs, G. Da Costa and J. Pierson, "Energy consumption of residential and professional switches", In Proc. of International Conference on Computational Science and Engineering (CSE '09), vol. 1, pp. 240–246, 2009.

[39] A. P. Jardosh, G. Iannaccone, K. Papagiannaki, and B. Vinnakota, "Towards an energy-star WLAN infrastructure", In Proc. of 8th IEEE Workshop on Mobile Computing Systems and Applications (HotMobile '07), pp. 85–90, 2007.

[40] A. Murabito, "A comparison of efficiency, throughput, and energy requirements of wireless access points," University of New Hampshire, InterOperability Laboratory, 2009.

[41] H. El Abdellaouy, D. Bernard, A. Pelov and L. Toutain, "Green home network requirements", In Proc. of IEEE International Energy Conference and Exhibition (ENERGYCON 2012), pp. 896–902, 2012.

[42] IEEE 802.3az-2010, "IEEE P802.3az Task Force: Energy Efficient Ethernet", 2010. Available online: http://ieee802.org/3/az/public/index.html. Last accessed: 01.12.2013.

[43] C. Gunaratne, K. Christensen, B. Nordman and S. Suen, "Reducing the energy consumption of Ethernet with adaptive link rate (ALR)", IEEE Trans. Comput., vol. 57, no. 4, pp. 448–461, 2008.

[44] K. He, Y. Wang, X. Wang, W. Meng and B. Liu, "GreenVLAN: An energy-efficient approach for VLAN design", In Proc. of International Conference on Computing, Networking and Communications (ICNC 2012), pp. 522–526, 2012.

[45] P. Morales, J. Ok, M. Minami and H. Morikawa, "Smart wireless aggregation for access network infrastructure power saving in the office environment," IEICE Technical Report, IN2008-209, 2009.

[46] C. Rossi, C. Casetti and C.-F. Chiasserini, "An energy efficient protocol for gateway-centric federated residential access networks", 2011. Available online: http://arxiv.org/abs/1105.3023. Last accessed: 01.12.2013.

[47] Y. Agarwal, R. Chandra, A. Wolman, P. Bahl, K. Chin, and R. Gupta, "Wireless wakeups revisited: energy management for VoIP over Wi-Fi smartphones", In Proc. of 5th International Conference on Mobile Systems, Applications and Services (MobiSys'07), pp. 179–191, 2007.

[48] W. Ye, J. Heidemann, and D. Estrin, "Medium access control with coordinated adaptive sleeping for wireless sensor networks," IEEE/ACM Trans. Netw., vol. 12, no. 3, pp. 493–506, 2004.

[49] B. Wang, "Coverage problems in sensor networks: a survey," ACM Comput. Surveys, vol. 43, no. 4, pp. 32:1–32:53, 2011.

[50] M. Panjwani, A. Abbott, and T. Rappaport, "Interactive computation of coverage regions for wireless comm. In multi floored indoor environments," IEEE J. Sel. Areas Commun., vol. 14, no. 3, pp. 420–430,1996.

[51] R. Whitaker, L. Raisanen, and S. Hurley, "A model for conflict resolution between coverage and cost in cellular wireless networks", In Proc. of 37th Hawaii Intl. Conf. on System Sciences (HICSS'04), 2004.

[52] F. Ganji, Ł. Budzisz, and A. Wolisz, "Assessment of the power saving potential in dense enterprise WLANs", TKN Group, TU Berlin, Tech. Rep. TKN-13-003, 2013. Available online: http://www.tkn.tu-berlin.de /fileadmin/fg112/Papers/papers_all/TR-final-20130426.pdf. Last accessed: 01.12.2013.

[53] F. Ganji, Ł. Budzisz, and A. Wolisz, "Assessment of the power saving potential in dense enterprise WLAN", In Proc. of IEEE 24th International Symposium on Personal Indoor and Mobile Radio Communications (PIMRC '13), pp. 2850–2855, 2013.

[54] M. Ajmone Marsan, L. Chiaraviglio, D. Ciullo, and M. Meo, "A simple analytical model for the energy-efficient activation of access points in dense WLANs", In Proc. of 1st International Conference on Energy-Efficient Computing and Networking, pp. 159–168, 2010.

[55] J. Lorincz, A. Capone, and M. Bogarelli, "Energy savings in wireless access networks through optimized network management", In Proc. of 5th International Symposium on Wireless Pervasive Computing (ISWPC '10), pp. 449–454, 2010.

[56] J. Lorincz, M. Bogarelli, A. Capone, and D. Begusic, "Heuristic approach for optimized energy savings in wireless access networks", In Proc. of International Conference on Software, Telecommunications and Computer Networks (SoftCOM '10), pp. 60–65, 2010.

[57] S. Lee, S. Seo, and N. Golmie, "An efficient power-saving mechanism for integration of WLAN and cellular networks", IEEE Commun. Lett., vol. 9, no. 12, pp. 1052–1054, 2005.

[58] J.-W. Yoo and K. H. Park, "A cooperative clustering protocol for energy saving of mobile devices with WLAN and Bluetooth interfaces", IEEE Trans. Mobile Comput., vol. 10, no. 5, pp. 491–504, 2011.

[59] N. Mishra, K. Chebrolu, B. Raman, and A. Pathak, "Wake-on-WLAN", In Proc. of 15th international conference on World Wide Web (WWW '06), pp. 761–769, 2006.

[60] Ł. Budzisz, F. Ganji, G. Rizzo, M. Ajmone Marsan, M.Meo, Y. Zhang, G. Koutitas, L. Tassiulas, S. Lambert, B. Lannoo, M. Pickavet, A. Conte, I. Haratcherev and A. Wolisz, "Dynamic resource provisioning for energy efficiency in wireless access networks: a survey and outlook," IEEE Commun. Surveys Tutorials, vol. 16, no. 4, pp. 2259–2285, 2014.

[61] J. Cullen, "Radioframe presentation", Femtocell Europe, 2008.

[62] X. Wang, A. Vasilakos, M. Chen, Y. Liu, T. Kwon, "A survey of green mobile networks: opportunities and challenges," Mobile Netw. Appl. vol. 17, no. 1, pp. 4–20, 2012.

[63] M. Yavuz, F. Meshkati, S. Nanda, A. Pokhariyal, N. Johnson, B. Roghothaman and A. Richardson, "Interference management and performance analysis of UMTS/HSPA+ femtocells", IEEE Commun. Mag., vol. 47, no. 9, pp. 102–109, 2009.

[64] T. Zahir, K. Arshad, A. Nakata and K. Moessner, "Interference management in femtocells", IEEE Commun. Surveys Tutorials, vol. 15, no. 1, pp. 293–311, 2013.

[65] F. Mhiri, K. Sethom and R. Bouallegue, "A survey on interference management techniques in femtocell self-organizing networks", J. Netw. Comput. Appl., vol. 36, no. 1, pp. 58–65, 2013.

[66] H. Claussen, L. Ho and L. Samuel, "Self-optimization of coverage for femtocell deployments", In Proc. of Wireless Telecommunications Symposium (WTS 2008), pp. 278–285, 2008.

10

Towards Delay-Tolerant Cognitive Cellular Networks

Bi Zhao and Vasilis Friderikos

Centre for Telecommunications Research, King's College London, London, UK

10.1   Introduction

Over the last few years, we are witnessing a significant increase in the aggregate traffic in mobile networks, which is due to the proliferation of smartphones and Internet applications for mobiles. In this environment, mobile users expect to enjoy ubiquitous wireless Internet experience, which boils down to providing high-capacity connectivity to them anywhere and at any time. For sustainability reasons, operational as well as capital expenditure for mobile operators will need to be reduced. Energy consumption plays a significant role in the overall operational expenditure of a mobile operator. To this end, significant efforts have been recently placed on reducing the overall energy consumption leading to the so-called green networks.

The problem of energy-efficient transmission under delay constraints over wireless networks has been studied extensively over the past few years. Traditionally, the problem of data scheduling has been considered at the medium access control—MAC (i.e., packet) level, which considers short time intervals of the order of few milliseconds at most. Significant volume of research efforts have been placed on energy-efficient data transmission for delay-tolerant applications, especially the trade-off between transmission cost and time delay over wireless networks [1–5]. The work in Ref. [1] deals with the problem of packet scheduling with deadlines within a predefined time window of length T. Based on that, the authors in Ref. [2] explore the energy-efficient packet transmission with individual packet delay constraints, in which a trade-off between flexible energy and delay is analyzed under various individual packet delay constraints and bandwidth efficiencies. In Ref. [3], the authors consider a delay constraint for each packet and reveal the relationship between reliable transmission rate and QoS requirements, while a dynamic programming based algorithm is introduced to acquire throughput maximization and energy minimization according to different channel qualities of a fading

Green Communications: Principles, Concepts and Practice, First Edition.
Edited by Konstantinos Samdanis, Peter Rost, Andreas Maeder, Michela Meo and Christos Verikoukis.
© 2015 John Wiley & Sons, Ltd. Published 2015 by John Wiley & Sons, Ltd.

channel with time constraints [4]. The work in Ref. [5] further investigates the problem of energy-delay trade-offs under dynamic traffic loads and user populations. The target-set selection problem has been studied in the emerging Mobile Social Networks for traffic offloading by delaying the delivery [6]. In Ref. [7], a framework is proposed to investigate the trade-off between the amount of offloaded traffic and the users' delay tolerance over a 3G network.

Meanwhile, many portions of the spectrum are not in use for a significant period of time, thereby implying the existence of plenty of spectrum opportunities that can still be potentially exploited. Hierarchical Cognitive Radio (CR) networks improve spectrum efficiency by allowing the low-priority Secondary Users (SUs) to temporarily seek the wireless spectrum that is licensed to different operators serving Primary Users (PUs) [8]. It should be noted that whenever a PU captures the channel, SUs should be able to sense that the channel is occupied and defer from transmitting or competing unnecessarily for the access to the channel. At the same time, SUs opportunistically use available channels and can be preempted by PUs; in other words, the SU connections would be interrupted by the stochastic nature of the PU traffic. Consequently, the SUs should firstly estimate the channel availability by probability analysis based on PUs' historical traffic information or spectrum sensing. A common technique utilizes traffic characteristics from available long-term observations/statistics [9], and in an alternative approach, an SU selects the operating channel according to the instantaneous sensing results from the channel pre-scanning [10]. In Ref. [11], authors design optimal sensing strategies via a model assuming that the PU transmissions are unslotted as a continuous-time Markov chain while the SUs are slotted to sense the frequency channels. On this basis, the work in Ref. [12] proposes a stochastic multichannel sensing scheme based on traffic information as well as sensing history.

As modern digital devices are equipped with multiple wireless interfaces (such as Wi-Fi, 3G, LTE, and TV White Spectrum TVWS interface), the energy cost on data transfer can potentially differ significantly due to the different operational characteristics of these wireless interface. A detailed study regarding energy consumption in 3G, GSM, and Wi-Fi (802.11b) has been reported in Ref. [13]. Wi-Fi Access Points (APs) are becoming significantly popular, and hereafter the assumption will be that there are a number of APs within the coverage area of a macrocell base station of a cellular network. In cases where the mobility of vehicles and pedestrians can be predicted, the roadside APs evidently improve the average wireless throughput for file delivery by estimating the signal strength of APs along a predicted route utilizing historical RF fingerprints data statistics [14].

In addition to the Aforementioned factors, and as the vacant TVWS spectrum are permitted to be used in several countries, this new available spectrum will unfold new possibilities in data transmission with strong potential to decrease further energy consumption. In Ref. [15], it has been previously shown that the percentage of observed territory with nonzero number of available frequency channels is 64.7% under the ECC rules. In other words, when an SU is in any given location, there is a high probability that at least one frequency channel in TVWS could be available for SU transmission. When the SUs are using 3G cellular networks within the coverage of a TVWS master, it is possible that the SUs would prefer the TVWS connection over the cellular networks in terms of cost, RF coverage, capabilities, and transmission algorithm. However, the cognitive users have to cease wireless transmission immediately and relocate to a new band as soon as a PU appears and requires access to the channel. For the purpose of avoiding PU transmission, the SU nodes could utilize spectrum sensing or query a database that maintains information about the available channels for the details of the local radio environment. The SU connection can utilize the means as sensing or contacting a trusted geospatial database that records the information regarding PUs occupation with a specific location and time duration, prior to message transmission, to determine available spectrum at a given location [16, 29]. In this scenario, to predict the future location and the path of mobility of wireless nodes is another challenging issue in White-Fi networks.

10.1.1   Device-to-Device Communications (D2D)

Device-to-Device (D2D) Communications is a feature that has been introduced in the 3GPP Release 12. D2D communication has been considered as an underlay to an LTE-A cellular network. Devices are allowed to be engaged in direct communication with the network having the control in terms of interference management and resources used [17]. In this context, both the cellular network and the D2D communication use the same LTE resources. Therefore, D2D communications can be considered as an enabler for delay-tolerant networking techniques within the cell, i.e., D2D communication can be used to inteligently delay transmissions and/or relay information to another device or to the base station.

10.1.2   5G Wireless Communications

Currently, mobile operators are trying to cope with the high demand of Internet applications in cellular networks that utilize carrier frequencies that range between 700 MHz and 2.6 GHz. Available spectrum at these carrier frequencies can be deemed as rather limited, and as a result, in order to increase aggregate transmission rates to cope with the ever-increasing demand, there is a need to move higher in the spectrum. To this end, wireless technologies for 5G, or Beyond 4G as it is also called, envision the use of the current very much underutilized millimeter-wave (mm-wave) frequency spectrum ,for example, the use of 28 GHz, 38 Ghz [18], or even the unlicensed 60 GHz as envisioned by the Wireless Gigabit Alliance [19, 20]. Clearly, at these frequencies, signal attenuation is significant, and as a result, high-speed broadband access to the Internet can be considered only for picocells with radius of up to 200 meters. In these scenarios, delaying transmission of elastic user traffic based on the proposed set of solutions can allow for better utilization of the very high speed mm-wave links because it can allow users to refrain transmission until they are within the coverage area. Another important benefit stemming from the use of delaying message transmission until the user is closer to the access point is that in mm-wave spectrum due to the significant path losses, the energy gains that can be achieved by delaying the transmission are even greater compared to current operating frequencies in cellular networks. By inspecting the Friis Law for free-space path loss, it can be seen that when moving from 3 GHz to 30 GHz, path loss increase by 20 dB. Also in these carrier frequencies, the PA has an efficiency of less than 10%; therefore, energy consumption is a key issue. Consequently, by utilizing the elasticity of Internet application and the inherent mobility of users delaying message transmission is well fitted to be utilized in envisioned 5G wireless networks that are based on the use of mm-wave frequency spectrum.

10.2 Scenarios and Applications

Hereafter, the focus is on Internet applications, which can tolerate significant delays without deteriorating the experience of the end user. To this end, the consideration is on highly delay-tolerant traffic, which can tolerate delays that can range from few seconds up to few minutes. Traffic with these characteristics are e-mails, ftp data transfers, updates of social networking portals, message/file exchanged via the File Transfer Protocol (FTP), Rich Site Summary (RSS) feeds, non-real-time video streaming and Operating System (OS) and firmware updates, and updates to social networks, to mention just a few. We can utilize their inherent characteristics to significantly save the energy cost for embedded systems of mobile nodes. Note at this point, and as have been mentioned previously, the increased usage of smartphones and the rich ecosystem of Internet applications are having a severe effect on the recharging cycles of devices due to the increased levels of energy consumption and limitations of battery technology. As the infrastructure of hotspots for Wi-Fi and TVWS (TV White Space) interfaces are becoming ubiquitously available in urban areas, and mobile devices are equipped with multiple air interfaces, they could switch among these networks to seek and use any licensed spectrum bands as long as they do not cause interference to the PUs. In this case, the energy usage in modern devices for transmitting a fixed amount of data could differ drastically due to the significant difference on the achievable data rates on these radios. In addition, channel conditions change according to user mobility and spatial characteristics of the channel. Therefore, predicting the future location and the mobility path of SUs is another challenging issue in White-Fi networks. Furthermore, apart from the availability of the primary channels, the mobile nodes have to compete with other SUs to seek an optimal time duration for wireless transmission. Consequently, a virtual queuing model based on an M/M/K/L system is designed to analyze the optimal population of SUs to be served in the system to minimize the energy consumption of message transmission for the delay-tolerant applications. Wireless nodes (SUs) gather PU traffic information from a historical database in order to predict over a short term the traffic pattern of PUs. In the database server, there are two types of information about primary channels. One is the 24-hour traffic characteristics of different channels across several months [9], and the other is the noise power level in different channels that is updated continuously by the SU devices. An SU has to send a query to the database server for the available channel to transmit. According to the available channel information at the same time slots in previous days and long-term statistics regarding channel availabilities, the database server will certificate the noise power levels by comparing to a threshold. Then, the set of best candidate channels for the inquired SU will be determined.

10.3 Previous Research

For delay-sensitive traffic, there has been an enormous previous research both applied and theoretical within the general area of optimal job scheduling with deadlines [21]. With respect to wireless packet transmission, proposed solutions focus on low-complexity algorithms (to allow real-time implementations) over small timescales, which are technology-specific and depend on the standardized Transmission Time Unit (TTI), which, for example, in UMTS Release-5 has been define to be 2 msec. There has also been significant volume of previous research work on energy-efficient scheduling for wireless packet transmission. The work in Ref. [5] further investigated the problem of energy-delay trade-offs under dynamic traffic loads and user populations. The target-set selection problem has been studied in the emerging Mobile Social Networks for traffic offloading by delaying the delivery of messages [6]. In Ref. [7], a framework is proposed to investigate the trade-off between the amount of offloaded traffic and the users' delay tolerance over a 3G network. Furthermore, a theoretical study of optimal delay-tolerant multi hop transmissions within the cell have been studied in [30].

In accordance with the information of channel utilization, mobile systems can decide which frequency channels to utilize without deteriorating the quality of service of experience of the Primary connections. To this end, queuing theory can be utilized to analyze the wireless transmission in the CR scenario described earlier. In this case, we assume that there are M/M/K/L queue systems for the wireless connection, in which the SU message could be buffered. A preemptive priority queuing system has been utilized to analyze the mean system dwelling time of the SU traffic and the blocking probability for real-time SU connections [22]. In Ref. [23], authors analyze the queue lengths and average queuing delay of the SUs based on Poisson distribution of the SUs. In Ref. [24], a dynamic strategy learning (DSL) algorithm relied on the priority queuing systems including the SUs and the PUs is proposed for the delay-sensitive multimedia applications in order to maximize the user's utility function.

A detailed study regarding energy consumption in 3G, GSM, and Wi-Fi (802.11b) has been reported in Ref. [13]. As Wi-Fi hotspots are becoming very common in recent years, the study in Ref. [25] examined large-size file transmission protocols for vehicle-to-vehicle utilize existing Wi-Fi APs and navigation systems. In the cases where the mobility of vehicles and pedestrians can be predicted, the roadside APs evidently improve the average wireless throughput for file delivery by estimating the signal strength of APs along a predicted route utilizing historical RF fingerprints data statistics [14]. In addition to the above, and as the vacant TVWS spectrum is permitted to be used in several countries, it is without doubt that this new available spectrum will unfold new possibilities in data transmission with strong potential to decrease further the overall energy consumption. In the United States, the unoccupied TVWS has already been filled up with unlicensed users without significant interference to TV viewers, while Ofcom is determined to permit TVWS for unlicensed use by checking with a database in the United Kingdom. Once mobile devices are equipped with multiple air-interfaces allowing them to connect to LTE, WiFi, and TVWS, they could switch among these networks to seek and use any licensed spectrum bands as long as they do not cause interference to the PUs. Prior to wireless transmission via TVWS, digital devices have to sense or contact a trusted geospatial database that records PUs' occupation within a specific location and time duration [16]. In this scenario, predicting the future location and the path of mobility of SUs is another challenging issue in White-Fi networks.

10.4   System Model and Energy Saving Schemes

The ability of performing Adaptive Modulation and Coding (AMC) based on signal quality allows cellular networks to dynamically use higher-order modulation, up to $64QAM$, within strong signal areas and lower-order modulation $QPSK$ in poor signal quality areas for the sake of better signal recovery. To this end, the cell can be separated into n concentric rings of radii $R_i, i = 1, \ldots, N$ according to the distance between the wireless device and its serving BS as shown in Figure 10.1. The set of available modulation and coding schemes is denoted by $\{M_{R_1}, M_{R_2}, \ldots, M_{R_N}\}$. Consequently, each circular region with distance R_i to the BS corresponds to a different constellation size and coding scheme.

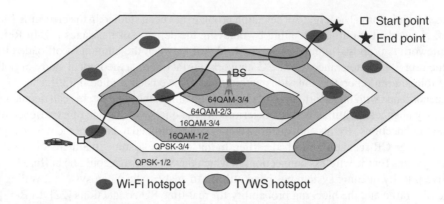

Figure 10.1 The bit rate of mobile users within the coverage of the base station (BS) ring range

Let r denote the coding rate, and r_{BS} be the (Euclidean) distance between the mobile node and the serving BS. Then, the spectral efficiency (bits/s/Hz) is given by Schoenen and Walke [26]:

$$\text{IEC}(r_{BS}) = r \cdot \log_2(M_{r_{BS}}) \quad \text{bits/s/Hz} \tag{10.1}$$

According to the sector bandwidth of cellular channels, the average sector throughput could be estimated so as to quantify the total capacity through the sector or site coverage. With the aggregate of the user data bit rates and the number of simultaneous concurrency users within the sector, we could determine the network throughput for individual user.

10.4.1 Storage Cost

Broadly speaking, the general architecture of a mobile device (such as smart phones, tablet computers, digital cameras, etc.) can be decomposed into the processing unit (CPU), the local Dynamic Random-Access Memory (DRAM), and flash/hard disk (HDD). When a wireless terminal prepares to transmit data to other nodes, the data has to be ready in the local DRAM. However, if the system decides to wait for a short period to transmit, the data stored in DRAM would be transferred to internal storage devices (e.g., internal NAND flash) or external storage units (e.g., SD card, HDD), which depends on the delay constraints and transmission strategies of mobile applications.

In the numerical investigations, the assumption is that the DRAM has three background power operating modes, namely, the Self Refresh, Precharge Fast Powerdown, and Precharge Standby. Figure 10.2 shows the DRAM power states in different states for wireless transmission. During the transfer process from the DRAM to NAND flash, except for the operation commands, the DRAM would be in the Precharge Standby state for the sake of short wake-up latency, which will consume most power. After the completion of the transfer, the DRAM will enter into a self-refresh state, consuming least power with significantly higher exit latencies. Once the data return to the DRAM for data transfer to the RF module, the DRAM would enter into the Precharge Standby state again. In the mean time, the NAND flash will remain in the idle state except the duration of data transfer between the DRAM and NAND flash.

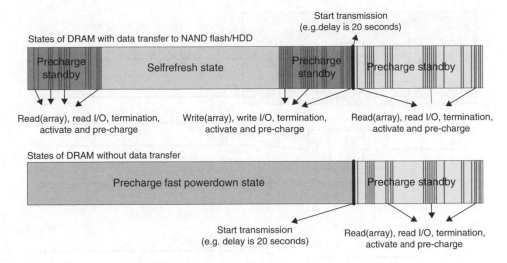

Figure 10.2 Operation and background power state of DRAM

10.4.2 Optimal Stopping Problem

Concerning the energy consumption for wireless data chunk transmission, we can identify the following main sources: (i) the energy required to operate the electronic circuits at the nodes; (ii) the energy consumed for wireless transmission of the data chunk; (iii) the energy consumed to receive the information; and finally, (iv) the energy consumed in DRAM and NAND flash/HDD devices. Based on the aforementioned classification, the inherent trade-offs regarding message delay and energy cost of nodes for wireless transmission can be modeled with an OSP formulation in order to seek optimal solutions for the message transmission scheduling by taking into account available spectrum opportunities for multiple radio interfaces (such as cellular networks, Wi-Fi, and TVWS).

The stopping decision (i.e., the time to transmit the information) would be made based on average channel conditions, delay constraints, and energy consumption. To be more specific, firstly, the policy calculates the data rate (bits/second) at each time slot and finds the optimal time duration that could be utilized for the data transmission of mobile nodes in this time slot, thereby fixing the data length to be transmitted in this time slot. Secondly, this policy calculates the energy cost at all of the candidate stopping time slots from the initial location along the route of vehicle. Finally, the policy should make sure that all of the messages will be transmitted to the BS before the (hard) time deadline. The policy compared all the schemes that launch the message transmission at different time slots, finding the scheme with optimal trade-off between energy cost and time delay under deadline constraints.

10.4.3 Optimal Number of Users

In this section, we turn our focus on the SU message competition without competing PU connections. Figure 10.3 gives an example of the physical queues for the case of K frequency channels and N concentric rings with different modulation and coding schemes. When the

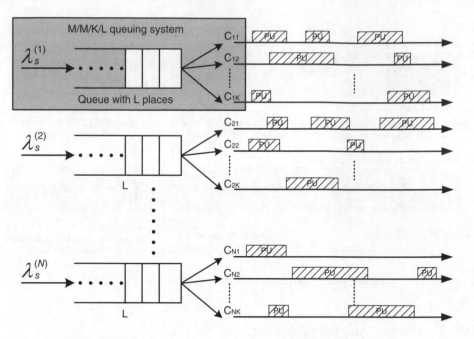

Figure 10.3 Access for SUs modeled as an M/M/K/L queuing system

traffic of the SUs needs to be transmitted in the system, it can be inputted to the queue for the SU connections. This proposed channel selection model could approximate the virtual SU message queue using an M/M/K/L queuing system. If the number of SUs is large, the input traffic of the virtual queue can be modeled as a Poisson process, where K is the number of servers and L is the finite number of waiting positions for each queue.

Let $\lambda_s^{(i)}$ denote the average number of the SUs per unit time in ith concentric ring of radii R_i and L denote the number of unit time as a kind of queue length. Therefore, each queue of this system, namely each ring, can accommodate $\lambda_s^{(i)}L$ number of the SUs. Given the set of candidate channels $\Omega = \{1, 2, \dots, K\}$ and the set of concentric rings $\Re = \{1, 2, \dots, N\}$, we denote C_{ij} to be the capability of the SUs in ring $i \in \Re$ within the channel $j \in \Omega$ and have

$$C_{ij} = \text{IEC}(r_{\text{BS}}) \cdot \frac{B}{K \cdot N} \quad \text{(bit/second)} \tag{10.2}$$

where F is the size of SU message and B represents the bandwidth available at the BS. Let μ_{ij} represent the service rate of SU connections using the frequency channel j in ith ring, we have,

$$\mu_{ij} = \frac{C_{ij}}{F} \tag{10.3}$$

Let ρ_i denote the occupation rate (offered traffic load), we have

$$\rho_i = \frac{\lambda_s^{(i)}}{K \cdot \mu_{ij}} = \frac{F \cdot \lambda_s^{(i)}}{K \cdot C_{ij}} \tag{10.4}$$

Assume that p_m is the probability that there are m SU messages in the system and p_{thres} is the blocked traffic rate threshold; therefore, we have

$$p_m = \begin{cases} \dfrac{\rho_i^m}{m!} \cdot p_0 & m \leq K \\[3mm] \dfrac{\rho_i^K}{K!} \left(\dfrac{\rho_i}{K}\right)^{m-K} \cdot p_0 & K < m \leq \lambda_s^{(i)} L + K \end{cases} \tag{10.5}$$

subject to:

$$\sum_{m=0}^{\lambda_s^{(i)} L + K} p_m = 1, \tag{10.6}$$

$$P_{(\lambda_s^{(i)} L + K)} \leq P_{thres} \tag{10.7}$$

Note that the constraint of (1.7) depicts that the blocking rate of SU connections in the virtual queue should be lower than the predetermined threshold p_{thres}.

10.4.4 Wireless Interface Switch

We assume that alternative wireless networks (Wi-Fi, TVWS) may be available at limited locations or time duration. The challenge is whether a mobile device that needs to transfer N MB of data should search for alternative networks to transmit the file and possibly delay the transmission in order to achieve minimal energy consumption cost.

In the case of multiple alternative interfaces, the system should choose the network with the most expected energy saving. In this work, the primary network is the cellular network and the alternative networks are TVWS and Wi-Fi. If the data size transmitted by TVWS L_{TV} or by Wi-Fi L_W is smaller than the entire data size L_{total}, which means the entire data cannot be transmitted within the current cell, the system has to determine whether to delay the transmission to the next wireless hotspot cell. Otherwise, it has to utilize LTE access to download the remaining data portion. Meanwhile, if all of the data could be transmitted within the high-speed access coverage like Wi-Fi and TVWS, it will not consider the handover for the remaining data. We categorize the wireless transmission strategies for mobile applications into several different categories as shown in Figure 10.4.

1. Large size with short delay constraint (YouTube-like applications, 10 seconds delay constraint): (i) in the only-LTE-coverage area, the system will predict whether the mobile user can move into the next hotspot under the delay constrains of the applications. If cannot, the user has to utilize LTE interface to download a proportion of file in order to meet the applications' requirements before moving into next hotspot; (ii) within the Wi-Fi and TVWS coverage area, if all the data cannot be transmitted, the system has to determine whether to delay the transmission to next hotspot or immediately transmit the remaining data via LTE. In order to provide better user experience, the system might have to start the wireless transmission immediately to secure enough downloaded data in local storages for video playback rather than waiting until next high-rate hotspot.

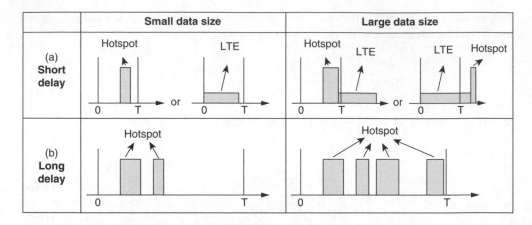

Figure 10.4 Transmission strategy for different data size under delay constraints

2. Small size with short delay constraint (podcast, audio files): the system will launch the file transfer immediately, no matter what kind of wireless radios it could access for transmission.

3. Delay-tolerant applications (e-mail, social network, and APPs updating, 1-minute delay constraint): if the mobile applications are delay-tolerant, the mobile users will have enough time to move into the next wireless hotspot, such as TVWS and Wi-Fi. Therefore, the wireless transmission could be always executed via the high throughput interfaces.

10.5 Numerical Investigations

10.5.1 Trade-Offs between Delay and Cost

In this section, numerical results are presented in general scenarios that a vehicle is moving toward a BS, in which the energy consumption is based on the embedded system of mobile nodes including RF, DRAM, and NAND flash. When the vehicle is approaching closer to the serving BS, then compared to the RF transmission power consumption, power consumption of storage devices will become non-negligible in the prediction of transmission energy consumption as observed from the distinction between Figure 10.5(a) with the distance of 400 meters to the BS and Figure 10.5(b) where the distance is assumed to be 800 meters. Due to the short distance between wireless terminals and the BS, the energy consumption of storage devices will have an increasingly overall effect on the energy cost, which is vividly shown in the curves of Figure 10.5(a).

Moreover, in order to evaluate the importance of time delay for the wireless transmission process, different values of γ are utilized for different applicable situations, such as the file transfer (delay-tolerant), video transmission (delay-sensitive). Figure 10.6 exhibits the difference of energy cost in different values of γ with 5, 10, and 20 second time delay constraints. The hyaline part of the bars stands for the time delay incorporated into the overall energy consumption. $\gamma = 0$ reveals the real value of energy consumption which delay cost has not been comprised. The cases of $\gamma = 1$ and $\gamma = 10$ clearly demonstrate the rising importance of time delay in energy cost.

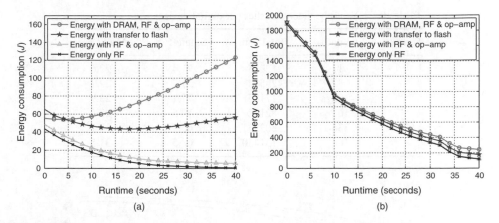

Figure 10.5 Energy cost from different distances to the BS (a) 400 meters to Base Station (b) 800 meters to Base Station

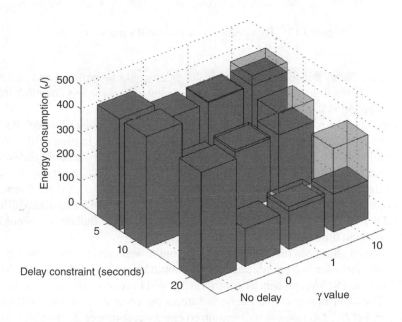

Figure 10.6 Importance of delay constraint in the overall energy cost

10.5.2 *Trade-Offs between Transmission and Storage Cost*

The research in Ref. [27] tracks the behavior of user requests from a campus network spanning an interval of 10 months. They use a commercial PC with a Data Acquisition and Generation (DAG) card to capture video information from YouTube server.

The values in Table 10.1 from Ref. [27] show the statistics regarding the videos requested during the aforementioned track period. The 4th row (Single) presents the percentage of clips requested by one PC only once, while 5th row (Multi) shows the percentage of videos requested

Table 10.1 YouTube video statistics per digital devices

Trace	Length(h)	Total num	Single(%)	Multi(%)
1	12	12955	77.4	22.6
2	72	23515	77.7	22.3
3	108	17183	77.1	22.9
4	162	82132	72.5	27.5
5	336	303331	65.9	34.1
6	168	131450	68.5	31.5

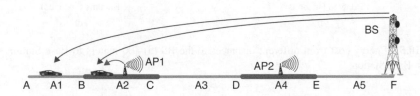

Figure 10.7   The selection from cellular BS and Wi-Fi AP

more than once. These statistics reveal that if the video could be buffered and cached in the local storages of digital devices, it will effectively decrease the wireless downlink traffic and energy consumption at the client end.

It is assumed that the road segment a vehicle is moving along is covered by a cellular BS with two Wi-Fi hotspots (segment BC and DE) in this domain as shown in Figure 10.7. We model the time duration until a popular video is requested again as an exponential distribution with mean time duration μ equal to 60 seconds. Let $\mathcal{A} = \{A_1, A_2, \ldots, A_M\}$ represent the segments along the entire route when the vehicle is moving toward the BS and X denote time interval that the mobile user would potentially re-load the same video. From the cumulative distribution function (CDF), we can calculate the probabilities $\Pr(A_k)$ that the mobile user would make a demand to consume the same content again.

Once a video clip has been downloaded from the multimedia servers and stored in the local DRAM already, there are two possible actions that could be taken. The first is that the content is deleted from the DRAM and hence future requests will have to be streamed again via wireless access. The other option is that the device stores the content in the local DRAM until a hard deadline. Let $E_{\min}(A_k)$ denote the minimized energy cost in area A_k from the storing and streaming schemes, which incorporate energy consumed for wireless transmission in one area, such as electronic circuits in mobile devices, and energy consumed from data transmission. $A_k \in \mathcal{A} = \{A_1, A_2, \ldots, A_M\}$. The strategies strive to balance between storing and transmission in the entire domain and explore the minimized energy cost across a long-term average, which is given as follows:

$$E_{\text{opt}} = \frac{\sum_{k=1}^{M} \{E_{\min}(A_k) \cdot \Pr(A_k)\}}{\sum_{k=1}^{M} \Pr(A_k)} \tag{10.8}$$

The proposed strategies combine transmission and selective storing aiming to achieve long-term energy efficiency according to the time interval distribution of user demand. The

Table 10.2 Probabilities and energy consumptions in different area

Area	Energy for rx (J)	Energy for store (J)	Probability %
A_1	45, 894.0	46.6088	0.2835
A_2	1509.8	151.4587	0.3408
A_3	2812.0	279.6088	0.3935
A_4	64.2	407.7587	0.3408
A_5	106.8	524.2587	0.3408

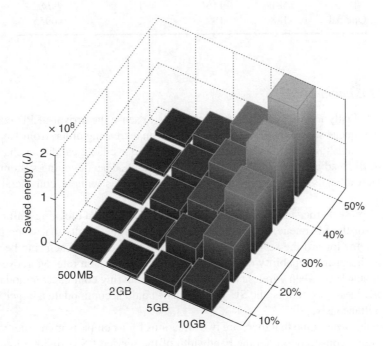

Figure 10.8 Saved energy consumption of the proposed over one month compared to the always streaming scheme

figures in Table 10.2 suggest that in the area A_1, A_2, and A_3, the device should store the downloaded content in local DRAM in case the mobile user would request this content again in a short time period. Although in the area A_2 with Wi-Fi coverage, the majority of transmission energy is less than the energy to store it in the local DRAM, the energy consumption of storing the content in the local DRAM performs better than the scheme that always downloads via wireless access on a long-term average.

Figure 10.8 depicts the energy saving for an individual mobile user in one month, compared with the always streaming scheme with different elastic percentage and cumulative content size. These results suggest that as mobile users frequently request the same video content, our proposed scheme could save a considerable portion of energy consumption, thus prolonging the lifetime of digital devices.

Table 10.3 Simulation results ($p_{\text{thres}} = 0.10$)

Num of ring	Num of time unit	Max length	Packet length F (Mbits)	Blocking probability
1	42	7	1	0.0490
2	23	11	1	0.0909
3	84	14	1	0.0477
4	77	22	1	0.0909
5	167	29	1	0.0805
6	188	33	1	0.0909
Total	581	116	1	N/A
Optimal	188	199	1	0.0955

10.5.3   Maximum of SU

The SUs must firstly monitor all frequency channels to sense the arrival of PU connections. The SU virtual queue senses the frequency channel in an increasing order, from 1th to the Mth channel. When finishing sensing channels and finding the available candidate channels, SUs regard PU traffic load as stable within a short time duration. Then, queuing system modeling can be utilized to analyze the competition of SU connections without consideration of PU connections.

Table 10.3 presents the simulation parameters and results with a blocking traffic threshold $p_{\text{thres}} = 0.10$ for the SU queue. It can be clearly seen from the results that when the mobile nodes are moving toward the service BS, the length of time units in each ring becomes less owing to the increasing capability of rings in terms of transmission rate. Moreover, with the decreasing distance between wireless nodes and the BS, the ring could accommodate a larger number of SU messages, that is, the SU virtual queue could accommodate a larger number of the SUs simultaneously.

In this model, we assume that there are two situations for a comparison in order to show the benefit of energy consumption: (i) the bandwidth of the service BS is divided into six equal parts, and each ring could only use one part for transmission; (ii) the bandwidth is occupied exclusively by the last ring, which is closest to the service BS. For instance, in the 7th line of Table 10.3, the SUs buffer all the messages and move into the area of the ring that is closest to the service BS. In this case, the SUs in the last ring occupy all of the bandwidth for message transmission, so that the capacity and throughput will be the maximum possible because this ring can support the higher constellation.

10.5.4   Battery Lifetime

It is assumed that the road segment a vehicle is moving along is covered by a cellular BS with two TVWS hotspots and two Wi-Fi hotspots in this domain as shown in Figure 10.9. Let $\Gamma = \{\tau_i, \tau_{i+1}, \ldots, \tau_j\}$ denote the decomposed time slots in which mobile users are moving along a

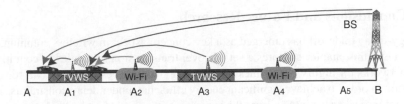

Figure 10.9 The selection from cellular BS, Wi-Fi, and TVWS AP

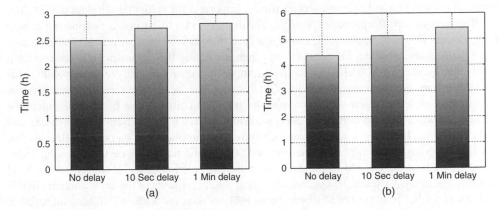

Figure 10.10 Battery life for different mobile applications (a) Video play back (b) Audio play back

road. Let $Q(t) = \{q_i, q_{i+1}, \ldots, q_j\}$ represent the energy consumed from embedded peripherals, such as CPU, Graphics, backlight, and storage devices.

Previous statistics have shown that the average multimedia video size is 10 Mb on YouTube, which is a nominal value for such video files [28]. Hence, we assume that the average length of a YouTube video is 4 minutes and 12 seconds, so in one cell, the mobile user will watch 250 seconds video, which translates to about 100 Mb downloading from wireless radios on average. Additionally, it is assumed that the mobile devices have a lithium-ion battery with a capacity of 1200 mAh, 3.7 V. Figure 10.10 shows a comparison of the battery lifetime of two mobile applications (video playback and audio playback) in three different scenarios. In the first scenario, when the mobile users make a request for the application, the system will start wireless transmission immediately without any delay. In the second and last scenarios, the mobile systems would delay the wireless transmission by 10 seconds and 1 minute, respectively. In the case of video playback, as seen in Figure 10.10a, the strategies that allow delay for mobile application can achieve drastic energy savings. As a result, the 10 second delay constraint strategy and 1 minute delay constraint could extend the battery lifetime by 9.3% and 12.7%, respectively.

As can be seen in Figure 10.10b, the proposed schemes, which allow delays on the transmission, perform better than the non-delay schemes due to the fact that in the audio playback scenario, the embedded system will have less CPU utilization with the backlight off, which takes a significant portion of overall energy consumption in the video playback scenario.

10.6   Conclusions and Future Research

The energy-delay trade-off has emerged as a key concern aspect in wireless communications. In this context, this chapter explores a set of novel transmission schedules in cognitive radio enabled networks for highly elastic messages aiming to select best time interval for message transmission in order to achieve significant energy efficiency under delay constraints. The proposed techniques take also into account the energy consumption of embedded storage memory at the terminal, which is used as a cost when message transmissions are delayed. Furthermore, when wireless nodes are competing for secondary access to the medium, the estimation of probability of PU arrival rate and service time is important for vehicular wireless device (SU) to effectively occupy the primary spectrum. The mobile nodes firstly contact a trusted database for historical information about PU traffic at a specific location and time duration so as to estimate the probability of the SU connections. Then, regarding the SU traffic, it is shown that it can be modeled as an M/M/K/L queuing system, which allows to analyze the capability that the system can serve users simultaneously. In this case, for the delay-tolerant applications, we recommend to delay the message transmission to the area close to the BS and maximize the throughput potential to minimize the energy consumption of message transmission under several constraints. In addition, for popular video streaming applications on portable devices that could be watched many times by one user, we studied the trade-offs between storing video content locally at the DRAM of the device or allowing deletion of the content from the local memory and relaying in wireless streaming in near-future requests of the same content. To this end, a scheme has been proposed where the mobility of the user is taken into account together with the probability of a user requesting the same content multiple times so that a decision is taken of whether or not the content should be stored locally. Finally, we analyze the problem of how to prolong the battery lifetime of mobile devices. This is especially important since the proliferation of always-on Internet applications has put significant strain on the battery capabilities and shortening the required re-charging time periods of the devices. Previous research has revealed that the data downloading via wireless radios is a dominant energy consumption factor in mobile devices. To avoid the drain of mobile device batteries, based on the delay tolerance of mobile Internet applications, we design the strategies that making selective use of the available high-rate wireless access such as roadside Wi-Fi/TVWS APs, by considering the energy cost, RF coverage, capabilities, and transmission algorithm.

In all of the aforementioned techniques in this chapter, the salient assumption was that energy savings can be achieved by capitalizing on the elasticity of Internet applications. This is an emerging area of research with significant potential to provide solutions for sustainability and better utilization of scarce wireless resources. Future avenues of research within this domain need to encompass a more architectural view on the wireless access network including overall system-required functionalities in order to allow for such mechanisms to be implemented in emerging and future wireless/cellular networks.

References

[1] E. Uysal-Biyikoglu, B. Prabhakar, and A. El Gamal, "Energy-efficient packet transmission over a wireless link," IEEE/ACM Trans. Networking, vol. 10, no. 4, pp. 487–499, 2002.
[2] W. Chen, M. J. Neely, and U. Mitra, "Energy-efficient transmissions with individual packet delay constraints," IEEE Trans. Inf. Theory, vol. 54, no. 5, pp. 2090–2109, 2008.
[3] X. Zhong and C.-Z. Xu, "Delay-constrained energy-efficient wireless packet scheduling with QoS guarantees," in GLOBECOM 2005, vol. 6, pp. 3336–3340, St. Louis, MO, 2005.

[4] A. Fu, E. Modiano, and J. Tsitsiklis, "Optimal energy allocation for delay-constrained data transmission over a time-varying channel," in INFOCOM 2003, vol. 2, pp. 1095–1105, San Francisco, CA, July 2003.

[5] H. Kim and G. de Veciana, "Leveraging dynamic spare capacity in wireless systems to conserve mobile terminals' energy," IEEE/ACM Trans. Networking (TON), vol. 18, no. 3, pp. 802–815, 2010.

[6] B. Han, et al., "Cellular traffic offloading through opportunistic communications: a case study," in Proceedings of the 5th ACM Workshop on Challenged Networks (CHANTS '10), New York, NY, 2010.

[7] X. Zhuo, W. Gao, G. Cao, and Y. Dai, "Win-Coupon: an incentive framework for 3G traffic offloading," in Network Protocols (ICNP), Vancouver, BC, pp. 206–215, October 2011.

[8] I. F. Akyildiz, W.-Y. Lee, M. C. Vuran, and S. Mohanty, "NeXt generation/dynamic spectrum access/cognitive radio wireless networks: a survey," Comput. Netw. Int. J. Comput. Telecommun. Networking, vol. 50, no. 13, pp. 2127–2159, 2006.

[9] X. Li and S. A. Zekavat, "Traffic pattern prediction and performance investigation for cognitive radio systems," in Wireless Communications and Networking Conference (WCNC), pp. 894–899, Las Vegas, NV, March 2008.

[10] L.-C. Wang, et al., "Load-balancing spectrum decision for cognitive radio networks," IEEE J. Sel. Areas Commun., vol. 29, no. 4, pp. 757–769, 2011.

[11] Q. Zhao, S. Geirhofer, L. Tong, and B. M. Sadler, "Opportunistic spectrum access via periodic channel sensing," IEEE Trans. Signal Process., vol. 56, no. 2, pp. 785–796, 2008.

[12] G. Noh, J. Lee, and D. Hong, "Stochastic multichannel sensing for cognitive radio systems: optimal channel selection for sensing with interference constraints," in Vehicular Technology Conference Fall (VTC 2009-Fall), pp. 1–5, September 2009.

[13] N. Balasubramanian, et al., "Energy consumption in mobile phones: a measurement study and implications for network applications," in IMC'09, Chicago, USA, pp. 280–293, November 2009.

[14] P. Deshpande, A. Kashyap, et al., "Predictive methods for improved vehicular WiFi access," in MobiSys'09, New York, pp. 263–276, 2009.

[15] P. Latkoski, J. Karamacoski, and L. Gavrilovska, "Availability assessment of TVWS for Wi-Fi-like secondary system: a case study," 7th International Conference on Cognitive Radio Oriented Wireless Networks CROWNCOM 2012, Stockholm, June 2012.

[16] S. Probasco and B. Patil, "Protocol to Waccess White Space database: PS, use cases and requirements," online at. http://tools.ietf.org/html/draft-ietf-paws-problem-stmt-usecases-rqmts-03. [Accessed 10 January 2015].

[17] L. Lei, Z. Zhong, C. Lin, and X. Shen, "Operator controlled device-to-device communications in LTE-advanced networks," IEEE Wireless Commun., vol. 19, no. 3, pp. 96–104, 2012.

[18] T. S. Rappaport, S. Shu, R. Mayzus, Z. H. Zhao, Y. Azar, K. Wang, G. N. Wong, J. K. Schulz, M. Samimi, and F. Gutierrez, "Millimeter wave mobile communications for 5G cellular: it will work!," IEEE Access, 1, pp. 335–349, 2013.

[19] Wireless Gigabit Alliance, Annual Report 2013 www.wirelessgigabitalliance.org. [Accessed 10 January 2015].

[20] Karl Stetson, Wi-fi alliance and wireless gigabit alliance finalize unification, Austin, TX, 2013, www.wi-fi.org /news-events/newsroom/wi-fi-alliance-and-wireless-gigabit-alliance-finalize-unification [Accessed 05 March 2015].

[21] P. H. Dave and H. B. Dave, Design and Analysis of Algorithms, Pearson Education, 2007, ISBN-10: 81-775-8595-9.

[22] J. Heo, J. Shin et al., "Ho-Shin Cho mathematical analysis of secondary user traffic in cognitive radio system," in Vehicular Technology Conference 2008 (VTC 2008-Fall), pp. 1–5, September 2008.

[23] S. Wang, J. Zhang, and L. Tong, "A characterization of delay performance of cognitive medium access," IEEE Trans. Wireless Commun., vol. 11, no. 2, pp. 800–809, 2012.

[24] H.-P. Shiang and M. van der Schaar, "Queuing-based dynamic channel selection for heterogeneous multimedia applications over cognitive radio networks," IEEE Trans. Multimedia, vol. 10, no. 5, pp. 896–909, 2008.

[25] I. Leontiadis, P. Costa, and C. Mascolo, "Extending access point connectivity through opportunistic routing in vehicular networks," in INFOCOM'10, Piscataway, NJ, pp. 486–490, 2010.

[26] R. Schoenen and B. H. Walke, "On PHY and MAC performance of 3G-LTE in a multi-hop cellular environment," in Wireless Communications, Networking and Mobile Computing (WiCom) 2007, Shanghai, China, pp. 926–929, September 2007.

[27] M. Zink, K. Suh, et al., "Characteristics of YouTube network traffic at a campus network - measurements, models, and implications," J. Comput. Networks, vol. 53, no. 4, pp. 501–514, 2009.

[28] www.websiteoptimization.com "Average Web Page Size Septuples Since 2003," May 2011, online at: http://www.websiteoptimization.com/speed/tweak/average-web-page/. [Accessed 10 January 2015].

[29] S. J. Shellhammer, A. K. Sadek, and W. Zhang, "Technical challenges for cognitive radio in the TV white space spectrum," in *Information Theory and Applications Workshop*, pp. 323–333, February 2009.

[30] P. Kolios, V. Friderikos, and K. Papadaki, "Ultra low energy store-carry and forward relaying within the cell," in IEEE Vehicular Technology Conference, pp. 1–5, Anchorage, AK, September 2009.

Further Reading

[31] M. Cha, H. Kwak, et al., "I tube, you tube, everybody tubes: analyzing the world's largest user generated content video system," in IMC 2007, California, pp. 1–14, October 2007.

[32] P. Gill, M. Arlittz, et al., "YouTube traffic characterization: a view from the edge," in IMC 2007, New York, pp. 15–28, October 2007.

[33] P. J. Smith, A. Firag, P. A. Dmochowski, and M. Shafi, "Analysis of the M/M/N/N queue with two types of arrival process: applications to future mobile radio systems," J. Appl. Math., vol. 2012, Article ID 123808, 14–p., 2012, doi: 10.1155/2012/123808.

[34] H. Ekstrom, A. Furuskar, J. Karlsson, et al., "Technical solutions for the 3G long-term evolution," IEEE Commun. Mag., vol. 44, no. 3, pp. 38–45, 2006.

[35] H. David, et al., "Memory power management via dynamic voltage/frequency scaling," in Proceedings of the 8th International Conference on Autonomic Computing (ICAC), Karlsruhe, Germany, June 2011.

[36] A. Carroll and G. Heiser, "An analysis of power consumption in a smartphone," in Proceedings of the 2010 USENIX Annual Technical Conference (USENIXATC'10), Berkeley, CA, June 2010.

[37] A. Hylick, et al., "An analysis of hard drive energy consumption," in Modeling, Analysis, and Simulation of Computer and Telecommunication Systems (MASCOTS), Baltimore, MD, pp. 1–10, 2008.

[38] D. Molaro, H. Payer, and D. Le Moal, "Tempo: disk drive power consumption characterization and modeling," in IEEE International Symposium on Consumer Electronics (ISCE '09), San Jose, CA, pp. 246–250, 2009.

[39] T. S. Ferguson, "Optimal Stopping and Applications Available," online at: http://www.math.ucla.edu/tom /Stopping/Contents.html. [Accessed 10 January 2015].

11

Green MTC, M2M, Internet of Things

Andres Laya[1], Luis Alonso[2], Jesus Alonso-Zarate[3] and Mischa Dohler[4]

[1]*KTH Royal Institute of Technology, Kista, Sweden*
[2]*Department of Signal Theory and Communications, Universitat Politècnica de Catalunya (UPC), Barcelona, Spain*
[3]*Centre Tecnològic de Telecomunicacions de Catalunya (CTTC), Barcelona, Spain*
[4]*Centre for Telecommunications Research, King's College London (KCL), London, UK*

11.1 Introduction

The capability of having any type of object interconnected and Internet-connected creates an unprecedented access and exchange of information that has been baptized as the Internet of Things (IoT) [1]. With the advancement of integrated technologies, improved batteries, and electronic miniaturization, everyday *things* will be equipped with sensors and microprocessors to collect information around them and execute smart applications. In addition, they will be able to communicate with each other. The IoT has the potential to revolutionize innovations; create new products, services, business; and reshape consumer's behavior. IoT represents a major player for the future of Information and Communications Technologies (ICT).

The development of the IoT must be environment friendly. ICT have shown to be a key contributor to global warming and environmental pollution; it is predicted that the global greenhouse gas (GHG) emissions from ICT will account for 12% of all emissions by 2020 at a growth rate of 6% per year [2]. Therefore, it is mandatory to develop environmentally friendly - or "green" - technologies for the IoT, and ICT in general.

Machine-to-Machine (M2M) communications constitute a fundamental part of the IoT. The term M2M refers to the exchange of data between two or more entities, objects, or machines

Green Communications: Principles, Concepts and Practice, First Edition.
Edited by Konstantinos Samdanis, Peter Rost, Andreas Maeder, Michela Meo and Christos Verikoukis.
© 2015 John Wiley & Sons, Ltd. Published 2015 by John Wiley & Sons, Ltd.

that do not necessarily need human interaction [3]. The envisioned market for such kind of communications is broader than the one traditional human-based communications in terms of number of users and variety of applications. Some forecasts [4] predict figures up to 50 billion machines that will be connecting to communication networks by 2020. This is a very big number compared to the entire world population of around 7 billion people. Therefore, there are some challenges that need to be addressed in order to fully support M2M services in current communication networks [3]. From the technical point of view, M2M communications are substantially different from Human-to-Human (H2H) communications. For example, network operators should provide communication services at low cost in order to face the low Average Revenue Per User (ARPU). Despite the large number of expected M2M connections, most of them will generate very little and infrequent data traffic. Communication networks shall also provide suitable congestion and overload control solutions in order to handle a huge number of simultaneous connections. Features such as low mobility, time-controlled data delivery, group-based policing and addressing, low connection delays, and a wide variety of Quality of Service requirements are among other challenges that need to be addressed. All of them must have the "green" concept embedded. In order to ensure that devices can operate autonomously for years or even decades without human intervention, it is necessary to provide networks with highly efficient communication protocols. This is the main focus of this chapter.

The European Telecommunications Standards Institute (ETSI) created in 2009 a dedicated technical committee to identify key M2M use cases [5–7], understand the service requirements [8], and promote standards for the complete end-to-end M2M functional architecture [9]. Later, in 2012, the global One M2M project was also established by ETSI, with other international standardization bodies, in order to define M2M standards that can accelerate the deployment and success of M2M applications. Figure 11.1 depicts the high-level architecture for M2M according to ETSI, which consists of a Device and Gateway Domain and a Network Domain. The Network Domain consists of the Access Network, the Core Network, and M2M Applications. The Access Network is used to interconnect the two domains, and it could be either a wired or wireless solutions, or a combination of both, to attain the best features of each alternative. Likewise, the Core Network could be of any kind, although it would be desirable

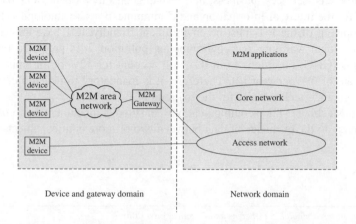

Figure 11.1   High-level architecture for M2M according to ETSI

to ensure that it has Internet Protocol (IP) connectivity. The set of M2M applications, which is referred to as the M2M Application Domain, can be divided into two layers: (i) the lower layer, which consists of the M2M Service Capabilities and constitutes an adaptation layer shared by different upper applications; and (ii) the upper layer, which consists of the M2M Applications and uses the M2M Service Capabilities to access the Core Network. The Device and Gateway Domain is composed of M2M Devices and M2M Gateways. An M2M Gateway can act as the Network Proxy between the devices and the Network Domain (see Figure 11.1), thus forming short-range M2M Area Networks.

In this chapter, the focus is on the Access Network, and in particular, on the pivotal role that cellular networks are deemed to play in the IoT. The benefits of cellular network above any other technology are mainly five:

1. **Ubiquitous coverage**: almost every populated inch of the planet has some level of cellular coverage.
2. **Mobility**: cellular networks enable handover and mobility around the planet.
3. **Infrastructure**: already deployed that can be leveraged to enable new services.
4. **Well-known and accepted technology** by users: there is no need to deploy new networks and infrastructure, and take an effort to make this new technology accepted by users.
5. **Controlled interference**: while other wireless systems operate in license-free bands, cellular networks operate in licensed bands. This provides better control of the interference, which may become a major issue when the number of simultaneous devices is very high.

That being said, and according to the vision of ETSI, cellular networks may be "extended" with capillary networks formed by other short-range technologies such as Low Power Wi-Fi based on the IEEE 802.11 Standard, ZigBee-like networks based on the new amendments of the IEEE 802.15.4, Bluetooth Low Energy (BLE), Radio Frequency Identification (RFID), Near Field Communications (NFC), and many others. Also, wired solutions, such as Ethernet of Power Line Communications (PLC), will very likely have their place in the IoT. However, the ubiquitousness, maturity and mobility capability of cellular networks are increasingly appealing for MTC. For this reason, the focus in this chapter is on the Green Communications that can be attained with current cellular networks. In particular, the focus is on LTE and LTE-Advance (Long-Term Evolution), which is the key technology being developed by the 3rd Generation Partnership Project (3GPP) to facilitate cellular networks. Even though the first releases of LTE did not include any optimization for Machine-Type Communications (MTC) and were mainly focused on achieving very high throughput and low delays, new releases of LTE already include architectural and functional optimizations for M2M applications.

Among other challenges, achieving very high energy efficiency is one of the primary targets. Low energy consumption is crucial for M2M devices because their energy sources are usually scarce, for example, devices rely on batteries or alternative harvesting technologies such as solar panels. In addition, M2M devices may be deployed in high numbers and in remote environments where it would be hard or even impossible to recharge or replace batteries. Not to mention that the fact that the total number of devices will be very high turns the replacement of batteries into a not scalable approach. Therefore, green communications must be embedded into cellular networks, and this is the main focus of this chapter.

The remainder of the chapter is organized as follows. Section 2 discusses different communication techniques that are being explored to improve the energy efficiency of cellular networks and make them suitable for M2M. In Section 3, a number of relevant M2M applications are presented and discussed. Finally, open research issues for green M2M are outlined in Section 4, leaving the final conclusions for Section 5.

11.2   Green M2M Solutions for M2M

The aim of this section is to introduce a number of improvements for wireless networks that have been proposed in the literature to date to reduce the energy consumption in M2M communications.

11.2.1   Discontinuous Reception (DRX)

In today's wireless networks, such as those based on LTE, devices continuously—or very frequently—listen to the downlink control channel to check whether there are pending downlink data transmissions intended for them or there are uplink resources available for transmission. To do so, the radio circuitry of devices is always on, thus consuming energy. While this approach is very efficient to reduce latency in highly demanding applications, it turns very inefficient for bursty packet data traffic patterns with long silent periods. For this reason, the concept of duty-cycling, also referred to as Discontinuous Reception (DRX), can be used for such traffic patterns. A generic DRX scheme is illustrated in Figure 11.2. Time is divided into DRX Cycles, comprising both On and Off Periods. During the Off Period, devices can switch off their receiver circuitries to save energy, and they are said to enter into sleeping state. During the On Period, devices can monitor the downlink control channel. As shown in [10], such operation can significantly improve the energy efficiency in LTE networks.

However, the execution of a DRX scheme inevitably increases latency and communication delay. Note that devices in sleeping state cannot immediately react to traffic changes [11]. For this reason, and in order to meet diverse Quality of Service (QoS) requirements, LTE defines

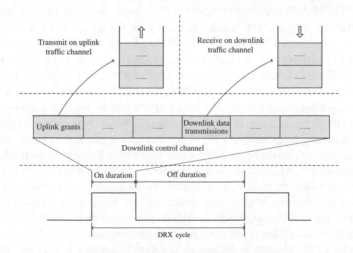

Figure 11.2   Example of DRX cycle

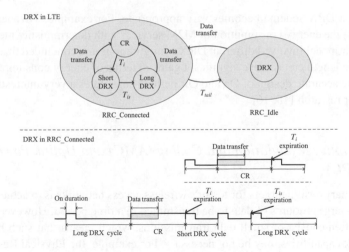

Figure 11.3   DRX mechanism in LTE

the DRX mechanism depicted in Figure 11.3 [12]. The upper part of Figure 11.3 shows the UE (User Equipment) modes in different Radio Resource Control (RRC) states and the transition criteria. There are two RRC states in LTE, RRC_Connected and RRC_Idle. When a UE is in RRC_Idle state, that is, without data to transmit, it operates in DRX mode, alternating on and off periods. However, when the UE needs to transfer data, it makes a transition to RRC_Connected state and operates in Continuous Reception (CR) mode. In this mode, it initiates an inactivity timer (T_i), which is reset every time a new data transfer is originated. When this timer expires, the UE enters into Short DRX mode. From this mode, the UE can go back into the CR mode whenever there is a new data transfer requirement. Whenever the short DRX cycle of duration T_{is} is completed and no data transfer has been conducted, the UE switches to Long DRX mode. Similarly, a UE will switch to CR mode if new data has to be transmitted. If there is no data transmission for the entire RRC_Connected inactivity timer (T_{tail}), the UE will switch back to RRC_Idle state. The lower part of Figure 11.3 shows details of LTE DRX in RRC_Connected state and relationship between the timers used for this DRX operation.

The three DRX modes (DRX in Idle mode and both Short and Long DRX in connected mode) have a similar operation but with different configuration parameters, mainly bound to the duration of the Off Period. The DRX in RRC_Idle state usually has a longer Off Period than those in RRC_Connected state because it is intended for UE with longer inactivity periods. Contrarily, the durations of these periods are shorter in connected state in order to reduce latency. This is vital for some applications. For example, for Voice-over-IP (VoIP), when a transmission starts (i.e. a conversation), periods of on and off activity will be regularly repeated over time.

The durations of On and Off Periods should be configured according to the service traffic pattern and QoS requirements [13]. For example, when the traffic model is stochastic, such as H2H service traffic model, long On and Off Periods will lead to considerable delays, thus degrading the overall user experience. However, some M2M applications are "time controlled" [3]. This means that devices can tolerate long access delays to send or receive data.

In such cases, a DRX scheme becomes very appropriate. For example, Tirronen et al. studied in Ref. [14] the energy consumption of M2M services with deterministic intervals in LTE networks. Their model involved different DRX configurations and concluded that maximizing the DRX Cycle length can achieve significant gains in terms of energy consumption.

All in all, the accurate design of On and Off periods constitutes a very interesting and open area for further research [15, 16].

11.2.2   Adaptive Modulation and Coding (AMC) and Uplink Power Control (UPC)

One of the primary design targets for today's wireless access networks is to achieve ultra-high data rates with large amounts of data to be transmitted in a time period. However, M2M communications often consist of small data transmissions, which means that such high wireless access network capabilities may be not necessary. For example, the Physical Resource Block pair, which is the minimum scheduling physical resource in LTE, can convey up to 712 bits of payload [17]. However, in many M2M applications, for example, sensors monitoring temperature, the amount of data to transmit could be of only a few bytes. This is typically referred to as the Small Data Transmission of M2M applications.

Most of today's wireless technologies provide AMC and UPC mechanisms to cope with the channel variations and adjust the transmission rate to the channel conditions. Indeed, these two techniques can be also used to reduce energy wastage for the transmission of very few bytes of information in M2M applications [18].

For some wireless networks with AMC scheme (such as LTE), the transport block size of a number of physical radio resources is adjustable. This means that the number of payload bits that can be transmitted on a given physical resource changes according to the Modulation and Coding Scheme (MCS) used. The higher the MCS, the more payload bits that can be transmitted per second. However, higher MCS implies that the total energy per bit is lower, and thus more prone to suffer from the errors induced by the wireless channel. In LTE, the AMC balances the data rate and the transmission robustness (probability of correctly decoding) by choosing a proper MCS [19]. From a different perspective, on the premise of receiving bits with a given outage probability, one can compute the capacity of a given channel by selecting the maximum MCS, which ensures the targeted outage probability.

In its turn, the mechanism to define the transmission power in the uplink, the UPC, can also combat channel fluctuations, so that the receiver can always perceive the same Signal to Interference plus Noise Ratio (SINR). In this case, the data rate of the channel can be maintained constant, but the sender needs to consume a variable amount of energy, which depends on the required transmission power. Therefore, if the channel is stable, a higher SINR can be attained by increasing the transmit power. With a higher SNIR, the transmitter can also select a more aggressive MCS to increase the data rate. Therefore, the capacity of a given link can be adjusted by changing the transmit power, in combination with the MCS.

In order to maximize energy efficiency, it is necessary to make sure that the AMC and the UPC are well-designed and optimized for the small data transmissions of M2M applications. For example, conservative MCS could be employed so that the channel capacity could be filled up with small payloads. At the same time, the transmit power could be decreased as the receiver would be able to decode the message with more redundancy bits even with a lower SINR. This approach somehow is counterintuitive to latest proposed mechanisms where the main target is to maximize capacity. For example, the UPC mechanism in LTE includes a parameter Δ_{MCS} [17], which is the MCS dependent power offset. The scheduler in LTE can utilize this parameter to inform the devices to reduce their transmit power when using conservative MCS. Therefore, devices can conserve energy when transmitting small M2M packets by using both conservative MCS and low transmit power.

11.2.3   Group-Based Strategies

M2M devices with either similar functions, or located in close geographical positions, communicating with the same gateway or server, can be grouped together. Then, group-based policies can be applied to reduce signaling and improve the energy efficiency.

Generally speaking, it is possible to define two types of MTC groups (or clusters): **static** and **dynamic**.

The devices connected to a common M2M Gateway could become a group, with the M2M Gateway acting as the group leader and relaying messages between the devices and the Network Domain. Considering that the M2M Gateway usually owns stable and unconstrained power supply, such as mains power, it can continuously act as the group leader, thus creating a static cluster.

In contrast, without a fixed M2M gateway, any device could temporary act as the cluster leader [20, 21]. Among other options, the selection of the device could be based on its battery level and the channel state with the corresponding Access Point (e.g., the eNodeB in LTE). In this case, the group members are dynamically assigned based on their connectivity with the group leader [22]. This means that both the group leader and the group members will change during the network operation when the selection metrics vary or the mobility of the network affects connectivity.

Once devices are grouped, either statically or dynamically, various group-based strategies can be applied. This is a non-exhaustive list of some examples that can be used to improve energy efficiency:

- **Relaying**: The group leader relays messages from the devices to the network, thus saving energy of the group members. Normally, the link distance from any group member to the network is larger than the one to the group leader, thus enabling transmission at lower transmission power and with higher reliability.
- **Offloading**: The establishment of short-range networks, such as a Wireless Local Area Network (WLAN) or a Wireless Personal Area Network (WPAN) among group members, can help reduce congestion and signaling overload in the cellular network. When a large number of devices access the network simultaneously, part of the devices will fail access because of congestion and high probability of collision. They will back off for some random time and re-attempt access again till success. This repetition of the access procedure will drain energy from devices. In such case, the use of an M2M Area Network can help reduce the number of simultaneous access requests through the cellular network and help the devices save energy.
- **Complexity reduction**: Complexity and intelligence can be gathered at some devices, thus letting other devices execute extremely simple functions at very low energy consumption. This could be applied, for example, when some devices are operating at critical battery levels

and need to wait for the harvester device to capture more energy. It is worth mentioning that the 3GPP is already working on the application of this approach to the network domain between the core network and the eNodeB [23, 24]. The main idea is to redistribute the work load of one entity to the other entities during off-peak times. For example, when one eNodeB serves very little number of UEs, it can redistribute such UEs to neighbor eNodeBs and enter into sleep mode to save energy. Another example: if one district is covered by several Radio Access Technologies (RAT), it is possible to shut down some of them if the traffic load is very low. In such case, the network shall transfer current users to other RATs (inter-RAT) or the same RAT but different cells (intra-RAT). If a simple low-complexity device is only equipped with a single RAT, it will lose the connection with the network. In such case, the device must rely on the group leader to act as a relay.

- **Data compression**: The M2M Gateway or group leader can perform packet integration and IP compression functions, which will decrease the volume of required data transmission between the Gateway and the Access network and will help improve the energy efficiency. Small packets received by M2M Gateway from all the devices in the group could be integrated into larger packets, which can better utilize the high capacity of the cellular access networks. In such case, IP Header Compression techniques would be necessary to avoid transmitting very long data packets packed with too many control bits.
- **Control signaling reduction**: Similar signaling messages from the devices could be integrated to decrease the control information load [25]. A solution to do so was presented in Ref. [25], where the signaling integration is done at the eNodeB, thus reducing the control load in the core network. These ideas could be applied by the group leader of a cluster in order to reduce the signaling load between the group leader and the eNodeB.

11.2.4 Low-Mobility-Based Optimizations

Some M2M applications feature low (or even no) mobility. In many applications M2M devices will rarely move, or do so in a very particular and known region. Among other examples, this is the case of Automatic Meter Readers (AMRs) or remote sensing devices deployed in specific locations of a warehouse to monitor some physical or environmental condition. In such cases, some of the mobility management procedures designed for traditional H2H communications could be removed, simplified, or optimized for M2M applications in order to reduce congestion and increase energy efficiency.

The low mobility feature may provide two kinds of relatively static information, which can be used to reduce the signaling load: **location** (according to the association to a cell) and **distance** to the serving eNodeB.

First, while in scenarios with high mobility, the network needs to track the position of devices at all time by using the Tracking Area Updates (TAUs) procedure in LTE, in low mobility scenarios, devices may not change the location frequently and the intervals between TAUs could be increased and thus signaling reduced.

Second, in the case of static devices with fixed distances to the serving eNodeB, the Timing Alignment (TA) parameter will be constant, thus avoiding its periodic update for the devices. This can also be used in the Random Access procedure in the following manner: devices can figure out that a Random Access Response (RAR) is intended for them through the TA information included in the RAR. This reduces the collision probability and the associated signaling load for the repeating access [26].

In the case of static devices with a fixed location and constant distance to the serving eNodeB, they may hold relatively stable path loss. Therefore, they can reduce the reporting period of the measured downlink signal strength.

11.2.5   Cooperative Communications

Cooperative communications are known by their capability to improve channel throughput and enhance the energy usage in wireless networks [27]. As it could be expected, their application to M2M is no exception. Cooperative communications exploit the broadcast nature of the wireless channel and are mainly based on one of the two following principles:

1. In the wireless medium, the distance between a transmitter and the receiver determines the path loss, that is, the loss of signal strength with the distance. This effect is typically modeled with an exponential function inversely proportional to such distance, elevated to the power of an exponent, referred to as path-loss exponent, which is generally comprised between 2 and 4 [28].
2. It is possible to achieve spatial diversity [29], also referred to as cooperative diversity. Relay devices in a cooperative link forward messages from source devices, thus providing the receiver with multiple copies of the messages that have been transmitted through independent propagation paths.

Figure 11.4 depicts the three basic cooperative strategies.

The first type of cooperative communication scheme is the Single-Input Single-Output (SISO) Multi-Hop scheme, based on the principle of the path loss and its relationship with the distance. A long distance link between a source and a destination can be split into short distance transmissions, wherein intermediate devices act as relays. By properly arranging the relays between the source and destination, the total energy consumption of cooperative transmissions will be lower than that of the direct transmission [30].

The second cooperative scheme represented in Figure 11.4 is the so-called cooperative relay. The operation of the scheme involves at least three devices: the source, the relay, and the

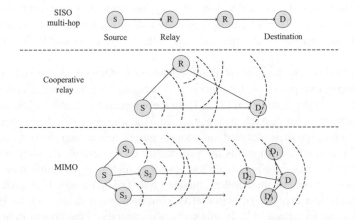

Figure 11.4   Examples of cooperative communications

destination. The transmission operates in a frame-by-frame basis. Each frame consists of two steps. In the first step, the source transmits a data packet to the destination. As the wireless transmission is of broadcast nature, the relay also receives this transmission. In the second step, the relay forwards the data packet to the destination. At last, the destination can combine the two received copies of data applying any combination technique. As the two copies of data are received through two independent wireless channels, the destination can get some spatial diversity gain, which can help decreasing the error rate of data transmissions. This cooperative schemes lets both the source and the relay transmit data with low power while achieving the same required data error rate.

In the case of energy-constrained MTC Devices acting as a source, they may always try to achieve the minimum transmission power by using the relay transmission in all cases. However, under some circumstances, for example, when the channel state between the source and the destination is suitable, the relay may not need to forward data packets for each transmission. In order to save energy, an incremental-relaying scheme such as the one presented in Ref. [31] can be applied. The relay is only active when necessary. In this incremental-relaying scheme, the relay monitors the feedback from the destination to the source after a data packet has been transmitted from the source. When the feedback is a negative acknowledgement (NACK), then the relay forwards the message to the destination.

The third cooperative scheme represented in Figure 11.4 is the Multiple-Input Multiple-Output (MIMO). This case involves a source and a destination, and multiple relays assisting the transmission (cooperative sources) and the reception (cooperative destinations). In this case, a transmission can be divided into three steps. In the first step, the source S transmits data to the cooperative sources S_x (they can be seen as relays). Then in the second step, all the cooperative sources (S and S_x) encode the same data into Space–Time Bloc Coding (STBC) symbols and transmit simultaneously to all the destinations. In the last step, all the cooperative destinations (D_x) forward the received data to the destination D, where a joint signal combination and data decoding is performed. As the MIMO scheme can achieve the diversity of several copies of the STBC, this mechanism allows the sources to transmit data with lower power compared to a transmission with a single hop. Generally, the cooperative sources (or the destinations) are geographically close to the source (destination) and can be grouped to form a cluster. The transmissions between the cooperative devices of the same cluster can be typically done at very low transmission power, thus leading to low power consumption. If there are no cooperative destinations but just a single destination device, the MIMO scheme is actually reduced to the Multiple-Input-Single-Output (MISO) scheme.

In addition to this schemes, where the cooperation is decided by the source, an amount of reactive cooperative protocols have been also presented wherein the decision to perform cooperation is left to the destination [32, 33]. In these reactive schemes, when the destination is not able to correctly decode a received packet from the source, it asks the cooperative devices in the cluster to provide their received data copies rather than asking the source to retransmit the packet. The destination improves the probability of successful decoding and avoids the long-distance data retransmissions from the source, which may consume more energy, and due to the coherence time of the wireless channel, may fail again with high probability.

The first theoretical analysis of cooperative communication dates back to 1979 with the paper by Cover and El Gamal [34]. In this work, the capacity of the cooperative relay system is computed, for an additive white Gaussian noise (AWGN) channel. Since then, a vast number of contributions related to cooperative communications have emerged. A simple Google search with the terms "cooperative communications" results in more than 40 millions results today (2013), just to give an idea. The current active areas of research include the relay selection, the retransmission strategy (Amplify and Forward, Decode and Forward, Compress and Forward, and all their variations), the transmission technique, the interference management, power control, and also research at higher layer of the protocol stack to handle multiple access at the Medium Access Control (MAC) layer, and to deal with cooperation delays at the transport and application layers.

Cooperative communications have been included in 3GPP's specifications for LTE-Advanced [35]. Through the inclusion of relays, LTE-Advanced provides two kinds of cooperative communication scheme:

- **SISO**: a fixed relay can forward messages between the eNodeB and UEs in order to extend coverage.
- **Cooperative Relay**: a fixed relay works with the eNodeB (or the UE) to achieve cooperative diversity and improve the system performance, which is known as the coordinated multipoint (CoMP) transmission or reception in LTE-Advanced.

These two cooperative approaches can benefit M2M applications to reduce energy. However, still a lot of research needs to be done in this area. In addition, particular attention deserves the possibility of enabling such cooperation between devices, leading to the concept of Device-to-Device communications, which are presented in the next subsection.

11.2.6   Device-to-Device (D2D) Communications

D2D communications allow devices to communicate directly with each other without having to route traffic through the base station. D2D communications include the possibility to use multi-hop routes, wherein some devices must collaborate by acting as relays. D2D communications can be exploited to reduce the transmission energy and extend the lifetime of M2M devices operating with cellular technologies, such as LTE. Figure 11.5 illustrates different forms of D2D communications applied to an LTE network.

In traditional wireless networks, devices communicate with each other by means of a common network infrastructure. The used frequency spectrum can be either licensed (e.g., cellular network) or unlicensed (e.g., WLAN). However, when the infrastructure becomes unavailable, for instance, due to a natural disaster or a power blackout, the devices are unable to communicate. Wireless ad hoc communications would be the solution for such cases; the devices in an ad hoc network communicate with each other directly or by using other devices as relays. Indeed, each device may become a router, which may forward the messages so they can reach its final destination. Even though ad hoc networks have received lots of attention from the scientific community in the past, existing deployments are based on proprietary solutions in short-range technologies, and have never reached the cellular domain. However, things are changing today. D2D communications are emerging as an appealing technology to create a revolution in cellular communications by providing very energy-efficient communications and a simple way of offloading core cellular links, thus establishing the ideal framework to deploy the IoT.

D2D communications that work without any network infrastructure can be referred to as non-network-assisted D2D schemes. The air interface can use either the licensed spectrum

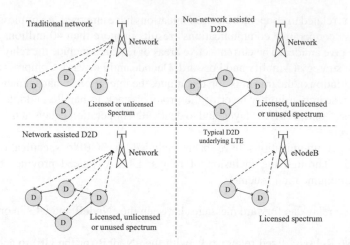

Figure 11.5   Four types of D2D communications

or the unlicensed spectrum. At the same time, some ad hoc networks with cognitive techniques exploit use the unused spectrum to improve the usage of the licensed bands. In contrast, network-assisted D2D communications work with the support of the network infrastructure. In general, the infrastructure may store some information about the ad hoc network and its devices (e.g., location and temporary network topology). By exchanging control information with the devices, the infrastructure helps establish, maintain, and terminate D2D transmissions more efficiently. More importantly, the infrastructure can help the devices establish different cooperative strategies as the ones described in Section 11.2.5), which can help the devices save energy.

In addition, D2D communications may act as an underlay for cellular networks (e.g., LTE as depicted in Figure 11.5) and this scheme has attracted a lot of attention from the scientific community. In such case, when the cellular network finds that the conditions to establish a D2D link are met, it will help the devices involved to establish the direct D2D link and the traffic between the devices will be offloaded from the one being served by the eNodeB. D2D as an underlay to cellular networks may hold several gains [36]; the first one is the **proximity gain**, where the proximity of devices helps reduce the energy consumption, and achieve both high data throughput and low communication delays. The second gain is the **hop gain**, because D2D communications only use one hop compared to traditional cellular communications where two hops (uplink and downlink) are needed, thus reducing latency and complexity, and improving spectrum efficiency at least in a factor of two. In addition, a D2D link saves energy on the eNodeB side as well. The third gain is the **frequency reuse gain**, where the spectrum for cellular network can be reused by D2D communications. In this case, efficient and dynamic **interference management** becomes fundamental piece to make D2D fully beneficial. Finally, **coverage extension** can be also facilitated by enabling the direct communication between devices.

The 3GPP is considering the inclusion of D2D communications as part of the evolution of LTE [37]. In particular, the technique is referred to as Proximity Services (ProSe), and there are two main areas of application:

- To improve the cellular network performance through network offloading, direct proximity transmission, and extended coverage.

- To deploy national security and public safety services on some allocated frequency bands when the cellular network is not available.

11.3   Green M2M Applications

M2M communications are the key cornerstone for realizing the IoT. A lot of innovative applications of IoT based on M2M can help achieve the green world with less greenhouse emissions [2]. This section introduces key applications that will enormously benefit from M2M communications:

- Automotive applications.
- Smart Metering for utility services.
- Smart Grids for electricity distribution networks.
- Smart Cities for sustainable and efficient use of resources in urban areas.

Even though other applications such as building automation, industrial automation, logistics, and remote e-Health, among many others, will also play an important role in the IoT, the focus in this chapter is on the four applications listed before.

11.3.1   Automotive Applications

Vehicles with embedded communication capabilities, for example, WLAN interface or SIM-enabled, can be interconnected to perform a variety of automotive applications that include automatic drive, improved safety, insurance or road pricing, emergency assistance, fleet management, electric car charging management, traffic optimization, or parking services, among many others. [6]. Figure 11.6 shows a simple network topology where M2M devices embedded in the vehicles report real-time location information to a data center. The data center will grasp the traffic information after analyzing all such vehicles location information together with other factors such as weather, accidents, an so on. The data center may broadcast

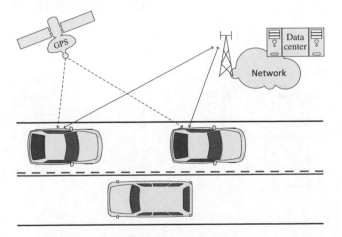

Figure 11.6   Example of automotive applications

a traffic congestion warning to the corresponding vehicles, so they can select other routes to avoid sinking into the traffic jam. In addition, vehicles can send route plan requirements to the data center, so that the data center can plan the optimized route based on the traffic information. Fluent traffic and optimized routes can help to save fuel and mitigate greenhouse gas emissions.

11.3.2 Smart Metering (Automatic Meter Reading)

Smart Metering, or Automatic Meter Reading (AMR), will facilitate the reading of instantaneous utility consumption (water, gas, electricity) by both costumers and utilities. This will benefit both users, who will be able to understand their consumption and adapt their demand to the price of the resource at each time of the day, and for the utility, who will be able to better manage their generation, distribution, and storage plants [7]. A better usage of the resources will have a benefit for both users and utilities, and will help reduce the dependency on natural resources.

Figure 11.7 depicts an example of a Smart Metering application, where the smart meters are connected to a data center through an M2M gateway. The data center can remotely read the meters, and pricing information and estimated bills can be sent to the in-home display for customers. Then, customers can "learn" how and when to use energy to reduce the total cost. In addition, Smart Metering can help suppliers collect customer usage information in order to provide better services. For the distributors, the Smart Metering helps them monitor and manage their networks effectively.

11.3.3 Smart Grids

A Smart Grid is an electricity supply network that relies on ICT to efficiently detect and react to changes in usage. This network is an economically efficient and sustainable power system

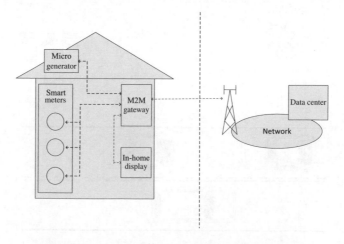

Figure 11.7 Example of smart metering application

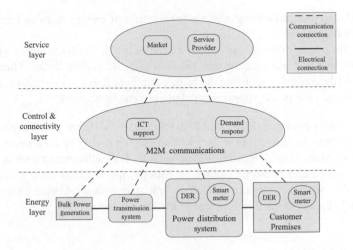

Figure 11.8   Example of smart grid architecture

with low losses and high safety, which also provides high levels of quality and security of supply [5]. In order to connect all users among them, Smart Grids need a telecommunication infrastructure with large-scale coverage, high quality, and security. In order to supply electricity efficiently, Smart Grids need to collect different types of information from the generators, the transmission system, the distributors, and the customers, which will rely on a great number of smart meters and sensors. The telecommunication infrastructure should support M2M communications efficiently.

Figure 11.8 describes the basic conceptual architecture of a Smart Grid, which can be divided into 3 main layers:

- The **energy layer** includes the energy generators, transmission systems, distribution systems, and the customer premises, where the Distributed Energy Resources (DERs) are local energy generators and storages. Between them, the electrical connections create the energy network.
- The **control connectivity** layer consists of the ICT and Demand Response (DR) entities, which are connected with all energy entities via M2M communications. The ICT entity collects, analyzes, and processes the information from all the energy entities. It also commits the strategies and commands from the market and service provider in the service layer.
- The **service** layer consists of the Market and Service Provider entities.

The DR entity is the key for smart grids to reduce the greenhouse gas emissions with the following functions:

- Reduce the power peak and peak time by properly arranging the power demands over time. DR provides mechanisms (such as dynamic prices) to encourage customers to shift their demand from the peak time to periods of lower power demands. As electrical grids are mainly sized by the peak demands, less peak demands decrease the total capital expenditure and operation expenditure of electrical grids. This will also lead to reduced need for energy storage and also reduced energy wastage (generation of energy that can be neither stored nor consumed).

- Balancing the power demand and supply in the network by dynamically optimizing the power flows, this will decrease the transmission and distribution losses. Through properly planning and dynamically adjusting the routes of power flows in electrical grids, the transmissions between the power supplies and demands can be reduced.

After the Obama's National Broadband Plan released in 2010, the push for Smart Grids has become very intense. Utilities all over the globe are trying to deploy such smart grid concept, and the inclusion of microgrids and the electric vehicle (EV) will create a shift in the mentality on how we generate and consume energy. In Europe, the Horizon 2020 program, funding research and development activities in Europe and also with worldwide collaboration, will give substantial support to the development of the Smart Grid.

11.3.4 Smart Cities

Today, more than half of the world population lives in cities. What is worse, it is predicted that this percentage will be about 70% by the middle of this century [38]. Therefore, cities are dominating the human society development and it is generally accepted that the prosperity of cities does not depend on their economic success. Actually, UN-Habitat believes that the prosperity of cites should include five dimensions [38]:

- Productivity.
- Urban Infrastructure.
- Quality of Life.
- Equity.
- Environmental Sustainability.

To some extent, the concept of smart city can represent this evaluation system of city prosperity, even though there is no consensus of its definition so far. Caragliu *et al.* in Ref. [39] believe a city to be smart when there is investment in human and social capital, transport, and ICT communication infrastructure; such investment fuels sustainable economic growth and improves the quality of life, with a wise management of natural resources, through participatory governance. As we can see, one main object of smart cities is the "green" city, which means that urban areas are able to manage appropriately the natural resources and modern infrastructures to improve the energy efficiency and reduce pollution, thereby achieving the sustainable environment. The modern communication infrastructure (especially the MTC or M2M communications) not only enables the other infrastructure wiser and greener, but also provides tools to monitor the environment and collect information for a more efficient management. From the technology point of view, M2M communications provide the interconnections between the information sources (such as monitors for electricity, natural gas and pollution), data centers (which store and analyze various data), and decision makers (which make decision according to the conclusions and results from the data centers.). Some key M2M applications for smart cities are as follows:

- Smart Parking provides the information of the adjacent parking space for drivers through monitoring the space usage. Then, people can save the time and gas to park, besides improving their well-being.

- Smart Litter Bins provide the bin usage information to the city councils. Then, they can arrange the optimal times and routes for garbage trucks, which can avoid the waste of excess garbage picking and can collect the garbage in time preventing the overfilling of bins.
- Pollution Monitoring utilizes sensors deployed widely in the cities to collect real-time environment information. People can produce immediate reactions for the environment destruction. The long-term information can be the basis for the authorities to make the city development strategies and policies considering the environment sustainability.

The number of possible applications is enormous, and the smart city concept is getting more and more popular around cities. As for the other applications outlined in this section, the opportunities are there, and the challenges as well. In order to make smart grids, smart automation, smart cities, and whatever smart application based on M2M communications a reality, still some political, societal, market, and technical challenges need to be faced. Even though many challenges have been already identified along this chapter, additional open challenges are summarized in the next section.

11.4 Open Research Topics

While the challenge of energy-efficient M2M communications has been extensively studied in the scientific community, there are many research topics still to be solved. Among others, it is possible to identify the following:

Optimization of Existing Wireless Networks for M2M: current communication systems have been designed to satisfy human-based applications demanding high capacity. However, the requirements of M2M applications are completely different, and thus adaptation and optimization of existing wireless networks are necessary. In addition, such an adaptation would be an economic, reliable, and fast solution to achieve efficient M2M communications without requiring to deploy new dedicated networks for M2M traffic. This adaptation and optimization must be done ensuring ultra-high energy efficiency so that real smart and autonomous applications can become feasible.

Cooperative Schemes for M2M with Network Coding: As discussed in Section 11.2.5, cooperative communications, embedding network coding techniques, are a promising technique to achieve ultra-high energy efficiency for M2M. M2M requires simplicity and low cost, thus the inclusion of network coding techniques—which are capable of distributing the complexity among cooperative relays—may be an interesting research path towards feasible M2M deployments. How to assign partners for this cooperation is still an important question, especially, when energy-constrained devices and non-energy-constrained UEs coexist in the same network. The two kinds of devices have different costs for energy consumption and different requirements for data rates. Designing algorithms to form cooperative groups ensuring that they will contain the two kinds of partners will help improve system performance. Furthermore, effective incentives must be designed for non-energy constrained devices assisting energy constrained devices. The reward for the relay may be higher data rate chances, for instance. Such kind of incentive should consider the potential selfish behavior, which may lead to the wrong cooperative partner assignment and lower system performance. Security is a key research topic in this area.

D2D Schemes for M2M: Enabling short-range communications using the main cellular air interface will leverage energy efficiency and enable scalable network deployments. To achieve all the benefits discussed in the previous section, it is still necessary to conduct research in a new technique that is in an exploration and maturation phase. Radio resource allocation (scheduling), power control, and interference management are key areas that need to be developed to enable D2D communications and thus facilitate the use of cellular networks for the efficient and scalable coexistence of humans and machines.

11.5 Conclusions

Even though there is no common agreement on the magnitude of the number of connected devices that emerge in the coming years, there is no doubt that the IoT is around the corner. Millions or billions of things will be equipped with sensor, microprocessors, and radio frequency transceivers to sense the environment, analyze the gathered data, communicate with each other, and enable smart applications. Smart driving, Smart cities, Smart Grids, and Smart Metering are just some key examples of emerging applications that will create a revolution in the very near future. However, still many challenges need to be faced before these applications can become commodities. One of them is ensuring almost zero-power operation of communications. The fact that these devices will be too many and possibly not reachable once deployed claims for autonomous operation. Even though it is improving, today's communication technologies are still power hungry, and this need to be changed. The design of green technologies that can enable long-lasting autonomous operation of communication devices constitutes one of the main challenges to be faced. In this chapter, we have discussed how cellular networks can become key players in providing the IoT with the requested connectivity. Of course, these networks will be extended and complemented with other short-range technologies, but cellular networks have inherent benefits that must be exploited. Even though emerging cellular standards (LTE-Advance) are very flexible and powerful, still optimization for M2M is necessary. DRX (duty-cycling), cooperative communications, and D2D communications have been identified as promising techniques to tailor the operation of cellular networks to the needs of M2M communications. The IoT is just around the corner and technology must be ready to face this new challenge; green communications are key to facilitate real smart applications.

Acknowledgements

The authors of this chapter would like to acknowledge and express their gratitude to the contribution made by Kun Wang during his affiliation to the Centre Tecnològic de Telecomunicacions de Catalunya (CTTC) in the organization and writing of this chapter.

The work presented in this paper has been partially funded by the research projects ADVANTAGE (FP7-607774), P2P-Smartest (H2020-646469), and by the Catalan Government under grant (2014-SGR-1551).

References

[1] International Telecommunication Union "ITU Internet Reports 2005: The Internet of Things; Executive Summary," November 2005.
[2] Smart 2020 "Smart 2020, Enabling the low-carbon economy in the information age," The Climate Group, London, June 2008, www.smart2020.org. [Accessed 10 January 2015].
[3] 3GPP TS 22.368, "Service requirements for Machine-Type Communications (MTC)," Stage 1, (Release 12), 2014.

[4] N. Lomas, "Online Gizmos Could Top 50 Billion in 2020," 29 June 2009, http://www.businessweek.com /globalbiz/content/jun2009/gb20090629/_492027.htm. [Accessed 10 January 2015].

[5] ETSI TS 102 935, "Machine-to-Machine communications (M2M); Applicability of M2M architecture to Smart Grid Networks; Impact of Smart Grids on M2M platform," V2.1.1, September 2012.

[6] ETSI TS 102 898, "Machine-to-Machine communications (M2M) M2M Use cases of Automotive Applications in M2M capable networks," V1.1.1, April 2013.

[7] ETSI TS 102 691, "Machine-to-Machine communications (M2M); Smart Metering Use Cases," V1.1.1, May 2010.

[8] ETSI TS 102 689, "Machine-to-Machine communications (M2M); M2M service requirements," V1.1.1, October 2010.

[9] ETSI TS 102 690, "Machine-to-Machine communications (M2M); Functional architecture," V1.1.1, November 2011.

[10] C. S. Bontu and E. Illidge, "DRX mechanism for power saving in LTE," IEEE Commun. Mag., vol. 47, no. 6, pp. 48–55, 2009.

[11] K. Zhou, N. Navid, and S. Thrasyvoulos, "LTE/LTE-A discontinuous reception modeling for machine type communications," IEEE Wireless Commun. Lett., vol. 2, no. 1, pp. 102–105, 2013.

[12] 3GPP TS 36.321, "Evolved Universal Terrestrial Radio Access (E-UTRA) and Evolved Universal Terrestrial Radio Access Network (E-UTRAN); Medium Access Control (MAC) protocol specification," 2012, (Release 11).

[13] Y. Yu and K. Feng, "Traffic-based DRX cycles adjustment scheme for 3GPP LTE systems," in IEEE 75th Vehicular Technology Conference (VTC Spring), Yokohama, May 2012.

[14] T. Tirronen, A. Larmo, J. Sachs, and B. Lindoff, "Reducing energy consumption of LTE devices for machine-to-machine communication," in IEEE Globecom Workshops, pp. 1650–1656, Anaheim, CA, December 2012.

[15] S. Jin and D. Qiao, "Numerical analysis of the power saving in 3GPP LTE advanced wireless networks," IEEE Trans. Veh. Technol., vol. 61, no. 4, pp. 1779–1785, 2012.

[16] J. Liang, J. Chen, H. Cheng, and Y. Tseng, "An energy-efficient sleep scheduling with QoS consideration in 3GPP LTE-advanced networks for internet of things," IEEE J. Emerging Sel. Top. Circuits Syst., vol. 3, no. 1, pp. 13–22–, 2013.

[17] 3GPP TS 36.213, "Evolved Universal Terrestrial Radio Access (E-UTRA); Physical layer procedures," 2012, (Release 11).

[18] K. Wang, J. Alonso-Zarate, and M. Dohler, "Energy-Efficiency of LTE for Small Data Machine-to-Machine Communications," IEEE ICC 2013, June 2013.

[19] 3GPP TS 36.211, "Evolved Universal Terrestrial Radio Access (E-UTRA) and Evolved Universal Terrestrial Radio Access Network (E-UTRAN); Physical Channels and Modulation," 2012, (Release 11).

[20] J. Alonso-Zarate, E. Kartsakli, L. Alonso, and Ch. Verikoukis, "Performance analysis of a cluster-based MAC protocol for wireless Ad Hoc networks," EURASIP J. Wireless Commun. Networking, vol. 2010, Article ID 625619, p. 16, 2010.

[21] J. Alonso-Zarate, L. Alonso, and Ch. Verikoukis, "Performance analysis of a persistent relay carrier sensing multiple access protocol," IEEE Trans. Wireless Commun., vol. 8, no. 12, pp. 5827–5831, 2009.

[22] R. Y. Kim, "Snoop based group communication scheme in cellular Machine-to-Machine communications," Information and Communication Technology Convergence (ICTC), pp. 380–381, Jeju, November 2010.

[23] 3GPP TS 36.927, "Potential solutions for energy saving for E-UTRAN," (Release 11), 2012.

[24] 3GPP TS 23.866, "Study on System Improvements for Energy Efficiency," (Release 12), 2012.

[25] T. Taleb and A. Kunz, "Machine type communications in 3GPP networks: potential, challenges, and solutions," IEEE Commun. Mag., vol. 50, no. 3, pp. 178–184, 2012.

[26] K. Ko, M. Kim, K. Bae, and D. Sung, "A novel random access for fixed-location machine-to-machine communications in OFDMA based systems," IEEE Commun. Lett., vol. 16, no. 9, pp. 1428–1431, 2012.

[27] T. Nguyen, O. Berder, and O. Sentieys, "Energy-efficient cooperative techniques for infrastructure-to-vehicle communications," IEEE Trans. Intell. Transp. Syst., vol. 12, no. 3, pp. 659–668, 2011.

[28] C. Pu, S. Lim, and P. Ooi, "Measurement arrangement for the estimation of path loss exponent in wireless sensor network," in Computing and Convergence Technology (ICCCT), pp. 807–812, Seoul, December 2012.

[29] H. Chen and M. H. Ahmed, "Performance analysis of cooperative-diversity wireless systems with adaptive modulation and imperfect channel state information," in Wireless Communications and Mobile Computing Conference (IWCMC), pp. 1698–1703, Istanbul, July 2011.

[30] T. Predojev, J. Alonso-Zarate, and M. Dohler, "Energy evaluation of a cooperative and duty-cycled ARQ scheme for machine-to-machine communications with shadowed links," in Proceedings of the IEEE International Symposium on Personal Indoor and Mobile Radio Communications (PIMRC), London, United Kingdom, 2013.

[31] S. S. Ikki and M. H. Ahmed, "Performance analysis of incremental-relaying cooperative-diversity networks over rayleigh fading channels," IET Commun., vol. 5, no. 3, pp. 337–349, 2011.

[32] G. Botter, J. Alonso-Zarate, L. Alonso, and F. Granelli, "Extending the lifetime of M2M wireless networks through cooperation," in IEEE ICC 2012, June 2012.

[33] A. Laya, K. Wang, L. Alonso, and J. Alonso-Zarate, "Multi-radio cooperative retransmission scheme for reliable machine-to-machine multicast services," in Proceedings of the IEEE International Symposium on Personal Indoor and Mobile Radio Communications (PIMRC), pp. 1–6, Sydney, NSW, September 2012.

[34] T. M. Cover and A. A. El Gamal, "Capacity theorems for the relay channel," IEEE Trans. Inf. Theory, vol. 25, no. 5, pp. 572–84, 1979.

[35] 3GPP TS 36.912, "Feasibility study for Further Advancements for E-UTRA (LTE-Advanced)," 2012, (Release 11).

[36] M. Belleschi, G. Fodor, and A. Abrardo, "Performance analysis of a distributed resource allocation scheme for D2D communications," IEEE GLOBECOM Workshops, pp. 358–362, Houston, TX, December 2011.

[37] 3GPP TR 22.803, "Feasibility study for Proximity Services (ProSe)," (Release 12), 2014.

[38] UN-HABITAT, "State of the World's Cities: Prosperity of Cities," 2012/2013.

[39] A. Caragliu, Ch. Bo, and P. Nijkamp, "Smart cities in Europe," 2009.

12

Energy Saving Standardisation in Mobile and Wireless Communication Systems

G. Punz, D. C. Mur and Konstantinos Samdanis
NEC Europe Ltd, Heidelberg, Germany

12.1 Introduction

As should have become apparent from the previous chapters, energy saving (ES) has to be considered as a system-wide optimization effort, and for distributed systems like wireless/mobile networks, where many components of potentially different vendors have to cooperate, functioning together in an interoperable way, standardization is essential. In this section, we focus mainly on ES-related efforts of the 3rd Generation Partnership Project (3GPP) and Institute of Electronic and Electrical Engineers (IEEEs) standardization bodies. 3GPP is a worldwide organization defining the most important and successful standard for mobile communication. The brand names of UMTS (also known as "the" 3G technology) and its "Long Term Evolution" (LTE, nowadays taken synonymously for 4G) are well known. They also maintain the specifications of predecessor technologies GSM and GPRS. Besides 3GPP, other complementary organizations among the most noticeable ones, the Next Generation Mobile Network (NGMN) Alliance and GSM Association (GSMA) aim to deliver the mobile operator perspective to other standardization bodies and into the telecommunications industry. Specifically, NGMN Alliance is an operator-driven forum concentrating on mobile broadband technologies and particularly on LTE and its evolution. GSMA is a mobile operator's interest group, which enables the advancement of mobile technology by coordinating with regulatory bodies, public policies, device and network infrastructure industry.

The IEEE 802.11 and the Wi-Fi Alliance (WFA) are the two standard organizations responsible for the definition of the Wi-Fi technology including ES-related mechanisms and operations. The two organizations work in a complementary way, where the IEEE

Green Communications: Principles, Concepts and Practice, First Edition.
Edited by Konstantinos Samdanis, Peter Rost, Andreas Maeder, Michela Meo and Christos Verikoukis.
© 2015 John Wiley & Sons, Ltd. Published 2015 by John Wiley & Sons, Ltd.

802.11 group is technology-oriented with the aim of developing standards, and the WFA is market-oriented and leverages the technical specifications defined by the IEEE 802.11 group in order to create certifications and test plans defined around market segments involving the Wi-Fi technology. Alongside standards bodies that shape ES technologies like 3GPP and IEEE, other standards organizations including the European Telecommunications Standards Institute (ETSI) and the Alliance for Telecommunication Industry Solutions (ATISs) specify measurements procedures and ES metrics for evaluating energy efficiency for telecommunication systems. ETSI introduced the Green Agenda to address ES in several sectors of Information and Communications Technologies (ICTs) including mobile, radio and broadcast, specifying measurements and metrics, which accommodate certain regulations from the European Commission. ATIS is a North American partner of the 3GPP, which created the Green Initiative that focuses on power consumption metrics and standards.

The remaining of this chapter is organized as follows. Initially, an overview of the early NGMN ES activities is provided and then a detailed overview of 3GPP is presented, analysing the main contributions across the different working groups. A summary of ES efforts in GSMA is then presented documented, followed by a brief analysis of the main activities on ETSI and ATIS. Finally, IEEE 802.11 and Wi-Fi Alliance are analysed before deriving the conclusions.

12.2   Next Generation Mobile Networks (NGMN)

NGMN launched one of first efforts for addressing ES in mobile networks as a part of the Self-Organizing Network (SON) study, which analysed business requirements focusing on network management for LTE. Such a study opened the horizons for 3GPP to specify technical requirements for SON functions and initiate ES work items, as analysed later in the corresponding 3GPP section. ES SON parameters and requirements are briefly introduced in Ref. [1], with the objective to provide a close match of the operator's offered capacity to the traffic demand, minimizing the energy consumption by powering off certain base station (or evolved Node B (eNB) in 3GPP terminology) components, for example, RF resources and circuits, pilot channels, and so on, or even power-off entire eNBs. The mechanism considered is based on a threshold policy, which activates and releases eNB resources enabling energy efficiency according to traffic load variations or QoS experienced by the end users.

A simple example that elaborates such threshold policy is illustrated in Figure 12.1 showing how the activation and release of resources is performed in relation with traffic load variations. When the network utilization satisfies a predetermined threshold for a sufficient time interval, then the appropriate resource release or activation takes place. Periodic checking, for example, every 15 minutes, or on per flow basis are the two options for examining the shortage or excess of resources.

The Operational Expenditure (OPEX) and network management aspects for enabling ES in LTE, providing a set of best practices is investigated in Ref. [2], assuming network elements can support a standby mode with the potential to be powered on and powered off remotely by a management system, which follows an autonomous capacity-driven ES optimization. A set of recommendations were introduced for network elements and for the element manager considering traffic reporting, neighbour sites updates, dynamic energy policy provision for configuring thresholds and policy rules, and interaction with other SON functions including alarm suppression, resiliency and heterogeneous network issues. As a result, potential standardization objectives were derived for physical layer broadcast channels, performance

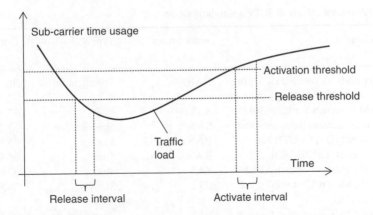

Figure 12.1   Energy efficiency triggering for releasing and activating resources at the eNB [1]

management reporting in terms of traffic or radio capabilities, cell coverage characteristics and configuration management for controlling network element functions, that is, power on/off, change tilt, control air conditioning units and power amplifier, and so on.

12.3 3rd Generation Partnership Project (3GPP)

3GPP is structured in many different work groups, which progress in parallel in a co-ordinated manner focusing on a particular aspect of the mobile network. An overview of the different working groups is depicted in Figure 12.2, highlighting the topics related to ES, which have been investigated so far.

3GPP integrates the work handled by different workgroups by issuing a set of parallel system-wide "releases," which provide a stable platform for implementation, while allowing the addition of new features in a coordinated manner. Table 12.1 provides an overview of ES-related activities in 3GPP standardization per release by listing the topic/goal, the work group and the documented results. It should be noted that for a comparative view between 3GPP releases, always the (latest) version of the specification/technical report per release should be considered. Technical reports indicated by "TR" are informative documents, which aim to provide use cases and technical content, while specifications represented by "TS" are

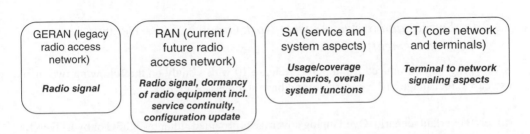

Figure 12.2   3GPP workgroups involved in energy saving (with the related topics)

Table 12.1 ES related efforts in 3GPP's standardisation

3GPP release	Topic/goal	Work group	Type	Documentation
9	Inter-eNB ES for E-UTRAN (by cross-eNB signalling)	RAN3	Specification	36.300 [3]
10	OAM scenarios for ES in RAN	SA5	Study	32.826 [4]
	ES management (requirements)	SA5	Specification	32.551 [5]
	ES in NB (UMTS/UTRAN)	RAN1, RAN3	Study	25.927 [6]
	ES in eNB (LTE/E-UTRAN)	RAN3, RAN2	Study	36.927 [7]
11	OAM aspects of inter-RAT ES	SA5	Study	32.834 [8]
	ES impacts on UE to NW signaling	CT1	Study	24.826 [9]
	Additional requirements on inter-RAT ES and probing	SA5	Specification	32.551 [5]
12	System improvements for ES	SA2	Study	23.866 [10]
	UE power Optimization	SA2	Study and specification	23.887 [11]; 23.401 [12]
	Network ES in E-UTRAN	RAN3	Specification	36.300 [3], 36.413 [13], 36.423 [14], 25.413 [15]
	ES Enhancements for E-UTRAN	RAN3	Study	36.887 [16]

normative documents. Typically, a technical report precedes a specification document, which is initiated once consensus is gained after the completion of a corresponding informative technical study.

In 3GPP, the work on ES was initiated mainly from the RAN workgroups and from the network management workgroup SA5. The motivation was based on observations that the permanent operation of the radio part of a radio base station consumes a considerable amount of the overall power [4], even during off-peak hours, where the utilization factor is quite low. Switching off equipment, especially radio parts, is thus a natural goal, but not a trivial task: the service should be guaranteed for the users remaining active with sufficient quality. Such operations were not foreseen in the original design.

12.3.1 Service and System Aspects Work Group 5 (SA5 – Network Management)

SA5 workgroup introduced ES in 3GPP Release 10 with a study on the following two main use cases, focusing on LTE radio cells and base stations (i.e. eNBs):

(a) eNB overlaid scenario: Overlapping coverage is provided either by 2G/3G or by LTE eNBs with different carrier frequency bands, as it would be the case in a combination of macro and pico/micro cells within the same area (see Figure 12.3(a)). Sub-variants are defined either as carrier restricted, that is, only some carrier frequencies of an eNB are subject to ES, or eNB restricted, that is, the eNB radio part is completely switched off.

(b) Capacity limited network: Here eNBs are classified into the ones that are subject to be switched off during off-peak periods and into the eNBs that provide coverage compensation when the former are switched off, that is, are in ES state (see Figure 12.3(b)). The focus of this use case is on homogeneous environments, that is, pure LTE, considering base stations with dynamic cell coverage capabilities. At peak times, base stations are assumed to be configured with a smaller coverage area per cell than the potential maximum one for providing increased capacity, while once the traffic is low they may exploit their full coverage for ES purposes allowing neighbouring base station to enter an ES state, that is, be switched off.

Following the initial ES study [4], the workgroup concluded that a clear differentiation regarding whether the ES mechanism is performed within a radio access technology (intra-RAT) or across RATs (inter-RAT) is essential. To address the latter, SA5 performed a detailed study and documented the results in 3GPP TS 32.834 [8]. The different combinations of RATs considered are given in Table 12.2, with the main focus concentrating on combinations 2 and 3. It is worth noting that CDMA2000 is defined by 3GPP2 and thus is not a genuine 3GPP RAT, but from Release 8 onwards it is treated with an especially high degree of integration into the 3GPP core network.

Inter- and intra-RAT ES methods can be performed alternatively or in combination. For instance, in the late evening, first intra-RAT ES for LTE (i.e. for eNBs) could be activated, reducing slightly the offered capacity, then – at night when user activity has dropped further

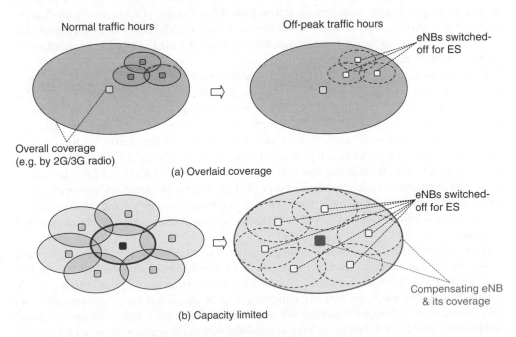

Figure 12.3   3GPP SA5 fundamental ES use cases

Table 12.2   Combinations of RATs for inter-RAT ES

RAT combination	RAT 1 (providing base coverage)	RAT 2 (target for ES)
1	GSM	UMTS
2	GSM	LTE
3	UMTS	LTE
4	CDMA2000	LTE

to a sufficiently low level – inter-RAT ES based on 3G overall coverage may follow, resulting in a complete switch-off of LTE eNBs. The reverse two-stage process could take place in the morning. The relative priority and optimal sequence of ES activation and deactivation is subject to an operator-defined policy, which typically aims to maximize the global network ES performance without compromising user's service quality. A crucial point to consider for deploying inter-RAT ES is that some UEs may not (i.e. not yet, or no longer) support a certain RAT, hence switching off a particular RAT may cause service loss for such UEs during the ES period.

The switch-off times of particular cells may in the simplest case be pre-scheduled. In a more advanced deployment, the switching on/off of cells and the modification of radio parameters for adapting cell's coverage should be dynamically determined and controlled, depending on traffic loads. These network management operations are desired to be autonomous and hence can naturally become a feature of SON (self-optimizing/organizing network) functions. Specific ES policies in terms of cell load thresholds (separately for intra and inter-RAT ES) need to be defined in order to trigger the activation and deactivation of ES in a network element. Additionally, related timers have to be determined, to regulate the ES activation/de-activation triggering based on stable load conditions, that is, when the load has been above/below the respective threshold for long enough.

The ES study also outlined three different alternatives for Energy Saving Management (ESM). All three alternatives rely on network-wide traffic statistics that enable the OAM to determine a stable low load period in order to enable/disable ESM procedures on particular parts of the network, ensuring that ES is profitably applied on network elements, avoiding local load minima and oscillations, that is, switching on/off cells frequently. With centralized ESM, the ES-related algorithms are performed in the (centralized) OAM system, whereas with distributed ESM this occurs at the network element level. In hybrid ESM, the ES-related algorithms are executed in a combined way on both OAM system level and on network element level, or only on network element level with a complementary centralized supervision, for example, for conflict resolution or prioritization using global information on certain network elements.

Obviously, with ES, some state modelling is required for base stations (and, in an aggregated view, also for the network); the two main states for configuration and control are "in ES" and "not in ES". The latter is equivalent to a conventional situation without introducing ES. For the purpose of coverage/capacity compensation, an additional state "compensating for ES" is needed, which applies only for intra-RAT ES. In ES state, a radio station's hardware components shall be switched off as long as possible, with the exception during an ES probing procedure (which is elaborated later), in where the radio transmitter is active for a minimum time. To ensure that eNBs can be switched on again, a minimum of hardware and related software needs to run constantly. A cell in ES state does not provide any service to UEs, so it is normally not "visible" to a UE, except during ES probing.

The procedure of "probing for ES" has been added in Release 11 for supporting further ES optimization. During this procedure, an eNB allows a particular cell to indicate its presence in order to collect UE measurements, but prevents idle mode UEs from camping on the cell and prohibits incoming handovers. In this way, the eNB can use such measurements to determine whether a cell in ES state has enough UEs within its reach to take over a certain amount of load by returning into the conventional "not in ES" state.

12.3.2   Radio Access Network Working Groups (RAN 1, RAN 2, RAN 3)

The RAN working groups initiated the first activities regarding ES in Release 9 concentrating on E-UTRAN, assuming network deployments with capacity booster cells, which can be clearly distinguished from the cells providing basic coverage. The rationale was to optimize energy consumption by allowing an eNB to switch off, that is, putting into dormant state, a so-called capacity-booster cell under its control, when its capacity is no longer needed and to reactivate it on demand (see Section 22.4.4 in Ref. [3]). Such a decision may be taken by the eNB based on cell load information and configuration data, or alternatively by the OAM system. The eNB is able to move out active mode UEs by enforcing handovers towards neighbouring cells, before the selected cell becomes dormant and can then inform all neighbouring eNBs by initiating an "eNB Configuration Update" procedure via X2 interface (see Section 8.3.5 in Ref. [14]). These peer eNBs keep the corresponding cell configuration data, that is, neighbour relationship, also during a cell's dormancy, and they can request the reactivation of a dormant cell via the cell activation procedure when the load increases.

In Release 10, ES problems, considerations and possible solutions were studied more intensively. First, the consumption characteristics of a UTRA base station (NodeB) were analysed in Ref. [6]; not as a surprise, the radio equipment and the base band unit have been found as the main power consumption sources. The following four solutions that address ES considering RAN characteristics have been analysed in Ref. [6]:

(a) Dormant mode (already utilized in Release 9, see above): this solution concentrates on switching off one or more carrier frequencies with the objective to reduce power or completely shut off a Node B. As long as UEs can still find carrier bands, the existing specifications would guarantee uninterrupted service, for example, by triggering handovers.
(b) Secondary antenna deactivation (in case of MIMO, usage of multiple antennas for the sake of higher transmission rates): It is obvious that ES is possible if parts of the antenna can be switched off depending on the load. MIMO usage in both terminals and base stations can be controlled fully, so no negative side effects are expected from the deployment of this solution. Effectively, this seems just a selective step back from the energy-hungry MIMO enhancements to a single antenna mode of transmission.
(c) Power control of common channels: In UMTS, the transmission power of common channels is fixed, configured according to the cell coverage. Hence, common channels are broadcast constantly all the way to the cell edge even when there are no UEs receiving them. Through power control, common channels may dynamically change, for example,

be reduced, ensuring though that all active UEs are in the coverage range excluding UEs in idle mode, which may be out of coverage. As a result, the total power consumption can be reduced. However, such an approach introduces problems with UEs around the cell edge especially in case of mobility or transition from idle to active mode.

(d) Cell DTX: Concentrates on deactivating the transmitter of the base station periodically, for example, on sub-second level. As the power amplifier consumes a considerable portion of energy, ES can be achieved on the order of the fraction inactivity time in relation to the total period. However, UEs would need to adjust to the DTX cycle; this is not possible in some of their sub-states, that is, in idle mode and not for legacy UEs (i.e. such a solution would create backwards compatibility issues).

Another, parallel study was performed for E-UTRAN in Ref. [7], focusing on the dormant mode for inter-RAT and inter-eNB ES, while for intra-eNB ES, an optimization on radio frame level was analysed. These efforts are further advanced in Release 12, detailed in the corresponding technical report TR 36.887 [16]. The target scenarios include ES in overlaid deployments and the LTE coverage layer scenario, which enables a cell switching off provided that coverage compensation is feasible by neighbour cells. The enhancements in the overlaid scenario are mainly QoS considerations, including the case where different overlaid cells may offer a distinct QoS, and cases where the UE QoS demand is considered when switching on/off particular cells. In the LTE coverage layer scenario, a set of additional goals were introduced to avoid compensation when not necessary, reduce interference and consider UE QoS, alongside specific coverage compensation scenarios including a single compensating eNB and multiple compensating eNBs.

12.3.3   Architecture Working Group 2 (SA2)

3GPP SA2 initiated a study on ES documented in TR 23.866 [10], regarding system-level architecture in Release 11 but completed only in early Release 12. The scope of the study was on deployment aspects (primarily in the Packet Switching (PS) domain), including potential system enhancements to support energy efficiency. As such, it was a first step into ES within the mobile core network, despite the fact that its current share of energy consumption is not major; this makes sense by anticipating that an increase of demands for resources in the mobile core network is forecasted over the next years, due to trends like machine-type communication and evolving social applications. In other words, the core network is expected to contribute a considerable share to the overall energy consumption unless the ES of the RAN is coordinated with the core network aligning the energy consumption on a user basis. The following four main deployment scenarios were considered in this study (but these have not yet found their way into normative specifications):

1. Pooled deployment of MMEs: When user plane network entities are deployed more distributed, that is, topologically closer to the RAN, it is possible to optimize the routing
   and reduce the required transmission resources. In contrast, control plane network entities, that is, MMEs, are deployed in centralized pools in order to take advantage of the
   pooling effects, which include better scaling and load sharing. Effectively, a drawback of
   such a pooled deployment is that considerable signalling traffic may occur between eNBs,
   S-GWs and MMEs. System enhancements to minimize the signalling drawback of the

pooled MMEs would be required, either enabling the S-GW to handle service requests without always the need of MME or enhance the management of the connected mode by the RAN, but such proposals are still early, on a conceptual basis.

2. Load redistribution during off-peak times: ES-favourable NW node utilization through load redistribution during off-peak times at the EPS assumes that the operator has configured pools of MMEs and/or S-GWs to serve the E-UTRAN. These resource pools are typically dimensioned for peak loads, that is, the number of S-GWs, P-GWs and MMEs is chosen according to the expected maximum number of UEs and their traffic demand. But most of the time, the load in the network is far from the dimensioned peak rate, thus operators can expect significant ES opportunities by setting unnecessary resources/nodes in the core network to an energy conservation mode. The net benefit depends here on the non-linearity of energy consumption with load, since a certain portion of energy is consumed already with zero load.

3. Network sharing: This is a standard way of reducing the operator's cost by cooperation. In rural areas with lower population density, mobile networks may need to provide coverage but already a minimum configuration could offer more resources than actually needed. Hence, network sharing could help realizing ES as a part of overall base station related CAPEX and OPEX. In urban, more densely populated areas, network sharing could also be performed dynamically, during low traffic hours; in this case, the focus is more on OPEX. Specifically, during peak times, the Public Land Mobile Networks (PLMNs) may use their own capacity and coverage, but at off-peak times, every PLMN may utilize only a low-capacity portion. Hence, network sharing could be used at these low traffic hours to improve energy efficiency by combining traffic of multiple PLMNs providing a lower variance than the traffic of each individual PLMN.

4. Scheduled communications: This feature aims to enhance inter-RAT ES developed in RAN addressing scenarios wherein devices cannot use another RAT. There may be devices (e.g. for Machine Type Communication (MTC)) that can support a single RAT alone and need to communicate only during regular intervals. The network could determine whether in particular areas there are only such single RAT devices that do not need to communicate constantly, or their communication has a low priority and schedule the activation/de-activation of RATs accordingly, while more mission-critical devices may be assumed to have multi-RAT capabilities.

In Release 12, optimizations for UE power consumption were also studied and documented in TR 23.887 [11] as a part of the mobile network improvements for MTC. Since the number of devices is expected to dramatically increase, their accumulated power consumption has to be taken seriously into consideration, apart from that fact that restrictions in power supply may apply in many cases. For MTC devices, SA2 has analysed the following two solutions:

1. Exploitation of (very) long DRX cycles: In this solution, the maximum DRX cycles in idle mode are extended, permitting the UE to save battery, as waking up and listening for a potential paging message is a major power consumption source. The paging transmission period should also be adjusted according to the extended DRX cycle applied to the UE. Indicating and applying extended DRX may be eNB or UE driven with MME assistance for configuring extended DRX and adapting paging. Hence, support of extended DRX from UE, RAN and core network is essential.

2. Addition of a new power saving state: UEs may support a "power saving state," a state wherein UEs enter based on a timer after transiting to idle state. During the "power saving state" UEs remain attached, that is, active PDP/PDN connections remain established, but halt cell/RAT/PLMN selection as well as NAS procedures. However, UEs still perform periodic registrations (RAU and TAU) following a timer value, which is provided by the network.

The system-level concrete specification of these ES features can be found in TS 23.401 [12].

12.3.4   User Equipment: Core Network Signalling Working Group (CT1)

3GPP's CT1 workgroup has performed studies on the impacts of ES on the signalling between UE and core network as documented in TR 24.826 [9], but note that they have not yet lead to normative specification. The UE to core network signalling relevant here are the ones associated to the idle state, that is, registration signalling, whereas active mode is managed by the RAN. In particular, when a base station is switched off, a UE needs to look for another cell to camp on, that is, perform a handover, by tuning to the corresponding radio frequency band and decoding the relevant broadcast information. If the new cell belongs to a different registration area, a re-registration has to be performed. This effect is less pronounced with intra-RAT ES, because it could potentially only happen at the borders of registration areas, and involve only a fraction of the UE population. However, re-registration could result in massive amounts of signalling with inter-RAT ES, that is, when UEs should change RAT, since it affects the registrations of all UEs in the target area, especially when UEs over a larger part of the network attempt to register at the same time. To avoid such a signalling peak towards the core network nodes, an obvious solution is to randomize the registration access or perform a stepwise activation of ES.

The most recent 3GPP system (since Release 8) supports also interworking with access networks of non-3GPP technology, for example, WLAN; if ES is applied to such access networks, handovers will be triggered for the UEs currently using them, either to the 3GPP RAT acting as coverage umbrella or to other non-3GPP access networks available to the UE. The resulting signalling is effectively an attachment in the target RAT and thus includes registration. Hence, depending on the amount of non-3GPP radio access points entering ES mode at roughly the same time, potentially similar issues, that is, peaks of signalling, can occur. Additionally, as handovers for UEs in non-3GPP access with ongoing data transmissions are now managed based on UE to core network signalling, the service experience may be affected.

12.3.5   GSM/EDGE Radio Access Network Working Group (GERAN)

For the legacy radio access GERAN, the activity towards ES solutions so far is limited to a study for intra-BTS ES [17]. The reference configurations for sectored cells of a BTS and load profiles are analysed based on the ES ETSI model documented in Ref. [18]. The proposed ES solution concentrates on dynamically adjusting power for some time slots on the BCCH carrier. Applying this ES scheme in simulations shows that the service quality decreases slightly, while still satisfying the performance target. The call drop rate also increases modestly, but remains in all cases below the 0.2% target. Unfortunately, no concrete values of power saving gains are yet published.

12.4 GSM Association (GSMA)

Since 2010, GSMA provides a Mobile Energy Efficiency Benchmarking service, intended as an aid for mobile network operators to measure and monitor the relative efficiency of their radio access networks. Underperforming networks that waste energy can be detected and potential ESs can be estimated. Meanwhile 35 of MNOs, representing 200 networks and more than half population of mobile subscribers, participate with the benefit that a wealth of data, which has already been collected, can be used for future benchmarking, although in an anonymized form.

In creating a benchmark, which represents a statement of relative performance for a particular type of networks, it is most useful to consider a set of networks with similar base characteristics. The energy efficiency is then tracked over time. Following on the benchmarking, GSMA also promotes an enhanced service for energy optimization since 2011; this is run in collaboration with a partner (vendor or system integrator). Aggregated results are compiled in GSMA's Green Manifesto 2012 [19] in the form of energy consumption per connection and per unit traffic. The Green Manifesto contains energy cost and CO_2 emission figures, including forecasts rates for 2020, while introducing also policy recommendations for governments exploiting the potential of mobile communications to enable reductions in global greenhouse gas emissions.

12.5 European Telecommunications Standards Institute (ETSI)

ETSI has been particularly active in developing standards, accommodating European regulations, for improving energy efficiency in the sector of mobile and wireless communications and introduced the Green Agenda as a strategic topic since 2008, which includes the following major categories:

(a) Definitions and reference models for energy efficiency, energy consumption measurements and Key Performance Indicators (KPIs), methods of measuring energy efficiency.
(b) Specifications, recommendations and best practice guidelines for energy efficiency.
(c) Life cycle assessment for telecommunication equipment.
(d) Environmental considerations for telecommunication equipment and installation, power supplies including alternative power sources.

Among ETSI technical committees, the main ones involved in the field of energy efficiency are the Environmental Engineering (EE), which mainly defines energy efficiency, measurement methods and indicators as well as the Access, Terminals, Transmission and Multiplexing (ATTM) that handles broadband technologies including wireless access radio and mobile networks. The EE technical committee has introduced energy efficiency considering wireless access equipment in TS 102.706 [18], providing the criteria for assessing the power consumption of GSM/EDGE, WCDMA, WiMaX and LTE. This specification primarily models the energy consumption of base stations, including integrated and distributed scenarios, base station sites and for the case of GSM the network-level energy efficiency. Additionally, it introduces power consumption measurement methods to address static and dynamic coverage and capacity measurements.

Extensions of such measurements are introduced in Ref. [20], which provides an analysis based on real network characteristics for GERAN, UTRAN and E-UTRAN. It also specifies as energy efficiency metrics: (i) the successfully transferred data volume per unit time over consumed power and (ii) the number of users over consumed power considering the perceived QoE. Further guidelines for performing energy measurements are also included, concentrating on area measurements, observation times and QoE estimations. Energy measurements regarding base stations are performed directly using a power meter or via supplying RF power measures to an estimation model. The user QoE is determined considering throughput measurements or via packet inspection; alternatively terminal probing or user reporting could inform the network about the perceived QoE.

Energy efficiency measurements for core network equipment are specified on ES 201.554 [21], considering UMTS and E-UTRAN. The specification defines the notion of power consumption and energy efficiency as well as the metrics that should be considered for MGW, HLR, AUC, EIR, MSC, GGSN, MME, SGW and PGW, while in later revisions the radio access control nodes and IMS core are planned to be added. The conditions and procedures for performing energy measurements are also introduced including the system configuration, KPIs and traffic parameters. The relation of energy efficiency subject to specific KPIs is further studied in Ref. [22] to enable consistent monitoring and network assessment. Such KPIs relate power with bandwidth and transmission distance considering also on-demand or varied services. For wireless access, [22] introduces the Energy Efficiency Factor (EEF) to relate the energy usage with the network capability for achieving a given data rate within a certain coverage area, while it also provides measurement recommendations for a broad range of wireless technologies including Bluetooth, Zigbee, WiFi, WiMaX, fixed wireless links, GSM, GPRS, UMTS and 4G.

ETSI also specified the European Standard EN 301.575 [23] concentrating on measurement methods for energy consumption of Customer Premises Equipment (CPE) and test conditions for end-user broadband equipment, that is, CPE (WLAN), under a series of power states/modes including disconnected mode, off mode, standby, idle state, low power state and on mode, conforming the EU regulation 1275/2008. Life cycle assessment considerations for base stations, network equipment and services are analysed in Ref. [24] taking into account equipment material, production and development, operation and support as well as the end of life treatment phases of a product life cycle. The use of alternative energy solutions in telecommunication installations is documented in [25], considering various alternative energy sources and paradigms to power on network equipment, analysing the life cycle of solar power system and diesel generator battery hybrid solutions for GSM base stations.

12.6 Alliance for Telecommunication Industry Solutions (ATIS)

ATIS launched the Green Initiative in 2008 and as a part of it the Exploratory Group on Green (EGG) explored environmental sustainability in wireless networks. Its fundamental contributions regarding energy efficiency concentrates on measurements practices and metrics. Specifically, ATIS has introduced in Ref. [26] the Telecommunications Energy Efficiency Ratio (TEER) metric to measure and report the radio base station energy efficiency. TEER is defined as a ratio of equipment performance to energy consumption with respect to data and voice, that is, it addresses base station throughput per Watt of input power. ATIS has also defined a methodology to determine a RF power efficiency ratio as the radiated RF power to the consumed power taking in account three different test locations, that is, near, middle and cell edge, considering a weighted factor that corresponds to a variable load level for each of the three test locations.

Besides the definition of energy efficiency metrics, [26] specifies methods, processes and system configurations for measuring the energy efficiency of LTE, CDMA, EV-DO, UMTS, GSM and WiMaX, with the intention to compare base stations with the same radio technology. ATIS has also published a study item, that is, not a specification, which captures the industry efforts for providing energy efficiency in wireless networks [27]. In particular, it documents base station and cell site energy efficiency considerations focusing on hardware optimizations, management mechanisms and empowering methods, including alternative power at cell sites. The study also analyses network design considerations regarding topology, access network technology, redundancy and network dimensioning and optimization.

12.7   IEEE 802.11/Wi-Fi

The Wi-Fi industry is currently steered by two major standards organizations that work in a cooperative way, namely: (i) the IEEE 802.11 Working Group, carrying out the task of defining technical specifications and (ii) the Wi-Fi Alliance (WFA), which takes the specifications developed by IEEE 802.11 and defines certification programs. Throughout this section, we refer to the technology developed by IEEE 802.11 and WFA simply as Wi-Fi technology. Unlike 3GPP, these standard organizations do not define the full system architecture but instead limit their efforts to the air interface between the Access Point (AP), that is, corresponding to eNB in 3GPP, and the station, that is, the device capable to use IEEE 802.11, equivalent to UE in 3GPP. Consequently, all ES-related extensions developed for Wi-Fi apply to the air interface only.

Currently, the main efforts to achieve ES in Wi-Fi have been oriented towards extending the battery life of stations, especially for smartphones or other battery handheld devices for which the Wi-Fi interface consumes a significant part of their total energy. For instance, a recent study demonstrates that when the Wi-Fi interface is continuously active, a modern smartphone has a battery lifetime for only 5–10 hours. Therefore, IEEE 802.11 has defined several power saving protocols that mitigate this problem by allowing the device to set its interface in a sleep state when no data needs to be transmitted. These protocols are described in Section 12.7.1.

Wi-Fi APs have traditionally been low-power base stations generally operated by independent entities, for example, private users. As operators start to integrate Wi-Fi hotspots as part of their mobile infrastructure, reducing the energy consumption of APs becomes a significant target. Section 12.7.2 introduces existent functions that can be used for this purpose. Finally, as already mentioned when discussing ES efforts in 3GPP, an emerging trend that requires further energy efficiency enhancements for Wi-Fi radios is MTC, where large number of small and battery-powered sensors/actuators are connected to an AP in order to communicate with a remote MTC server. Improvements to enable MTC, currently under discussion in the IEEE 802.11 working group, are described in Section 12.7.3.

12.7.1   Mechanisms to Extend the Station's Battery Life

The baseline 802.11 specification [28] introduces a legacy power save mode that was designed to save power for stations with non-time-sensitive traffic. The 802.11e amendment developed a protocol that was designed to address the needs of stations with QoS requirements, while

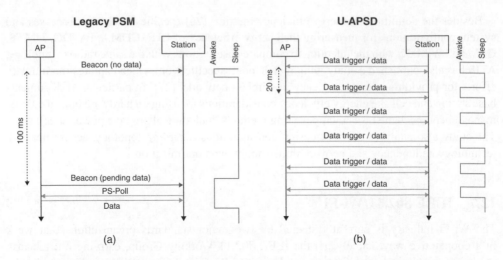

Figure 12.4 Detailed operation of (a) the legacy PSM and (b) the U-APSD protocols

recent amendments like 802.11v [29] contain mechanisms that further improve energy efficiency. In this section, we provide an overview of these mechanisms, highlighting their core ES contributions.

12.7.1.1 Legacy Power Save Mode (PSM)

The legacy PSM enables a Wi-Fi station to save energy by informing the AP. Specifically, a Wi-Fi station enters and leaves the PSM, by sending a frame to the AP with the power management bit set to 1 or 0, respectively. Stations in PSM are free to sleep, while the AP buffers any incoming data addressed to them. Consequently, the AP inserts into the Beacon frame, which is usually transmitted every 100 milliseconds, a signalling element that indicates if there is pending data for the stations in PSM. Stations wake up periodically to receive such a Beacon frame, and upon observing a pending data indication, transmit a so-called Power Save Poll (PS-Poll) frame to the AP that triggers the delivery of the pending data from the AP. The operation of legacy PSM is depicted in Figure 12.4(a).

The buffering delay introduced by the AP, which is up to 100 milliseconds, can impact the performance of regular data applications and especially the performance of applications with specific QoS demands. To address such an issue, state-of-the-art approaches like [30] have proposed extensions to the basic protocol that improve the performance of traditional data applications like Web traffic and long file transfers.

12.7.1.2 Unscheduled Automatic Power Save Delivery (U-APSD)

As already mentioned, the performance of the legacy PSM is critical, especially with applications having tight QoS requirements. For instance, considering Voice over IP (VoIP), the downlink leg of a conversation for a station operating in legacy PSM could potentially introduce a delay in the Wi-Fi link of up to 100 milliseconds. To reduce such downlink delay, the U-APSD mechanism allows stations to proactively poll the AP for data, instead of reactively triggering the AP as in legacy PSM. The operation of the U-APSD protocol is depicted in Figure 12.4(b). It should be noted that in U-APSD it is important for the station to have an accurate knowledge of the instants where data frames should arrive at the AP, as polling when no data is available may penalize energy consumption. To address such an issue, estimations of the inter-arrival times between frames in the multimedia stream could prove helpful. A comprehensive comparison and analysis between the legacy PSM and U-APSD is documented in Ref. [31].

12.7.1.3   802.11v Extensions

The 802.11v amendment [29] specifies a toolbox of mechanisms s for network management that can be used to further improve energy efficiency. The following ones are the most relevant:

(a) Proxy Address Resolution Protocol (ARP): The ARP [32] allows a station to discover the MAC address of another station through its IP address. For a station in power saving mode, waking up to respond to ARP requests sent by other stations could prove to be an energy-intensive process. To resolve this issue, the Proxy-ARP allows a station to register its IP address with the AP and let the AP reply to ARP requests on behalf of the station.
(b) Basic Service Set (BSS) Max Idle Period: The BSS in IEEE 802.11 is defined as a network composed by an AP and a set of associated stations. A common problem in traditional Wi-Fi implementations is that stations are forced to transmit periodic keep-alive messages to the AP in order to avoid being disassociated. The BSS Max Idle Period mechanism can resolve such an energy-inefficient issue, by allowing an AP to advertise towards the associated stations the duration they can sleep without being disconnected, which allows stations to sleep for longer periods.
(c) Traffic Filtering Service: Allows a station to convey a traffic template to the AP in order to inform it about the traffic that is interested to receive. Traffic addressed to the station that is not matching such a template is simply dropped by the AP. Consequently, a station is wakening up only by important traffic, saving energy at all other times.
(d) Flexible Multicast Service: In legacy PSM, APs deliver all broadcast and multicast traffic only at given periodic times known as Delivery Traffic Indication Message (DTIM) intervals, during which all stations must be awake. To provide ES, 802.11v defined the Flexible Multicast Service that allows stations of a given multicast group to request the AP to transmit data addressed to them only at a multiple time of the DTIM interval, permitting longer sleeping.

12.7.2   Reducing the Power Consumption of APs

In traditional Wi-Fi deployments, APs are powered by the electrical grid and their power consumption has not been considered as a significant issue. Therefore, the original 802.11 protocols were designed with the assumption that APs are always operational, that is, awake or powered-on, listening to transmissions coming from the associated stations. However, such an assumption is challenged by the adoption of AP functionality in battery constrained devices,

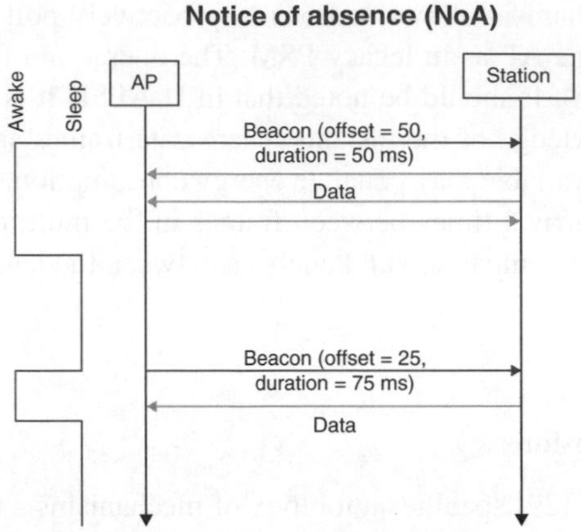

Notice of absence (NoA)

Figure 12.5 Detailed operation of the Notice of Absence protocol in Wi-Fi Direct

that is, smartphones, which is already standardized as documented in the Wi-Fi Direct spec-ification. In addition, large enterprise Wi-Fi deployments create a further challenge wherein aggregate power consumption becomes a relevant issue. This section presents emerging solu-tions to resolve these challenges.

12.7.2.1   Wi-Fi Direct: Enabling Battery-Enabled Devices to Act as APs

Wi-Fi Direct (WFD) is a technology specified by the Wi-Fi Alliance in 2010 [33], which aims to allow direct device-to-device connectivity without the presence of a traditional AP. WFD configures one device to emulate a traditional AP and permits other WFD devices to discover and connect to it. Considering ES, WFD needs to specify energy-efficiency operations for devices that act as APs. To address this issue, the WFD specification defined the Notice of Absence (NoA) protocol that allows an AP to specify certain time intervals where stations are not permitted to transmit, and thus the AP can switch off its radio in order to save power. Figure 12.5 depicts the behaviour of the NoA protocol.

The WFD specification does not specify though how an AP needs to configure its awake and sleep intervals, which can affect the performance of the applications run by the devices connected to such an AP. The interested reader is referred to Ref. [34] for a thorough evaluation of the WFD technology.

12.7.2.2   Energy Efficient Enterprise Wi-Fi Deployments

Recent Wi-Fi enterprise networks are typically dense deployments with a significant impact on energy consumption. To address energy efficiency, the 802.11v specification introduced the following two mechanisms to reduce the power consumption of APs via network management means:

(a) BSS Termination Notification: This mechanism allows an AP to notify its connected sta-tions that the AP is about to shut down, hence enabling the connected stations to gracefully transition to a new AP. The APs to be shut down are determined and controlled on-demand by a network management system.

(b) BSS Transition Management: This mechanism complements the BSS Termination Notification one by allowing an AP to indicate to its connected stations the target APs they could potentially handover. This mechanism can be used both for energy consumption and load balancing.

12.7.3   MTC Energy Saving Enhancements

MTC-specific enhancements for ES are introduced in 802.11ah amendment [35], focusing on MAC mechanisms MTC for low-power devices in the ISM 90 MHz band. Operating below 1 GHz, 802.11ah technology enables APs with very large coverage. Thus, a single AP could assist many MTC devices, in the order of thousands, for example, for remote metering purposes. The support of such a huge number of devices raises several ES issues especially at the MAC layer. The following mechanisms are discussed in the 802.11ah Working Group to address these issues:

(a) Grouping of stations: This mechanism allows an AP to create groups of stations, and assign to each group a specific time interval, wherein the stations of the group are allowed to transmit. This mechanism improves the network efficiency, including ES, by decreasing the contention overhead, or hidden nodes, and also enables stations to sleep when their group is not active.

(b) Target Wake Time (TWT): This mechanism is an extension of the previous one, which allows an AP to communicate to a specific station the time instants, wherein this specific station is allowed to transmit. Like in the grouping case the station can efficiently sleep when it is not transmitting. The TWT parameters are typically negotiated between the station and the AP.

(c) Two-Hop Relay Function: MTC communications can be limited in the uplink by the scarce transmission power of the sensor devices. Thus, 802.11ah allows sensor devices to communicate with the AP through a relay node.

It should be noted that since the 802.11ah specification is still under discussion (at the time when this chapter was compiled), the previously discussed mechanisms are subject to change.

12.8   Conclusions

This chapter provides an overview of the main standardization efforts regarding ES considering both licensed and unlicensed spectrum concentrating on 3GPP and IEEE 802.11/ Wi-Fi Alliance. NGMN created an early momentum for ES in LTE considering network management requirements and OPEX reduction. 3GPP explored the technical details considering network architecture and management, analysing also radio and protocol issues with the fundamental scenarios concentrating on overlaid, capacity limited, that is, pure LTE, and mixed scenarios including different RATs. To facilitate ES, power states were introduced for base stations mainly by the RAN and network management groups. OAM requirements and management processes were also introduced, while the RAN groups explored DTX and MIMO, with current efforts focusing on heterogeneous deployments and on QoS considerations. The architecture group provided an ES study concentrating on core network elements, while the core network signaling analysed the handover processes when ES is applied. Further efforts for ES concentrate on MTC, while efforts influencing ES include RAN sharing and Proximity Services. GSMA provides energy efficiency benchmarking services, with the Mobile's Green Manifesto contributing the latest details. ETSI and ATIS mainly specify ES measurements and metrics, with the EEF and TEER being the most significant ones. IEEE 802.11/Wi-Fi Alliance efforts concentrate on the radio interface introducing power saving methods to conserve the station's battery via PSM, U-APSD and 802.11v extensions and mechanisms to reduce AP power via Wi-Fi Direct and for enterprise environments. Current efforts concentrate on enhancements for MTC, which are still ongoing.

References

[1] NGMN Alliance, NGMN Informative List of SON Use Cases, Deliverable Apr. 2007.
[2] NGMN Alliance, NGMN Top OPE Recommendations, Deliverable Sep. 2010.
[3] 3GPP TS 36.300, Evolved Universal Terrestrial Radio Access (E-UTRA) and Evolved Universal Terrestrial Radio Access Network (E-UTRAN); Overall description; Stage 2, Rel.12, Sep. 2015.
[4] 3GPP TR 32.826, Telecommunication management; Study on Energy Savings Management (ESM), Rel.10, Mar. 2010.
[5] 3GPP TS 32.551, Telecommunication management; Energy Saving Management (ESM); Concepts and requirements, Rel.12, Oct. 2014.
[6] 3GPP TR 25.927, Solutions for energy saving within UTRA Node B, Rel.12, Sep. 2014.
[7] 3GPP TR 36.927, Evolved Universal Terrestrial Radio Access (E-UTRA); Potential solutions for energy saving for E-UTRAN, Rel.12, Sep. 2014.
[8] 3GPP TR 32.834, Study on Operations, Administration and Maintenance (OAM) aspects of inter-Radio-Access-Technology (RAT) energy saving, Rel.11, Jan. 2012.
[9] 3GPP TR 24.826, Study on impacts on signalling between User Equipment (UE) and core network from energy saving, Rel.11, Jun. 2011.
[10] 3GPP TR 23.866, Study on System Improvements for Energy Efficiency, Rel.12, Jun. 2012.
[11] 3GPP TR 23.887, Study on Machine-Type Communications (MTC) and other mobile data applications communications enhancements, Rel.12, Dec. 2013.
[12] 3GPP TS 23.401, General Packet Radio Service (GPRS) enhancements for Evolved Universal Terrestrial Radio Access Network (E-UTRAN) access, Rel.13, Mar. 2015.
[13] 3GPP TS 36.413, Evolved Universal Terrestrial Radio Access Network (E-UTRAN); S1 Application Protocol (S1AP), Rel.12, Mar. 2015.
[14] 3GPP TS 36.423, Evolved Universal Terrestrial Radio Access Network (E-UTRAN); X2 Application Protocol (X2AP), Rel.12, Mar. 2015.
[15] 3GPP TS 25.413, UTRAN Iu interface Radio Access Network Application Part (RANAP) signalling, Rel.12, Mar. 2015.
[16] 3GPP TR 36.887, Study on Energy Saving Enhancement for E-UTRAN, Rel.12, Jun. 2014.
[17] 3GPP TR 45.926, Solutions for GSM/EDGE BTS Energy Saving, Rel.13, Mar. 2015.
[18] ETSI TS 102.706, "Environmental Engineering (EE); Measurement Method for Energy Efficiency of wireless Access Network equipment", v1.2.1, Oct. 2011.
[19] GSMA, Mobile's Green Manifesto, 2nd Edition, Jun. 2012.
[20] ETSI TR 103.117, Environmental Engineering (EE); Principles for Mobile network level energy efficiency, v1.1.1, Nov. 2012.
[21] ETSI ES 201.554, Environmental Engineering (EE); Measurement methods for Energy efficiency of Core network equipment, v1.1.1, Apr. 2012.
[22] ETSI TR 105.174-4, Access, Terminals, Transmission and Multiplexing (ATTM); Broadband Deployment – Energy Efficiency and Key performance Indicators; Part4: Access networks, v1.1.1, Oct. 2010
[23] ETSI EN 301.575, Environmental Engineering (EE); Measurement method for energy consumption of Customer Premises Equipment (CPE), v1.1.1, May 2012.

[24] ETSI TS 103.199, Environmental Engineering (EE); Life Cycle Assessment (LCA) of ICT equipment, networks and services; General methodology and common requirements, v1.1.1, Nov. 2011.
[25] ETSI TR 102.532, Environmental Engineering (EE); The use of alternative energy solutions in telecommunication installations, v1.1.1, Jun. 2009.
[26] ATIS-0600015.06.2011, Energy Efficiency for Telecommunication Equipment, Methodology for Measurement and Reporting of Radio Base Station Metrics.
[27] ATIS Exploration Group on Green, ATIS Report on Wireless Network Energy Efficiency, Jan. 2010.
[28] IEEE Standard for Information technology–Telecommunications and information exchange between systems Local and metropolitan area networks--Specific requirements Part 11: Wireless LAN Medium Access Control (MAC) and Physical Layer (PHY) Specifications
[29] IEEE WG, Part 11: Wireless LAN Medium Access Control (MAC) and physical layer (PHY) specifications: Wireless Network Management, IEEE 802.11v, 2011.
[30] D. Camps-Mur, S. Sallent-Ribes, "Enhancing the Performance of TCP over Wi-Fi Power Saving Mechanisms", Wireless Networks, vol. 18, no. 8, 2012893–914.
[31] X. Pérez-Costa, D. Camps-Mur, "IEEE 802.11 E QoS and power saving features overview and analysis of combined performance", IEEE Wireless Communications, vol. 17, no. 4, 201088–96.
[32] D. C. Plummer, An Ethernet Address Resolution Protocol. IETF RFC 826, 1982.
[33] Wi-Fi Alliance, P2P Technical Group, Wi-Fi Peer-to-Peer (P2P) Technical Specification v1.0, Dec. 2009.
[34] D. Camps-Mur, A. Garcia-Saavedra, P. Serrano, "Device to Device Communications with WiFi Direct: Overview and Experimentation", IEEE Wireless Communications Magazine, vol. 20, no. 3 201396–104.
[35] IEEE WG, Part 11: Wireless LAN Medium Access Control (MAC) and physical layer (PHY) specifications: Amendment 6: Sub 1 GHz License Exempt Operation, 802.11ah D0.2, Sep. 2013

13

Green Routing/Switching and Transport

Luca Chiaraviglio, Antonio Cianfrani, Angelo Coiro, Marco Listanti and Marco Polverini
DIET Department, University of Rome "Sapienza", Rome, Italy

13.1 Energy-Saving Strategies for Backbone Networks

In the last decade, the energy consumption of Internet has been hugely fuelled by an increase of the number of connected devices causing an explosion of traffic volume. Consequently, energy-restraint strategies have become a crucial element to prevent the Internet from being throttled by an energy bottleneck. The challenge is to make the increase of the Internet operational efficiency faster than the rate of traffic growth.

Several studies [3] tried to foresee the future Internet energy consumption over the next 15 years (2010–2025). These studies divide the network in two segments: the access segment, realized with heterogeneous technologies, and a backbone segment, composed of high-capacity IP routers and transmission devices. It is foreseen that, at the end of the considered time interval, the overall carried traffic will achieve a total increment of 2–3 orders of magnitude, while the energy per bit consumed by the networking equipment will decrease in the range of 15–20% per year. Basic results of these studies show that, though at present (2013) the access network is about two orders of magnitude higher power than the backbone segment, in the near future, due to the foreseen traffic increase, the overall energy consumption of switching and transport equipment will progressively equal that of the access segment to become the dominant component after the 2020. Moreover, studies are unanimous to indicate that hardware technologies are expected to improve energy efficiency over the 2010–2020 decade, but the rate of the overall network efficiency improvement is expected to be slower than the traffic growth rate.

Green Communications: Principles, Concepts and Practice, First Edition.
Edited by Konstantinos Samdanis, Peter Rost, Andreas Maeder, Michela Meo and Christos Verikoukis.
© 2015 John Wiley & Sons, Ltd. Published 2015 by John Wiley & Sons, Ltd.

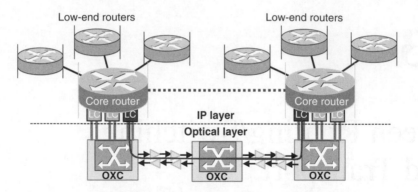

Figure 13.1 IP-over-WDM network architecture

Summarizing, the future design of IP networks oblige to reach an improvement of the energy efficiency of backbone network elements since they will be the dominant components of the overall energy consumption in the core network.

13.1.1 Backbone Networks and Energy Consumption

A typical ISP backbone network architecture is usually based on an IP-over-WDM model (Figure 13.1). The electronic IP routers make use of the huge capacity circuits, called light-paths, provided by the optical layer. Each IP link corresponds to a dedicated WDM lightpath. A specific IP interface (NIC: Network Interface Card) is associated to each logical IP link originated or terminated by the node; a NIC interacts with Optical Layer by means of an E/O converter. More details regarding energy efficiency in optical network technologies are provided in Chapter 15.

Focusing on power consumption requirements, recent studies (see e.g. Ref. [4]) have shown that the power consumption of IP devices is significant, thus an effective energy-saving strategy must act on IP routers, trying to reduce their power consumption.

An IP router is a very complex device, having an internal structure that depends on the specific vendor. However, it is possible to define a high-level model, reported in Figure 13.2, composed of three main blocks: a router processor, a set of line cards and a switching fabric.

The Router Processor (RP) manages the whole system and implements control plane functionalities, such as routing protocol execution and forwarding table management. Line cards interconnect the router with external networks receiving and sending IP packets; in Figure 13.3 a simple scheme of a line card is shown. Basically, there is an I/O interface and an internal interface connecting the line card with the switching fabric. The core of a line card is the Layer 3 processor that elaborates the header of incoming/outgoing IP packets and executes routing table lookup to perform IP forwarding. In terms of energy consumption, the line cards consume about 45% of the overall router power consumption, and the Layer 3 processor is responsible for about the 60% of line card power consumption [5, 6]. In the rest of this chapter, when referring to the power consumption of an IP link, we specifically refer to the power consumption of the line cards at the two ends of the link. Similarly, switching off a link means here switching off the related line cards.

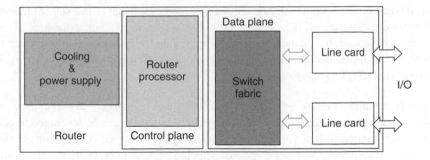

Figure 13.2   Simplified model of an IP router

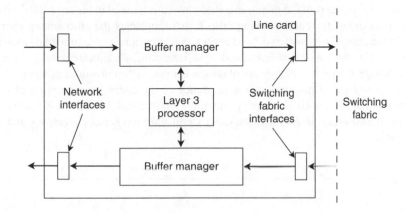

Figure 13.3   General scheme of an IP line card

13.1.2   Energy-Saving Strategies: Switch Off versus Energy Proportional

Traditional Internet design follows two main guidelines: use of equipment redundancy and over-provisioning in processing and transmission capacity normally dimensioned to support much more that the traffic peak values. In this way, network resources are usually largely under-loaded during off-peak periods. However, traffic variations are not followed by a corresponding decrease in the amount of energy consumed by the network because of the nearly independent power consumption of network devices. As a result, a key goal is to provide network energy consumption nearly proportional to the amount of carried traffic. In particular, two possible classes of strategies have been widely discussed in literature to bring the network energy consumption nearly proportional to the amount of carried traffic:

- *Energy proportional strategies*: They work on the individual network devices and try to achieve energy consumption proportionality by adapting the speed and capacity of the devices to the actual load, over relatively short timescales [7].
- *Switch-off strategies*: They involve the network as a whole and approximate load proportionality by carefully distributing the traffic in the network so that some devices are highly

utilized while others become idle and are put in sleep modes, that is, in a state of hibernation or completely switched off [8]. As this class of strategies need an enhancement of the network control procedures, namely routing, to aggregate traffic flows so as to reduce the set of active network resources, these strategies are also often named *Energy-Saving Traffic Engineering* (ES-TE) strategies.

Obviously, solutions belonging to these two classes can be merged, so that energy proportional devices are present, that is, powered-on, while sleep mode can be leveraged to possibly save additional energy. In Ref. [9], an evaluation study is performed considering the two different classes of solutions. As ES-TE enables selected devices to be switched off by steering traffic towards powered-on devices, increasing their power consumption due to higher traffic volumes sustained by them, this study brings light on identifying the scenarios where the switch-off approach is still beneficial.

The basic achievements of the mentioned study are summarized in Figure 13.4, derived from [9]. The curves shows the energy-saving ratio E that represents the ratio among energy consumption in the case of link switch off and energy consumption in case of load proportionality, as a function of p, the fraction of links (nodes) switched off. Results are averaged for a set of network topologies. The two curves are obtained for two different values of the parameter v, called *equivalent load*; it measures the dependency of the energy consumption of a link (or a node) on the load: if $v = 10$, the energy consumption of a link is practically independent of the load; on the contrary, if $v = 0.01$, the energy consumption rapidly increases and the load becomes high.

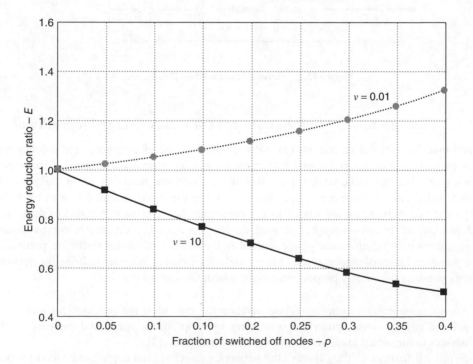

Figure 13.4 E versus p

When $v = 10$, the introduction of sleep modes saves energy in the network, that is, $E < 1$. Instead, if $v = 0.01$, because $E > 1$, the network consumes a higher amount of energy when devices are switched off. Anyway, in Ref. [9] it is also proved that link switch off is always convenient if $v > 0.2$, that is, even if a significant fraction of proportionality is assumed in link operation.

Summarizing, (i) switch-off methods are more convenient when the network is large with a high number of devices, whereas switch-off strategies are not convenient in small networks; (ii) if the device consumption is independent of the load, (or nearly independent, i.e. lightly variable), the use of sleep modes is very effective; on the other hand, if efficient power management techniques are implemented in the network devices, that is, devices are more energy proportional, the effectiveness of sleep mode approaches decreases.

Switch-off strategies aim at keeping the network-wide energy consumption nearly proportional to the amount of carried traffic. Traffic fluctuations are monitored in order to provide traffic aggregation over a subset of the network elements, allowing other equipment to enter power-saving mode and to wake them up only when needed.

In principle, network elements that can be switched off are both links (specifically, the network line cards associated to each IP link originated or terminated by a node) and entire nodes. Of course, an entire node can enter a sleep mode only if all its line cards are switched off. Nodes in sleep mode are essentially removed from the network topology, not able to receive, process and forward packets. Consequently, the node switch off is possible only if the node itself is neither a source nor a destination of traffic. This is the reason why most of studies exclusively concentrate on pure link switch-off strategies.

A particular Switch-Off strategy is possible when Link Aggregation functionality is enabled: A network link, referred to as a bundle, is composed of several physical links, referred to as "member" links. During low utilization times, traffic could be aggregated onto a subset of member links, providing the opportunity for the unused member links to be powered down, thus reducing the energy consumption of the overall bundle.

A classification of the *Switch-Off* strategies can be based on the constraints considered for path computation of packet traffic, that is, the constraints that are applied to determine the new routes that packets have to follow in order to put in sleep mode the highest number of network elements. Two main classes of *Switch-Off* strategies can be defined: (i) *flow-based* solutions and (ii) *destination-based* solutions. In the flow-based solutions, each packet is classified as belonging to a flow, implicitly identified by the source-destination addresses or explicitly denoted by a label or a set of rules, while the routing decision depends on the specific flow a packet belongs to. These approaches may require a connection-oriented layer 2 IP routing, typically implemented by Multi Protocol Label Switching (MPLS) or Open Flow. Hence, energy saving is obtained in a different protocol layer than IP; for this reason such strategies can be either called *MPLS* or Open Flow *solutions*. In the destination-based solutions, the forwarding decisions at each router are taken considering uniquely the destination IP address carried by the incoming packets. Specifically, the destination-based solutions maintain the traditional principles of IP routing.

Alternative strategies that aim to define low-power states avoiding switching-off nodes are based on deactivating a subset of router functionalities, concentrating on those regarding IP packets processing and routing. The selection of these functions is made since they are the most power hungry [4] and a considerable energy saving can be obtained through their deactivation. These strategies are classified as *Forwarding Freezing* approaches and can achieve significant energy saving without switching off entire hardware components. In this way, they are able to limit the impact on network performance, stability, and reliability levels. In Section 13.4, we discuss a specific *Forwarding Freezing* solution, referred to as Table Lookup Bypass (TLB).

13.1.3 Energy-Saving Strategies: Deployment Issues

One of the main limitations for fast deployment of energy-saving techniques in real networks is their integration in actual network protocols. Both Switch-Off and TLB strategies require the implementation of two main features: (i) an extension of the control protocol for the computation of new energy-aware paths and (ii) a device-level functionality to enable the specific power state (switch off or TLB) on the device (i.e. link or node).

The control protocol extension consists in integrating the new energy-aware routing algorithm to be executed during low traffic levels. The routing algorithms can be classified in two main categories: centralized algorithms and distributed algorithms. A centralized algorithm is executed on a single network element and the results (links to power down or node in TLB state) are then applied in the network by specific control-level messages. A distributed algorithm is executed on each network node, so avoiding the generation of control-level messages and then allowing each node to independently manage forwarding functionality. The implementation choice strongly depends on the network scenario considered: a destination-based or a flow-based scenario. In the case of a destination-based scenario, a distributed energy-saving routing algorithm is needed to maintain the principles of IP routing [10]. In the case of flow-based scenario, both centralized and distributed approaches can be used; in the first case, each router executes a distributed algorithm [11]; in the second case a centralized path computation element, such as Path Computation Element (PCE) for GMPLS networks, executes the energy-saving algorithm and manages the routing configurations of network routers [12].

The second step in the deployment of energy-aware solutions is the enabling of low-power state on network device. In the case of actual devices, where no advanced power-saving capabilities are available, the only possible solution is to power off the specific device (i.e. node or link); in this case, the TLB solution cannot be deployed. This solution has a drawback: the generation of a considerable amount of routing protocol-related messages and the network convergence. For instance, in the case of a link-state routing protocol such as OSPF, the removal of a link leads to the generation of a Link State Advertisement message that reaches all network nodes; each node performs path re-computation and updates its routing table. During the convergence phase, the network can experience performance degradation due to temporary unfeasible paths, that is, loops. New energy-aware devices, available in the near future [13], implement advanced power management techniques: (i) the standby of a line card with periodical wake-up allows to maintain physical layer synchronization and to generate routing protocol periodical messages (such as Hello messages for OSPF), making possible a more effective implementation of Switch Off; (ii) the management of line-card functionalities for the implementation of TLB techniques. The coordination among control-level algorithms and advanced hardware capabilities reduces the impact of energy-saving solutions on existing protocol and network performance.

In the case of energy-proportional devices, the implementation issues are limited: The modification of network paths is not needed and so each device acts independently, modifying its processing capability as a function of traffic load. The only aspect to carefully consider is the ability for the device to follow in real time the traffic behaviour and properly react to sudden traffic peaks.

13.2   Switch-Off ILP Formulations

The reduction of energy consumption in backbone networks can be obtained modifying network paths so as to use a subset of network resources; the computation of new paths and the detection of network devices to power off will be the result of energy-saving strategies. In this section, we define this problem by means of an Integer Linear Programming (ILP) formulation.

A network can be modelled as a graph $G(V,E)$, where V is the set of nodes and E is the set of directed edges (or links). The nodes represent the network routers and $N=|V|$ is the total number of router in the networks; the edges represent the network links and $L=|E|$ is the total number of directed edges in the network. The generic node i has a power consumption P_i. The generic edge from node i to node j, indicated with notation (i,j), has a power consumption P_{ij} and a capacity C_{ij}. We consider directed links to have a more general scenario, since the modification to an undirected links scenario can be easily derived. To represents switch-off operation in the formulation, two variables must be introduced:

- x_{ij}, a binary variable indicating if link (i,j) is active ($x_{ij}=1$) or it is switched off ($x_{ij}=0$);
- y_i, a binary variable indicating if node i is active ($y_i=1$) or it is switched off ($y_i=0$).

The energy minimization problem can be formulated by means of the following objective function:

$$\min \left[\sum_i \sum_j x_{ij}P_{ij} + \sum_i y_iP_i \right] \tag{13.1}$$

The scope is to minimize the overall consumption of nodes and links.

The traffic demand among network nodes can be represented by means of an $N \times N$ traffic matrix $T=[t_{sd}]$, where the element t_{sd} represents the amount of traffic originated from node s and directed to node d.

The complete formulation of the energy minimization problem for an IP network requires the introduction of several constraints: In particular, traffic-related constraints are needed to determine the network routing able to minimize the overall energy consumption, while still satisfying traffic requirements. These constraints are thus related to the network scenario considered, that is the flow-based or the destination-based one. For the sake of simplicity in the problem formulation, we introduce an assumption regarding paths computation: For each source-destination pair, a single path is computed. The extension to the multi-path case can be easily done but since it will lead to a more complex formulation we limit our analysis to the single path case.

13.2.1   Flow-Based Routing Formulation

In a flow-based network scenario, each traffic relationship t_{sd} is routed independently by means of a dedicated flow, that is, a dedicated path from the source to the destination. To take this into account, the binary variable f_{ij}^{sd} is introduced: f_{ij}^{sd} is equal to 1 only if the flow from s to d is routed through edge (i,j).

The first constraint to be considered is the classical flow conservation constraint:

$$\sum_{j=1}^{N} f_{ij}^{sd} t^{sd} - \sum_{j=1}^{N} f_{ji}^{sd} t^{sd} = \begin{cases} t^{sd} & \text{if } i = s \\ -t^{sd} & \text{if } i = d \\ 0 & \text{if } i \neq s, d \end{cases} \quad \forall s, d \in V \tag{13.2}$$

The second one is the bandwidth constraint: allowing a maximum utilization value α_{ij} for each edge (i,j), the amount of traffic routed on each link is limited by the following equation:

$$\sum_{s=1}^{N} \sum_{d=1}^{N} f_{ij}^{sd} t^{sd} \leq x_{ij} \alpha_{ij} C_{ij} \quad \forall (i,j) \in E \tag{13.3}$$

Another effect of equation (13.3) is that only unused links can be switched off ($x_{ij} = 0$).

The last constraint to be considered regards the node switch-off condition, and use the big-M notation to make possible the switch off of a node only if all its links are switched off:

$$\sum_{j=1}^{N} x_{ij} + \sum_{j=1}^{N} x_{ji} \leq My_i \quad \forall i \in V \tag{13.4}$$

The M value is an integer constant assuming a "big" value so that even when all x_{ij} variables are equal to 1 the first term of Eq. (13.4) is smaller than M; on the other side, a node can be switched off ($y_i = 0$) only when all x_{ij} variables are equal to zero.

The objective function (13.1) and the constraints (13.2), (13.3) and (13.4) represent the general energy minimization problem in a flow-based network scenario.

The general formulation can be adapted to specific cases introducing further constraints. One of the most common implements a physical requirement of router interfaces: It is possible to switch off a directed link only if also the link on the other direction is switched off:

$$x_{ij} = x_{ji} \quad \forall (i,j) \in E \tag{13.5}$$

13.2.2 Destination-Based Routing Formulation

The destination-based formulation is introduced to model the conventional data plane of IP networks and in particular the forwarding action performed by IP routers. A router selects the outgoing interface for an incoming IP packet exclusively based on the destination IP address; so, all the packets entering a router and directed to the same destination will be forwarded on the same output link, regardless of the source that generated them or the communication flow they belong to. Therefore, packets emitted by the network nodes directed to a given destination node are constrained to follow a *Reverse Path Tree* (RPT) having as root the destination node itself.

To formulate the energy minimization problem in such a scenario, it is necessary to modify the traffic assumption provided for the flow-based case. Traffic demands cannot any longer be split in source-destination flows, which are treated independently, since different flows having the same destination node must "share" respective parts of their route towards such common destination. To take this into account, a new binary variable n_{ij}^d must be introduced in the MILP formulation: n_{ij}^d is equal to 1 only if node i uses edge (i,j) to route traffic directed to destination d; in other words, using the IP terminology, j is the next-hop router of i to reach destination d. Otherwise, it remains equal to 0.

Starting from the flow-based formulation, two further constraints must be introduced to define the destination-based problem. The first one is needed to relate the new n_{ij}^d variables with the flow variables:

$$\sum_{s=1}^{N} f_{ij}^{sd} \leq Mn_{ij}^d \quad \forall i, d \in V \tag{13.6}$$

The second one is needed to maintain the single path assumption, that is, the use of a single next hop for each destination:

$$\sum_{j=1}^{N} n_{ij}^d = 1 \quad \forall i, d \in V \tag{13.7}$$

13.2.3   Comparison of Flow-Based and Destination-Based Formulations

The MILP formulations proposed in the previous section highlight that the solutions space of the energy minimization problem in the destination-based scenario is included in the solution space of the same problem in the flow-based scenario. A more rigorous analysis on the optimal solution in both cases is not possible since the problems are both NP-hard and can only be solved for small network scenarios. In Figure 13.5, we provide the solution of both formulations for a 12-node full mesh network considering a reference peak traffic matrix scaled to obtain nine different traffic scenarios.

Figure 13.5 shows that when the traffic is lower than 60% of peak values, the two problems have the same optimal solution, while in other cases, the flow-based optimal solution is slightly better than the equivalent destination-based one.

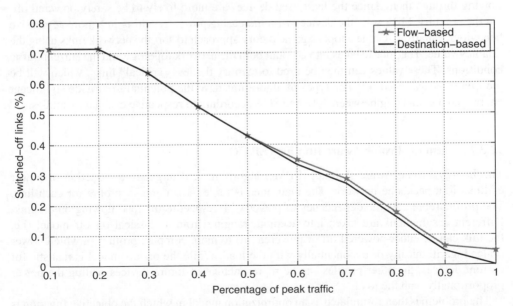

Figure 13.5   Optimal solution of flow-based (FLOW) and destination-based formulations (DBF) in a 12-node network scenario

As mentioned before, the problem formulations analyzed can be solved only for small networks. In real network scenarios, a low-complexity energy-saving algorithm must be considered, in order to compute the set of devices to switch off and the new routing configuration in a reasonable time. The next section provides a brief overview of the most common flow-based and destination-based energy-saving algorithms.

13.3 Switch-Off Algorithms

In this section, we provide a brief description of energy-saving algorithms for backbone networks proposed in literature classifying the solutions into flow and destination-based.

13.3.1 Flow-Based Algorithms

13.3.1.1 Least Flow Algorithm (LFA), Most Power Algorithm (MPA) and L-Game

The Least Flow Algorithm (LFA) [14], the Most Power Algorithm (MPA)[14] and L-Game [15] are different algorithms based on the same operative modality: the set of device (either a node or a link) is first sorted according to a given rule, and then the devices are selectively powered off. For each link powered off, the heuristics re-compute the routing, and check the connectivity and the maximum link utilization constraints. If such constraints are satisfied, the current device is left powered off; otherwise it is powered on again. The heuristics then try to power off another device, until all the devices are considered. The variations among the different heuristics rely on the rules adopted to sort list the network devices. In particular, LFA sorts the devices according to the load switching off first devices with the least amount of traffic flowing through them, since the least used devices are more likely to be safely powered off. On the contrary, MPA sorts the devices with increasing power, resulting in a more aggressive behaviour. Finally, L-Game adopts a game theory approach to sort the network links using the Shapley value. This metric takes into account both the network topology and the current traffic conditions. These values can then be used to identify the less "critical" links, which will be the first to be switched off. The types of algorithms have the potential to provide significant energy savings ranging between 30% and 80% according to respective evaluation studies.

13.3.1.2 Energy Profile Aware Routing (EPAR)

The Energy Profile Aware Routing (EPAR) algorithm [16] targets the reduction of power of links in a backbone network. The main idea is to define an energy profile for each link. The profile represents the dependency of power with respect to the load flowing on the link. Different assumptions are taken into account, ranging from a classical on–off model (i.e. the link can be either powered off or powered on) to more complex profiles in which power scales logarithmically or exponentially with the load. While the on–off model is realistic for current devices, the other profiles can be representative of future devices scaling the power proportionally with the load.

The problem is then formulated as an optimization model in which the objective function is the minimization of the total link power, still guaranteeing maximum link load and connectivity among nodes. The problem is then optimally solved over a realistic case study, showing that

EPAR can save more than 35% of energy compared to a classical formulation adopting shortest path routing and not targeting explicitly the power of links.

13.3.1.3 Green Distributed Algorithm (GRiDA)

The Green Distributed Algorithm (GRiDA) [17] aims at reducing the power consumption by switching off links while guaranteeing Service Level Agreements (SLAs). Different from LFA/MPA/L-Game approaches, switch-off decisions are completely decentralized. In particular, each node continuously monitors the utilization and the power status of the incident links. Based on this information, the GRiDA algorithm (i) switches off links when they are underutilized and (ii) switches on certain links back when traffic increases. The switch-on and switch-off decisions are taken at random intervals. Such decisions are based on the minimization of a utility function that integrates the power consumption of links with history information regarding past decisions. Moreover, the GRiDA algorithm assumes periodic Link State Advertisements (LSAs) exchanged among nodes reporting the status of the network considering the following two values: normal operation and congestion. When link congestion is detected, this information is flooded in the network towards other nodes, which power on the appropriate incident links. The GRiDA algorithm is able to achieve up to 50% energy savings with a limited impact on the network SLAs.

13.3.1.4 Distributed and Adaptive Interface Switch Off for Internet Energy (DAISIES)

Distributed and Adaptive Interface Switch off for Internet Energy Saving (DAISIES) [11] is an algorithm that exploits the control mechanisms provided by MPLS to switch-off network links. The actual amount of traffic carried by each Label Switched Path (LSP) is monitored by the ingress node on a fine granular observation period (e.g. tens of seconds). Then, whenever traffic goes beyond a prefixed threshold, the ingress node re-computes the path of the LSP and reroutes it updating both the path and the reserved bandwidth (make-before-break). The information about available (unreserved) bandwidth advertised by the Traffic Engineering (TE) routing protocol is used by DAISIES to properly compute link weights, and in turn aggregate the traffic on a reduced set of links. With this solution, energy savings between 35% and 50% are possible.

13.3.1.5 Green Traffic Engineering (GreenTE)

The Green Traffic Engineering (GreenTE) [18] is an intra-domain traffic engineering mechanism, which focuses on switching off links, considering the maximum link utilization and the network delay constraints. Rather than exploring the full space of paths among each source and destination, the authors consider a heuristic solution in which only a subset of paths is considered as input to the problem. Each candidate path has to meet two properties: (i) the candidate path cannot be larger than the diameter of the network, and (ii) the candidate path length cannot be larger than twice the shortest path for a given source destination node pair.

GreenTE is run by a centralized controller which continuously collects information from network routers, solves the traffic engineering problem and then notifies the corresponding routers about the links that should be powered-off and/or powered-on. Such a process is suggested to be run every 5–15 minutes to prevent frequent routing disruptions. The network topology, the link load and the state of the links (power on and power off) are distributed via LSAs adopting the Traffic Engineering Metric of the Open Shortest Path First protocol (TE-OSPF). Results show potential power saving between 20% and 40% can be obtained, while keeping both link delay and queuing delay close to the conventional case without GreenTE.

13.3.1.6   Energy-Aware Traffic Engineering (EAT)

Energy-aware traffic engineering (EAT) [19] considers the energy consumption of links while maintaining the same traffic rates between source and destination pairs as the traditional non-energy-aware approaches. The main assumption of this approach is the capability of links to operate at different transmission rates, while the basic idea is to spread the load across different paths with the objective to increase energy saving. In particular, the proposed algorithm shifts traffic towards links that can forward it without increasing their operation rate, provided that this would reduce the operation rate of other links. Every source node makes an independent decision based on the information collected from paths used towards certain destinations using a drop margin parameter. If the link utilization is below the drop margin, traffic is shifted from the link to other links. Hence, the drop margin plays a crucial role to select the number of links that will be considered to increase energy saving. In particular, the higher is the drop margin, the higher will be the number of links operating in low-power mode.

The EAT algorithm collects information regarding the candidate links considered to shift away their traffic and uses a metric to specify the distance from the lower operating rate. This metric is normalized by the utilization and represents the fraction of traffic that needs to be removed in order to enable the link to operate in a lower energy state. Since decisions are taken at the source nodes, the destination simply reports the amount of traffic that needs to be shifted and then the source checks the feasibility of such suggestion before performing the appropriate traffic alternations. EAT is able to switch off between 15% and 31% of the links and between 10% and 24% of the nodes.

13.3.1.7   Greening Backbone Networks with Bundled Links (GBNB)

Green backbone networks with bundled links [20] aim to switch off links that compose a link aggregation bundle at off-peak times when the traffic load is low, in order to save as much energy as possible. In particular, links are first sorted considering different rules and then are selectively powered off until no further link can be removed without causing service disruption. The heuristic initially computes the amount of flow on each link. In particular, an ILP model is formalized. The objective function is the minimization of the amount of flow on each link. Then, as a second step, the algorithm checks the amount of spare capacity on each link. Based on this information, the algorithm then turns off some of the links composing the bundle. Energy savings between 35% and 75% are achievable.

13.3.1.8   Green MPLS Traffic Engineering (GMTE)

Green MPLS Traffic Engineering [21] employs a traffic engineering means to switch off routers in a backbone network. The proposed heuristic takes as input the traffic demands and

Table 13.1 Main features of the flow-based algorithms

Algorithm	Computation	Targeted devices	Constraints	Complexity
LFA/MPA	Centralized	Routers/links	Connectivity, maximum link utilization	Central controller computing the set of devices to be powered off
L-Game	Centralized	Links	Connectivity, maximum link utilization	Central controller computing the set of links to be powered off
EPAR	Centralized	Links	Connectivity, maximum link utilization	Central controller computing the set of links to be powered off
Grida	Distributed	Links	Connectivity, maximum link utilization	Single node maximizing a utility function
DAISIES	Distributed	Links	Connectivity, maximum link utilization	Ingress nodes monitoring a traffic threshold
GreenTE	Centralized	Links	Connectivity, maximum link utilization, link delay	Central controller computing the set of links to be powered off
EAT	Distributed	Routers/links	Connectivity, maximum link utilization	Source node that shifts traffic the paths to the destinations
GBNB	Centralized	Bundled Links	Connectivity, maximum link utilization	Central controller computing the part of bundle to be powered off
GMTE	Centralized	Routers	Connectivity, maximum link utilization	Central controller computing the set of routers to be powered off

the routing information, and produces as output the set of powered off routers and the set of alternate label-switched paths with energy saving for re-routing traffic. The heuristic works as follows: Routers are first sorted according to the number of LSPs that they carry. Then, the routers are selectively considered starting from the ones with the least number of LSPs. For each router, traffic is shifted on new LSPs that are created. If the maximum link utilization does not exceed a threshold, the router is powered off. Otherwise, the router is kept powered on and the initial LSPs are restored. Results show that energy savings between 18% and 30% are achievable with the proposed solution.

Table 13.1 summarizes the main features of each algorithm.

13.3.2   Destination-Based Algorithms

13.3.2.1   Energy Saving IP Routing Strategy (ESIR)

The Energy-Saving IP Routing Strategy (ESIR) [10] is designed to be integrated into the OSPF routing protocol. The path computation strategy, realized by means of a modified Dijkstra algorithm, is distributed and fully compatible with OSPF mechanisms. At the same time ESIR is able to satisfy QoS requirements by maintaining traffic load on all the network links under fixed configurable values. The modified version of the Dijkstra algorithm is able to select a subset of paths to route the traffic, leaving the unused interfaces to enter low-power mode. The main advantage of this approach is that the IP topology does not change, and consequently no exchange of LSAs packets is needed. To achieve this goal, the set of routers is divided in importers and exporters. The main idea is the exportation mechanism, which allows to share a Shortest Path Tree between neighbours routers, so that the overall set of active links is minimized. Only the importer routers modify their shortest path tree starting from the tree of the exporter routers. This allows to avoid triggering an entire path re-computation inside the network. In order to work properly, the exporter has to be the neighbour of the importer. Moreover, an importer can be object of a single exportation, that is, it can receive the shortest path tree from a single exporter. Finally, once a node has become an exporter, its shortest path tree cannot be modified any more. The obtained results show that it is possible to reduce a percentage equal to about 40% the number of active links.

13.3.2.2 Energy Saving Based on Algebraic Connectivity (ESACON)

Energy Saving based on Algebraic Connectivity (ESACON) [22] adopts the algebraic connectivity to identify network links to be powered off. The algebraic connectivity is a metric developed in the context of complex networks, while in this work it is used as a parameter to keep the network connectivity above a suitable level. In particular, the algorithm works in two steps: It first creates a set of ordered links and then a subset of links is powered off. The links are ordered based on the value of the algebraic connectivity, since the aim is to switch off those links that have a low impact on the network connectivity. In the second step, a set of links is powered off in order to maintain the network connectivity above a given threshold. Despite not considering directly traffic flowing in the network, the ESACON algorithm reveals comparable performance with respect to traffic-aware solutions. Result show that the connectivity threshold plays a crucial role in determining the fraction of links powered off. Therefore, this parameter needs to be carefully set. Results show that up to 73% of links can be switched off in a realistic scenario, while maintaining the same link utilization as the LFA algorithm.

13.3.2.3 Ant Colony-Based Self-Adaptive Energy Saving Routing for Energy-Efficient Internet

Ant colony based energy-aware routing [23] relies on the concept of traffic centrality to switch off links, maximizing the energy saving. In particular, the traffic centrality is a measure of traffic volume on a link connecting a node to a neighbour. The authors proved that when this metric is maximized the energy consumption is minimized. Therefore, they propose a heuristic solution to maximize the traffic centrality. In particular, the heuristic, called A-ESR, is based on the ant colony technique, in which a number of artificial individual ants in the employed colony explore the network in real time. There are two types of artificial ants: the forward ant and the backward ant. At regular intervals, forward ants are launched from the nodes to other randomly chosen nodes. While moving to the destination, forward ants gather network information, including the arrival time and the identifier of each node. Once the forward ant

Table 13.2 Main features of the destination-based algorithms

Algorithm	Computation	Targeted devices	Constraints	Complexity
ESIR	Centralized	Links	Connectivity, maximum link utilization	Central controller computing the list of exporters and importers
ESACON	Centralized	Links	Connectivity	Central controller computing the link to be powered off
A-ESR	Distributed	Links	Connectivity, maximum link utilization	Single nodes computing the links to be powered off considering the information injected by "ants" in the network

has received the destination, a backward ant is generated. While moving to the original source, the backward ant applies the decision of the energy-aware algorithm for all nodes along the path. Results show an energy efficiency up to 71% in a realistic scenario.

Table 13.2 summarizes the main features of each algorithm.

13.4   Table Lookup Bypass

The "table lookup bypass" (TLB) is an alternative strategy to improve the energy efficiency of IP backbone networks [22] by avoiding using entirely or partly the forwarding capacity of a router, allowing a low-power mode for the hardware specifically dedicated to this function, that is, the L3 Processor. Since a relevant part of the power consumed by an IP router is wasted in performing layer 3 processing and, specifically, in forwarding packets, such an approach is very promising and could be a valid alternative to link switch off.

The main idea under TLB is to deterministically forward some or all packets entering a node to a specific and fixed next-hop router regardless of their destination address. By doing so, the lookup operation of the routing table, which is the main operation performed by the L3 Processor of IP routers, is bypassed and, consequently, the related hardware can be frozen.

Similarly to link switch off, this general mode of operation may take place during low traffic load periods during which the actual routing operation that should be performed by a specific router is delegated to some other routers within the network. Moreover, as TLB does not determine a reduction of the transmission capacity of the network, it may also take place during medium/high traffic load periods. In more detail, the TLB strategy may lead to changing the routing of some traffic flows within the network as elaborated in detail in the next section, which will generally experience a path length increase. Consequently, some spare capacity is generally needed within the network in order to support the increased load. In contrast, link switch off leads to both a path length increase of traffic flows and a reduction of the active transmission capacity. Therefore, link switch off generally needs higher reductions of the offered load in order to allow a significant number of links to be powered off.

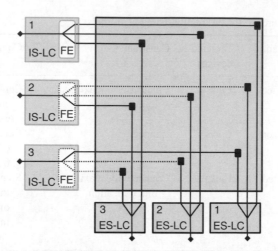

Figure 13.6   Graphical representation of a router with TLB enabled on all line cards

13.4.1   General Model and Implementation Aspects of TLB

Let us consider the router model presented in Figures 13.2 and 13.3, wherein each line card implements its own L3 Processor which is responsible for forwarding the packets received by the line card. This is performed by inspecting the forwarding table in order to find the correct next-hop router to forward packets and consequently the correct egress line card.

In such a model, it is possible to freeze the L3 Processor of a line card regardless of the state of the other line cards of the same router. Line card whose L3 Processor has been frozen cannot perform TL operation and must forward packets to a predetermined egress LC. LCs in this state are said to be in TLB state.

Figure 13.6 shows a graphical representation of a node having TLB-enabled LCs. IS-LCs are depicted on the left side of the figure whilst ES-LCs on the bottom. The node is represented as a set of cross points, which identify all possible switching operations between each IS-LC and each ES-LC. These switching operations are actually allowed only if there exists a path to reach the related cross point. According to the state of each IS-LC, only a subset of these paths can be actually used to forward packets, which are represented by continuous lines, whilst other ones, represented by dotted lines, cannot. The first IS-LC (top-left of the figure) is working normally forwarding received packets to any ES-LC. The second IS-LC (middle-left of the figure) is in TLB state towards the third ES-LC. Finally, the third IS-LC (bottom-left of the figure) is in TLB state towards the first ES-LC. Note that, if an LC is in TLB state, the related IS-LC can forward packets to only one ES-LC, whilst the related ES-LC can receive packets from any IS-LC. This mode of operation leads to change the routing of some traffic flows but allows to still exploit all the capacity deployed in the network to carry traffic demands.

Figure 13.7 shows a simple example of a TLB operation applied on the LC related to link (A,B) on node B, which is put in TLB state towards the outgoing interface on link (B,C). In the depicted scenario, the traffic demand t_{AC} follows its original path whilst the traffic demand t_{AD} is deviated from its original path $A \to B \to D$ when enters node B (the dashed line in Figure 13.7) and routed on the path $A \to B \to C \to D$. This operation allows the LC of B on (A,B) to enable the TLB and so to reduce the power consumption.

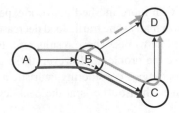

Figure 13.7   Example of paths modification in a network with a LC in TLB mode

A key point concerns how to implement the TLB functionality in current routers LCs. With reference to the LC scheme presented in Figure 13.3, we can note that, when an LC is in TLB state, packet redirection can be easily performed outside the IP packet processor and the lookup engine. For instance, this operation may be included in the hardware chip performing buffer management operations, since this module usually drives I/O operations towards other line cards and local ports. In this way, the L3 Processor represented in Figure 13.3 can be switched off (or put in a low-power state).

13.4.2   Network-Wide Solution

The application of TLB strategy foresees that each specific LC may be either in TLB state or work as usual (FULL state) regardless of the state of any other LC, including those belonging to the same router. Moreover, for each LC in TLB state, a specific egress LC (or equivalently a next-hop router) must be chosen.

A network-wide solution for TLB strategy may provide the state of each LC and the specific predetermined next-hop router for each LC in TLB state. The solution must consider routing and capacity constraints and should maximize the number of LCs put in TLB state.

Such a problem has been modelled as ILP [22], whilst sub-optimal heuristic solutions have been proposed in Ref. [23].

The ILP formulation includes typical flow conservation and capacity constraints of multi-commodity flow problems in addition to a set of specific constraints related to the unique state of each LC. The complexity of the general TLB problem turns out to be higher than the link switch off problem making the problem intractable even for small networks (i.e. even ten nodes).

In order to give an estimation of the energy savings achieved by the TLB approach in comparison to the link switch off, tone needs to consider (i) how much energy can be saved by putting in TLB state a single LC and (ii) how many LCs can be put in TLB state.

Concerning the first point, we have to specifically refer to the amount of power wasted by the L3 processor with respect to the total power consumed by an LC. As outlined in Section 13.4.1, L3 functionalities are responsible of about 60% of the power consumed by an LC. Thus, we expect that the power consumption of an LC in TLB state is reduced of about 50%.

The actual energy saving that can be achieved via TLB use depends on the amount of LCs that may enter TLB state according to the offered load. In this respect, it has been shown that a significant percentage of LCs can be put in TLB state even during peak traffic load periods. Specifically, in Ref. [22] it is shown to be between 30% and 80% depending on the average nodal degree of the network, with more meshed network experiencing higher performance. Moreover, it has also been shown that, as the traffic load decreases, up to 80% of LCs can enter TLB state even in little meshed networks. These results refer to small networks for which it was possible to optimally solve the problem. However, these general results have also been confirmed for large real ISP topologies considering heuristic approaches [23], which lead to about 40% of LCs in TLB state when traffic is at the peak level and about 90% when it is scaled down to 10% of the peak.

13.5 Conclusion

In this chapter, we proposed a classification of energy-aware techniques for wired backbone networks. We focused our attention on Switch-Off strategies, since they represent the preferred choice in the near future from a technological point of view. We first provided two different formulations for the energy minimization problem: the flow-based and the destination-based formulations; then we described the most known heuristics proposed in literature, showing that a considerable energy saving is possible in actual backbone networks exploiting both network overprovisioning and daily traffic behaviour. Finally we introduced an advanced power saving techniques for backbone networks: the TLB; this technique allows to manage the forwarding capabilities of a network device so that to enable low-power state during low-level traffic periods. The TLB technique represents a first step towards mid-term energy-aware techniques that will allow energy consumption of network devices "to closely follow" the traffic levels.

References

[1] C. Lange, D. Kosiankowski, R. Weidmann, and A. Gladisch: "Energy consumption of telecommunication networks and related improvement options", IEEE J. Select. Top. Quantum Electron., vol. 17, no. 2, pp. 285–295, 2011.

[2] F. Idzikowski, L. Chiaraviglio, and F. Portoso, "Optimal Design of Green Multi-layer Core Networks", Third International Conference on Future Energy Systems (e-Energy 2012), Madrid, Spain, 2012.

[3] R. Tucker, R. Parthiban, J. Baliga, K. Hinton, R. Ayre, and W. Sorin, "Evolution of WDM optical IP networks: A cost and energy perspective", J. Lightwave Technol., vol. 27, no. 3, pp. 243–252, 2009.

[4] S. Aleksic, "Analysis of power consumption in future high-capacity network nodes", IEEE/OSA J. Opt. Commun. Netw., vol. 1, no. 3, pp. 245–258, 2009.

[5] L. A. Barroso and U. Holzle, "The case for energy-proportional computing", IEEE Comput., vol. 40, no. 12, pp. 33–37, 2007.

[6] R. Bolla, R. Bruschi, A. Cianfrani, and M. Listanti, "Enabling backbone networks to sleep", vol. 25, no. 2, pp. 26–31, 2011.

[7] L. Chiaraviglio, D. Ciullo, M. Mellia, and M. Meo, "Modeling Sleep Mode Gains in Energy-Aware Networks", Comput. Netw., vol. 57, no. 15, pp. 3051–3066, 2013.

[8] A. Cianfrani, V. Eramo, M. Listanti, and M. Polverini, "An OSPF-integrated routing strategy for QoS-aware energy saving in IP backbone networks", IEEE Transactions on Network and Service Management, vol. 9, no. 3, pp. 254–267, September 2012.

[9] A. Coiro, M. Listanti, A. Valenti, and F. Matera, "Energy-aware traffic engineering: A routing-based distributed solution for connection-oriented IP networks", Comput. Netw., vol. 57, no. 9, pp. 2004–2020, 2013.

[10] A. Tzanakaki et al., "Energy Efficiency in integrated IT and optical network infrastructures: The GEYSERS approach," INFOCOM Workshop, 2011.

[11] R. Bolla, R. Bruschi, F. Davoli, L. Di Gregorio, P. Donadio, L. Fialho, M. Collier, A. Lombardo, D. Reforgiato, and T. Szemethy, "The green abstraction layer: a standard power management interface for next-generation network device," IEEE Internet Comput., vol. 17, no. 2, pp. 82–86. 2013.

[12] L. Chiaraviglio, M. Mellia, and F. Neri, Minimizing ISP Network Energy Cost: Formulation and Solutions, EEE/ACM Transactions on Networking, vol. 20, no. 2, pp. 463–476, April 2012.

[13] A. P. Bianzino, C. Chaudet, S. Moretti, J.-L. Rougier, L. Chiaraviglio, and E. Le Rouzic, Enabling Sleep Mode in Backbone IP-Networks: a Criticality-Driven Tradeoff, IEEE ICC'12 Workshop on Green Communications and Networking, Ottawa, Canada, 2012.

[14] J. Restrepo, C. Gruber, and C. Mas Machuca, "Energy profile aware routing", Communications Workshops, 2009. ICC Workshops 2009. IEEE International Conference on. IEEE, 2009, Dresden, Germany, 2009.

[15] A. P. Bianzino, L. Chiaraviglio, M. Mellia, and J.-L. Rougier, GRiDA: green distributed algorithm for energy-efficient IP backbone networks, Comput. Netw., vol. 56, no. 14, pp. 3219–3232, 2012.

[16] M. Zhang, C. Yi, B. Liu, and B. Zhang, "GreenTE: Power-Aware Traffic Engineering", IEEE ICNP 2010, Kyoto, Japan.

[17] N. Vasić and D. Kostić. "Energy-aware traffic engineering". Proceedings of the 1st International Conference on Energy-Efficient Computing and Networking. ACM, Passau, Germany, 2010.

[18] W. Fisher, M. Suchara, and J. Rexford, "Greening backbone networks: Reducing energy consumption by shutting off cables in bundled links," in Proc. 1st ACM SIGCOMM Workshop Green Netw., New Delhi, India, pp. 29–34, 2010.

[19] H.W. Chu, C.C. Cheung, K.H. Ho, and N. Wang, "Green MPLS traffic engineering", Australasian Telecommunication Networks and Applications Conference (ATNAC), IEEE, pp. 1–4, 2011.

[20] F. Cuomo, A. Abbagnale, A. Cianfrani, and M. Polverini: "Keeping the Connectivity and Saving the Energy in the Internet", IEEE INFOCOM 2011 Workshop on Green Communications and Networking, Shanghai (China), 2011.

[21] Y.-M. Kim, E.-J. Lee, H.-S. Park, J.-K. Choi, and H.-S. Park, "Ant colony based self-adaptive energy saving routing for energy efficient internet", Comput. Netw., vol. 56 no. 10, pp. 2343–2354, 2012.

[22] A. Coiro, M. Polverini, A. Cianfrani, and M. Listanti, "Energy saving improvements in IP networks through table lookup bypass in router line cards," Computing, Networking and Communications (ICNC), 2013 International Conference on, pp. 560–566, 2013.

[23] A. Cianfrani, A. Coiro, M. Listanti, and M. Polverini, "A Heuristic Approach to Solve the Table Lookup Bypass Problem", The 24th Tyrrhenian International Workshop on Digital Communications, Genoa (Italy), 2013.

14

Energy Efficiency in Ethernet

Pedro Reviriego[1], Ken Christensen[2], Michael Bennett[3], Bruce Nordman[3]
and Juan Antonio Maestro[1]

[1] Universidad Antonio de Nebrija, Madrid, Spain
[2] University of South Florida, Florida, USA
[3] Lawrence Berkeley National Laboratory, Vallejo, USA

14.1 Introduction to Ethernet

Ethernet is the dominant technology for wired local area networks (LANs) used to connect servers and desktop computers. Ethernet has also been adopted for industrial and automotive applications, to connect audio and video equipment and in wireless and wireline access networks. This wide adoption has fostered the continuous development of Ethernet since it was first introduced in the 1970s. In the 1980s, the IEEE 802.3 Ethernet working group was formed to develop Ethernet standards within the IEEE 802 LAN/MAN standards committee. Since then, many Ethernet standards have been produced to cover different transmission media, increased data rates, link aggregation, virtual LANs, Power over Ethernet, congestion management, and other capabilities.

The Ethernet standards support a wide variety of transmission media that include coaxial cables, unshielded twisted pair (UTP), optical fibers, and backplanes [1]. Typically, optical fiber is used for high-speed links between network equipment, whereas UTP is used to provide edge device connectivity. Therefore, the majority of the Ethernet links use UTP as the transmission media. For each medium, different rates are supported, and the rates have traditionally followed a 10× increase from one standard to the next. The rate is clearly identified in the standard name. For example, for UTP, the following rates are supported 10 Mb/s (10BASE-T), 100 Mb/s (100BASE-TX), 1 Gb/s (1000BASE-T), and 10 Gb/s (10GBASE-T). In some of the latest standards, a 4× increase in rate is targeted. That is the case of the next standard for UTP that will target 40 Gb/s and for which a study group has been formed [2]. Standards for optical fibers commonly support higher data rates. For example, the IEEE formed a study group recently to determine whether or not to start a project to specify 400 Gb/s Ethernet. As the

Green Communications: Principles, Concepts and Practice, First Edition.
Edited by Konstantinos Samdanis, Peter Rost, Andreas Maeder, Michela Meo and Christos Verikoukis.
© 2015 John Wiley & Sons, Ltd. Published 2015 by John Wiley & Sons, Ltd.

rate increases, so does the complexity of the transceivers. High-speed Ethernet transceivers use advanced modulation and coding techniques and are very complex devices. This results in high power consumption. To facilitate interoperability when different rates are supported, the standard defines a procedure known as auto-negotiation that allows link partners to negotiate the link rate and other parameters.

Ethernet networks are built around switches and routers that connect the different elements in the network [3]. There are different types of switches ranging from a simple four- or eight-port switch to a modular switch (in which line cards and ports can be added or removed) that can host hundreds of ports and provide advanced features such as virtual LANs or link aggregation. In all cases, the switch has a number of ports that may support different rates and media, and a fabric and logic to forward incoming frames to the appropriate output port. The number of ports is typically a multiple of four or eight as the physical layer transceivers (PHYs) are commonly grouped in quad or octal devices to reduce cost. In switches that aggregate traffic from a set of users or servers, it is common to have a large number of lower speed ports and a reduced number of higher speed ports known as uplinks over which the aggregated traffic is sent. The fabric and control logic is simple for small switches that provide only a basic set of features but can be very complex for high-end switches.

The elements that are connected by Ethernet are typically desktop computers and servers. In both cases, Ethernet is commonly implemented on the motherboard, which is known as LAN on Motherboard (LOM). Ethernet can also be implemented in a Network Interface Controller (NIC) card. This is less frequent and NIC cards are mostly used to provide additional Ethernet ports when needed. For servers it is common that the LOM provides multiple ports, usually a dual or quad port implementation. The Ethernet controller can implement advanced features such as receive side coalescing or TCP offload to reduce the processing required in the CPU [4]. The vast majority of the switches and computers in the market support at least 1 Gb/s data rates with an increasing number supporting 10 Gb/s.

The wide adoption and low cost of Ethernet has motivated its use in applications for which it was not originally designed. Examples include industrial applications, in which latency and reliability are critical [5], automotive applications [6], access networks, with 1 Gb/s and 10 Gb/s Passive Optical Networks (PONs), and audio and video equipment (which need synchronization and other features). There are also now Ethernet standards specifically for data center connectivity, where a need for higher speeds and low power consumption enables larger system integration. This can be achieved by designing transceivers optimized for reduced link lengths in data centers as the links are often short and predictable. The NGBASE-T standard, for example, will likely target only short reach channels [2].

Despite the wide adoption of Ethernet and its continuous evolution, the issue of energy efficiency was not formally considered in the standards until 2005 when a tutorial on energy efficiency was presented to the IEEE 802 community [7]. This led eventually to the creation of a study group and a subsequent task force that produced the IEEE 802.3az "Energy-Efficient Ethernet" standard [8]. The 802.3az standard was a milestone in terms of addressing energy efficiency not only for Ethernet but also for wireline communications generally.

In the rest of this chapter, first the key features of the Energy-Efficient Ethernet standard are presented. Then energy savings estimates are presented showing the benefits of the standard. Finally, future directions for energy efficiency are discussed covering current standardization efforts related to energy efficiency within IEEE 802.3 as well as other standards that have been influenced by Energy-Efficient Ethernet.

14.2   Energy-Efficient Ethernet (IEEE 802.3az)

The first Ethernet standards, such as 10BASE-T, had simple transceivers that transmitted only when there were frames to send (except for short periodic pulses to keep the link alive). Ethernet has evolved since then to support higher data rates using more complex transceivers. To achieve those rates, most subsequent Ethernet standards have specified a continuous transmission at the physical layer. When there is no user data to send, a control signal known as IDLE is transmitted to keep all the transceiver elements well aligned. This ensures that when user data arrives for transmission, the link is ready to send the data. Transmitting the IDLE signal typically requires the same power consumption as transmitting frames (i.e., sending user data), so the transceiver power consumption is roughly constant and independent of traffic load. Link utilization is usually low as shown in the measurements presented during the standardization process [9]. Therefore, making power consumption more proportional to the actual traffic load would result in large energy savings given the widespread use of Ethernet.

With this objective in mind, a tutorial was presented to IEEE 802 in July 2005 [7]. This presentation triggered substantial interest in the IEEE 802.3 community and led to a Call for Interest (CFI) in November 2006 [9]. A CFI is the first step to get a project started in IEEE 802.3. The CFI was approved and a study group was formed to determine whether or not to request a project to create a standard for Energy-Efficient Ethernet (EEE). The goal of the standard was to improve the energy efficiency of Ethernet transceivers. The study group phase included six meetings and 43 presentations supporting the creation of a standard for EEE. Contributions covered technical and economic feasibility, indications of broad market support, and compatibility with existing Ethernet devices. These are needed to meet the "five criteria" requirement to start a project for a new standard. The study group wrote a Project Authorization Request (PAR) that defined the project's purpose, need, and scope. The study group also produced objectives, which defined what the task force would work on, such as the media types. The project was authorized in September 2007, and the P802.3az Task Force was formed.

Two methods were proposed to improve energy efficiency: Adaptive Link Rate (ALR), which switches a link to a lower speed during periods of low utilization, and Low Power Idle (LPI), which creates a low-power sleep mode for use when there is no user data to send. ALR was proposed first and LPI later. Both methods save energy and have advantages and disadvantages. These were reviewed, and debated until LPI was eventually selected. The deciding factor was that ALR introduces a latency of hundreds of milliseconds which is not acceptable in many cases. This is so because the link has to be established at a lower/higher speed and therefore all the elements in the PHY have to be adjusted. On the other hand, LPI does not change the link speed and therefore only minor adjustments are needed in the PHY to enter/exit the low power mode. Those can be made in microseconds thus ensuring a low impact for most applications. In addition, it was thought that LPI might be somewhat easier to write in the standard and to implement.

The task force produced its first draft of the standard in October 2008 and completed selection of baseline proposals in March 2009. The EEE standard covers several media types including twisted pairs, backplanes, the XGMII Extender Sublayer (XGXS), and the 10-Gigabit Attachment Unit Interface (XAUI). Optical fiber is not covered in the EEE standard, though not for lack of interest. It is worth noting that at the very end of the study group phase, a proposal to include some fiber-optic transceivers was made; however, much more work was needed at the time to build the necessary consensus to add the work to the PAR. The study

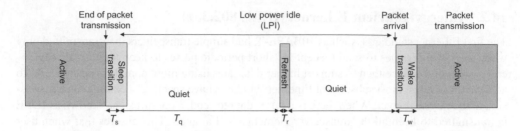

Figure 14.1 Illustration of the low power idle mode defined in energy-efficient Ethernet

Table 14.1 Minimum wake, sleep, frame transmission times, and single frame efficiencies for different link speeds

Protocol	Min T_w (μs)	Min T_s (μs)	Frame size (bytes)	T_{Frame} (μs)	Single frame efficiency (%)
100BASE-TX	30.5	200	1518	120	34.2
1000BASE-T	16.5	182	1518	12	5.7
10GBASE-T	4.48	2.88	1518	1.2	14.0

group had to choose between accepting the proposal and delay submitting the project request, which would add months to the overall timeline, or reject the proposal. They chose the latter. The EEE standard was approved as IEEE 802.3az [8] in September 2010.

The EEE standard defines an LPI mode to allow a transceiver to minimize transmission when there is no user data to send. Instead of transmitting continuously, transmission can be stopped with only short periodic refreshes sent to keep the transmitters and receivers aligned. As illustrated in Figure 14.1, sleep and wake transitions are needed to enter and exit LPI. The sleep transition ensures that the transceivers save relevant state while the wake transition ensures that they are aligned for reliable data transmission. In LPI, transmission is stopped except for short refresh intervals that keep a coarse alignment of the transmitters and receivers. Transceiver energy consumption in LPI is much smaller than in the active mode. However, packets can only be sent in the active mode and therefore a wake transition is needed before a packet can be sent.

Minimum transition times are specified in the standard and vary across the transmission media and interface rates. Table 14.1 shows the twisted pair transceiver values and the time required to transmit a maximum-length 1518-byte frame. Transitions times are larger than frame transmission times so that when packets arrive spaced in time and are transmitted isolated, one per active period, more time is spent in transitions than spent in actually sending the data. As power consumption during transitions can be similar to that of the active mode, isolated packets require considerably more energy to send than those sent in bursts. The transition times also impact latency as a frame that arrives for transmission to an interface that is in LPI mode has to wait until the link is back in active mode to be transmitted. Since transition times are on the order of microseconds, this is only an issue for latency critical applications. For example, in some datacenter applications, latency values of a few hundred microseconds

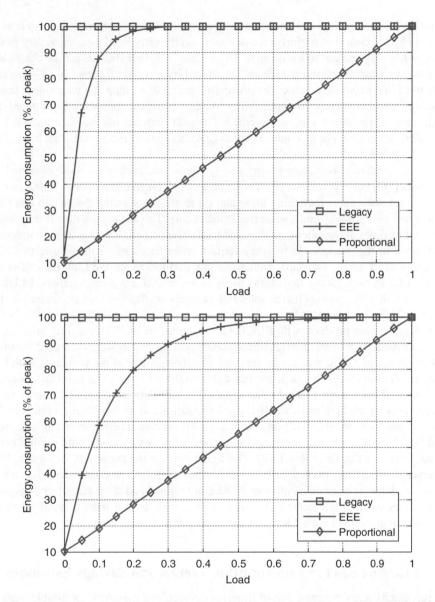

Figure 14.2 Energy consumption versus link load for 1000BASE-T (a) and 10GBASE-T (b)

are assumed [10]. In such cases, EEE could have an unacceptable impact on end-to-end delay, especially when a packet finds several links in LPI mode along its route.

Energy-Efficient Ethernet performance has been evaluated by simulation [11] and measurements [12]. Several analytical models have also been proposed and validated against simulations [13–15]. In all cases, the results confirmed that transitions limit the effectiveness of EEE when frames are transmitted isolated. Figure 14.2 shows the performance of a 1000BASE-T (a) and a 10GBASE-T (b) transceiver sending 1250-byte frames that arrive with a Poisson distribution, assuming that consumption in LPI mode is 10% that of the active mode as done in Ref. [11]. It is clear that performance deviates significantly from proportionality in which power draw is proportional to traffic load. To mitigate this issue, frames can be coalesced for transmission so that transition overheads are shared by multiple frames [16]. Coalescing is already used in computers to reduce the processor burden of sending and receiving frames. As coalescing increases latency, its applicability to latency critical applications is limited. However, in many applications, coalescing will not impact performance and is a natural feature of the information sent. For example, in video transmission, all the packets that belong to the same video frame can be coalesced for transmission [17].

The EEE standard also enables energy savings opportunities beyond those obtained in the transceivers [18]. For example, when all ports on a switch are in LPI mode, no traffic can arrive in the next microseconds. Switching logic and other elements within the switch can then be put into a low power mode until a wake transition occurs. This is only possible if the switching element wake times are smaller than those of EEE. The potential savings at system level can exceed those in the transceivers but may require wake times larger than those of EEE. To address this situation, the standard enables the link partners to use the Link Layer Discovery Protocol (LLDP) to negotiate transitions times larger than the minimum values. LLDP is an IEEE 802.1 link layer protocol that enables link partners to advertise parameters and exchange capabilities [19].

Complete network products with EEE began to appear in late 2010, shortly after the standard was ratified. By the end of 2012, EEE-enabled port shipments were measured in hundreds of millions/year, and most new computers and switches supported the standard. With typical equipment renewal cycles, it can be expected that in three to five years from 2013 most Ethernet links will implement EEE. At that point EEE will deliver substantial energy savings. Measurements on the first available EEE implementations show significant reductions in the power consumption of NICs [12] and switches [20]. The overall reduction in LPI mode is around 70% for NICs, whereas for the switches the results vary by model with savings between 50% and 75% (see details in Ref. [20]). The figures are for the entire NIC or switch. For the transceiver only, savings exceed 90% in some cases. In future products, the relative energy savings may increase as implementations of EEE are refined and the potential savings in other systems elements are implemented. Absolute savings may drop as reductions of active mode power consumption level are achieved in each new product generation.

14.3   Ethernet Energy Consumption Trends and Savings Estimates

The aggregated energy consumption of Ethernet devices, and the savings available from EEE, both depend on a number of factors:

1. The type and number (stock) of installed devices and the presence of Ethernet links between them.
2. The power consumption levels for each of them.
3. Link usage patterns.

All of these factors vary over time; hence, creating accurate estimates for each variable is difficult and they cannot be credibly forecast with reliability. So, rather than create a forecast that is only one of many plausible scenarios, we instead present a savings estimate based only

Table 14.2   1000BASE-T link counts by device type (millions; United States only, 2010)

	Device stock	Link-years
Residential		
Routers	53	53
Desktop PCs	101	40.4
Game consoles	109	38.2
Printers	113	21.7
Set-top boxes	45	11.3
AV receivers	99	7.4
TVs	353	7.1
Notebook PCs	132	6.3
VOIP adaptors	4.7	4.7
Switches	1	1
Blu-Ray players	12	0.8
Total	1023	191.8
Commercial		
VOIP phones	50	50.0
Desktop PCs	72.7	43.6
Printers	25	22.5
Notebook PCs	57.0	8.5
Total	205	124.7
TOTAL	1227	316.5

on current conditions. This provides an estimate of the order-of-magnitude for energy savings. We first present such an estimate and then delve into complications.

14.3.1   Number of Links

Table 14.2 shows our estimates for the number of active 1000BASE-T links today. A link-year is defined as one link, being active all the time, or ten links each being active for only 10% of the year. Interfaces without active links should consume little or no power. Stock data for network equipment are taken from Ref. [21], whereas other residential stock data are from Ref. [22]. The assumptions made in elaborating the table are described in the following. Set-top boxes cover only standalone boxes, not service provider equipment. Printers include multifunction devices. Many links between certain devices are constantly powered on and hence the link is always active; network equipment uplinks are an example. Some edge devices (e.g., many PCs) are powered down at certain times and so the corresponding links to these devices contribute to the active link count only fractionally. For some device types (e.g., TVs), only a portion of the stock support Ethernet at all, and out of such devices that support Ethernet, only a portion of the stock actually use it. All these factors are taken into account for estimating the average number of active links based on the amount of devices. In many cases, this relies on informed judgment of usage patterns in the absence of good data. Hence, while TVs are about a third of

Table 14.3   10GBASE-T link counts by device type (millions;
United States only, 2010)

	Device stock	Links
Servers		
Volume	11.3	22.6
Mid range	0.3	1.2
High end	0.035	0.28
Storage	2.9	2.9
Switch uplinks	8.3	16.7
Total	1250	31.2

all relevant devices, they contribute less than 4% of the residential EEE savings. About 80% of residential EEE savings derive from routers, desktop PCs, game consoles, and printers.

For commercial buildings, a similar but simpler process is employed. Stock data is taken from U.S. EPA, Energy Star program, except the data for VoIP phones, which is estimated by the authors. Not all notebook PCs and imaging equipment have Ethernet. Some devices are powered down some of the time for example at night. Desktop PCs and VOIP phones dominate the savings; some phones have two ports with PC traffic combined with phone traffic in a single uplink to the switch. This does not change the link count from the case of separate switch ports to each device as in both cases there are two links.

Table 14.3 shows our estimates for the number of active 10GBASE-T links today. Server stock data are taken from Ref. [23], with servers having multiple ports (2, 4, and 8 for volume, mid-range, and high-end, respectively). Data storage equipment is assumed to contribute 0.25 ports/server on average (not all storage equipment uses Ethernet). Switch uplinks are mostly in commercial buildings as the number of PCs and VOIP phones far exceeds the number of servers. We assume 400 million ports, 48 ports/switch, and 2 uplinks per switch. We examined market research data on shipments of network equipment ports and concluded that are broadly consistent with our estimations, considering also that on most switches, some ports are not connected. As we count only links, not NICs, switch downlinks are accounted by the end device count. All data about equipment is based on data for 2010.

Network links other than edge links and switch uplinks that may also use EEE, but their number is relatively small in comparison, whereas their utilization rates are likely higher. Hence, our estimation is not taking into account these links. Similarly, there are many other applications of Ethernet including industrial purposes, cameras, displays, and niche applications, but these are modest in number compared to the categories in Table 14.2. Backplane Ethernet could be a significant application in the future, but our estimates concentrate on reporting from 2010.

14.3.2   Power per Link

Power consumption typically increases with the operational rate of the link. Considering Ethernet, the power consumption of successive product generations decreases due to deployment optimizations of the implementations and also because of the use of more advanced microelectronic technologies. For example, the first generations of 1000BASE-T transceivers consumed approximately 6 W, whereas current implementations need less than 0.5 W [24, 25]. Similarly for 10GBASE-T, there is a reduction from 10–12 W to 2.5–4 W between the first and second product generations [25]. All power values presented are AC-equivalent and hence are larger than low-voltage DC values since they account for conversion losses from AC/DC to DC/DC conversion.

At the outset of EEE work, a typical power for a 1000BASE-T NIC was 1 W. Data for a typical 1000BASE-T NIC shipping in 2012 from a major manufacturer shows a maximum power of about 0.45 W for a link under 30 m in length, whereas for a 100 m cable consumption is about 0.55 W but most cables are shorter so we use the lower figure.

Data on 10GBASE-T PHYs can be found in Ref. [26], which notes the high consumption of early hardware, that is, 10–12 W. New 40 nm 10GBASE-T PHYs consume 2.5–4 W for a 100 m cable, and even newer designs reduce power around 10% for 30–50 m cables, and over 25% for 10 m cables. Consumption is supposed to be less than 2.5–4 W due to the use of shorter cables, but then an increase is expected due to the power supply loss factor. Accordingly, we assume a 3 W per 10GBASE-T PHY for the 2012 technology version.

14.3.3   Usage Patterns

The calculations above of number of average links accounts for the fraction of time that links are established. The usage in question here is how much data is transmitted (and its burstiness). With the use of EEE, link utilization plays a fundamental role in determining energy savings. In most cases, communication networks will continue to be underutilized [27], as concluded in the study and measurements that supported the initiation and development of the EEE standard [9]. However, Figure 14.2 shows that for 10GBASE-T, EEE can consume about 60% of the peak power with a link load of only 10%, so in certain cases even low loads may affect energy consumption substantially. For 1000BASE-T links, the average load is expected to remain very low (around 1% or less), which simply means a sustained load of 10 Mb/s. For datacenter links, utilization is much higher [28] with values that exceed 10%. Since Internet service providers consider data utilization statistics as sensitive information, they are reluctant to share it. To keep the estimates simple an average load of 10% for 10GBASE-T links and 1% for 1000BASE-T links are assumed, respectively.

Consistent with Figure 14.2, we assume that LPI mode uses 10% of the peak power, whereas the corresponding savings with "No Coalescing" are around 79% for 1000BASE-T links and 42% for 10GBASE-T links, respectively.

To summarize, in our energy savings calculations we used 2010 active link estimations and consider the increased data rates based on today's use. We also considered power levels corresponding to 2012 data utilization statistics, assuming a potential for underutilized network equipment and resources.

14.3.4   Results

Table 14.4 summarizes the estimated total energy saving using the active Ethernet link data introduced in Tables 14.2 and 14.3, considering also power per link values previously discussed (note that total savings sums are rounded). Simulation results, and specifically the ones illustrated in Figure 14.2, were used to convert link utilizations into durations of low power

Table 14.4 Assumptions and results for EEE savings (United States only, 2010 stock)

	1000BASE-T	10GBASE-T	Total
Assumptions			
Link-years	316.5	31.2	347.7
Peak power/link (W)	0.9	6.0	
Link utilization (%)	1%	10%	
No coalescing power	21%	58%	
Ideal savings (no transition time)			
Percent	89%	80%	
Per link (W)	0.80	4.80	
Total (MW)	254	150	403
Total (TW h/year)	2.2	1.3	3.5
Total ($million/year)	222	131	353
No coalescing savings			
Percent	79%	42%	
Per link (W)	0.71	2.52	
Total (MW)	225	79	304
Total (TW h/year)	2.0	0.7	2.7
Total ($million/year)	197	69	266
Coalescing opportunity			
Percent	10%	38%	
Per link (W)	0.09	2.28	
Total (MW)	28	71	100
Total (TW h/year)	0.2	0.6	0.9
Total ($million/year)	25	62	87
Moderate coalescing savings			
Percent	84%	61%	
Per link (W)	0.76	3.66	
Total (MW)	239	114	353
Total (TW h/year)	2.1	1.0	3.1
Total ($million/year)	210	100	310

mode. For low utilization levels, most frames require transitions from LPI to active mode, which significantly reduces low power mode time. Considering the power savings per link and the total number of active links in the United States, the cost savings is estimated at least $260 million/year, and over $310 million/year assuming moderate use of packet coalescing, whereas global savings are expected to be higher (an electricity rate of $0.10 is used, which approximates typical rates paid in the United States in 2010). Additional savings may also accrue from reductions in power and cooling energy in conditioned spaces such as data centers, and by using the Link Layer Discovery Protocol (LLDP) to negotiate longer wake transitions that enable savings beyond the PHY. In May 2007, a savings estimate [29] presented to the IEEE EEE Study Group was 7.5 TW h/year; it had substantially higher per-link savings but lower link counts. A savings estimate in [16] was 5 TW h/year that had higher numbers of 10 Gb/s links than this estimate. These savings take the number of Ethernet links from those in use in 2008. A recent study [30] estimates 2.8 TW h/year for 1000BASE-T alone for 2012.

The "Ideal Savings" section in Table 14.4 shows the potential energy savings via employing EEE if there was no energy consumption in transitions to and from the LPI mode; that is, each PHY uses the LPI power level plus the full rate level only for the percent of time when sending traffic. The "No Coalescing" section factors the energy consumed in the transitions to and from LPI mode, with fewer than two packets per active time. It shows substantial savings from EEE even without coalescing. The use of coalescing can ensure most of the additional savings an ideal link with zero EEE transition times would achieve. The "Moderate Coalescing" section shows savings considering double amount of packets in each active period, less than four considering both speeds. In computing the overhead, many factors are involved including packet sizes, packet interarrival patterns, utilization variation with time, and PHY power consumption during transitions. These estimates are for large frames, independent arrivals, no load variations, and 100% PHY power consumption during transitions. The frame size used tends to underestimate overhead, whereas using no utilization variations and transition power of 100% have the opposite effect. Therefore, the results presented are a reasonable first-order approximation to EEE overhead.

Further savings that are expected to occur over time depend on the number of links, which could rise substantially since Ethernet is currently continuously adopted widely in distributing audio and video, in homes and elsewhere. On the other hand, as wireless technologies increasingly enhance the offered capacity, reliability, and security, they could replace a significant segment of the current wired technologies such as Ethernet. In the near future, it is very likely for edge links in residential and commercial buildings to convert to 10 Gb/s Ethernet, with higher savings per link. Considering the equipment operational power consumption, some additional reduction is expected with the introduction of smaller semiconductor technologies, but the pace of reduction is likely to be significantly less than in the past. On the other hand, as traffic increases over time the actual utilization of Ethernet links goes up, and EEE savings will fall.

Global savings will be several times that in the United States. According to Ref. [23], the United States provides about 36% of the global stock of servers, and the International Energy Agency (IEA) [31] estimates that for all consumer electronics and ICT, North America consumes around 30% of the global total. Thus, global savings by our approach should be about $1 billion/year.

14.4 Future Directions of Energy Efficiency in Ethernet

The Energy-Efficient Ethernet standard is a milestone for improving energy efficiency in wire-line communications and will result in substantial energy savings in the coming years. EEE has also influenced other standards. The Fibre Channel community is working on a standard to add energy efficiency mechanisms to Fibre Channel transceivers similar to the LPI mode defined in EEE [32]. The same has occurred in the VDE 0885–763 standard for high-speed communication over plastic optical fibers [33]. In this case, the energy efficiency mechanisms have been designed as an integral part of the physical layer specifications. In IEEE 1904, Service Interoperability in Ethernet Passive Optical Networks (SIEPONs), a protocol similar to LPI is being developed to enable energy savings. The purpose of the SIEPON standard is to provide a system-level and network-level standard based on IEEE 802.3 EPON specifications, making it the first *system-level* specification for energy-efficient operation of Ethernet equipment [34]. IEEE 1904.1 refers to EPON standards developed in

802.3 for ways to save energy in the PHY and explains that further reduction in energy use may be achieved when combined with the use IEEE 802.3az interfaces. As of early 2013, six Energy Star specifications require that products supporting EEE must be tested connected to a device that also supports EEE, as this is necessary to have the measured power level reflect EEE savings. This encourages manufacturers to include EEE components. The Energy Star specification for small network equipment [35] goes further and takes into account savings at the other end of the link to provide more incentive to include EEE. The upcoming specification for large network equipment will most likely also incorporate EEE.

In addition to influencing standards beyond IEEE 802.3, EEE has also inspired incorporating energy efficiency in other Ethernet standards. Most IEEE 802.3 physical layer standards that followed the standardization of EEE specify energy efficiency mechanisms or they are being discussed as part of standards projects. One of the most compelling reasons to include an energy-efficient mode in new Ethernet projects at the start is that it is significantly easier to design the features at the beginning of the project, rather than at the end of the project where the risk of introducing new problems in the design is much greater. Such is the case in the IEEE P802.3bj project to define a 100 Gb/s standard for backplane and copper data center links. The cost of LPI mode transitions is very important in data center links as the load is significant and frame transmission times at 100 Gb/s are very small. This prompted the task force to define two LPI modes, an EEE-like mode in which physical layer transmission is stopped and a fast mode in which physical layer signaling continues, but savings can be achieved in other system elements [36]. The fast mode enables fast transition times that are beneficial when load is high or latency is critical. Energy efficiency mechanisms are also being incorporated in the latest IEEE 802.3 standards for optical fibers [37]. The IEEE P802.3bm standard for 40/100 Gb/s communication over optical fibers adopted the fast LPI mode from IEEE P802.3bj. This capability enables system energy savings while avoiding issues associated with stopping optical transmission [38]. The introduction of this new fast LPI mode increases the necessity for network equipment manufacturers to develop methods to take advantage of the signal to creatively reduce energy use in the system.

A key evolution in the acceptance of energy efficiency as a requirement for future projects in IEEE 802.3 was evident in the CFI for Reduced Twisted Pair Gigabit Ethernet (RTPGE) standards project, P802.3bp, which is the first project stating the need for energy efficiency explicitly. RTPGE needs a very low power standby mode in automotive applications [39], which may lead to having two low-power modes, one similar to LPI and another with larger energy savings and transition times. For the NGBASE-T project, there is a concern that power consumption will be high and an LPI mode will provide only limited savings due to the link speed (40 Gb/s) and the expected load in data centers [10]. Therefore, efforts are concentrating on optimizing the power use versus link length [26, 40]. Network equipment manufacturers have used this idea previously as an enhancement to the standards in their Ethernet products [41]. Incorporating it to the standard design opens a new avenue for energy efficiency in which the transceiver is designed to adapt its functionality and power consumption to the requirements of the channel. This results in significant energy savings that do not depend on the traffic load.

Clearly Energy-Efficient Ethernet has had an impact on improving energy efficiency in wire-line communications. Energy efficiency is now considered at the beginning of new project proposals in IEEE 802.3 standards development efforts. EEE has influenced, and will continue to influence, development of energy-efficient communications in other standards as well. With the development of new modes to achieve energy proportionality while mitigating operational impact, such as with fast LPI, comes the need to develop ways to accomplish the systems-level savings they enable. This need may drive the next cycle of innovation in energy efficiency for Ethernet as EEE becomes more routinely accepted.

14.5 Conclusions

In the last decade, energy efficiency in wire-line communication systems has changed from being practically disregarded to being a primary design objective in new communications standards. One example of this evolution is Ethernet for which the development of the EEE standard marked a milestone. This chapter presented first an overview of both Ethernet and the EEE standard. Then estimates and trends of energy consumption in Ethernet were discussed showing that the adoption of the new IEEE 802.3az Energy-Efficient Ethernet standard will result in large energy and economic savings likely exceeding $250 million per year in the United States alone. Finally, future directions and efforts to improve the energy efficiency in the Ethernet standards that are being developed were summarized. For the reader who wants to know more detailed information can be found in the references many of which are available online.

References

[1] C. Spurgeon, "Ethernet: The Definitive Guide," O'Reilly Media, 2000.
[2] "IEEE 802.3 Next Generation BASE-T Study Group," available online at http://www.ieee802.org/3/NGBASET/index.html.
[3] R. Seifert and J. Edwards "The All New Switch Book: The Complete Guide to LAN Switching Technology," Wiley, 2008.
[4] S. Makineni, R. Iyer, P. Sarangam, D. Newell, L. Zhao, R. Illikkal and J. Moses, "Receive side coalescing for accelerating TCP/IP processing," Proceedings of International Conference on High Performance Computing (HiPC), pp. 289–300, 2006.
[5] J. D. Decotignie "Ethernet-based real-time and industrial communications" Proc. IEEE, vol. 93, no. 6, pp. 1102–1117, June 2005.
[6] "IEEE 802.3 Reduced Twisted Pair Gigabit Ethernet (RTPGE) Study Group" available online at http://www.ieee802.org/3/RTPGE/index.html.
[7] K. Christensen and B. Nordman, "Reducing the Energy Consumption of Networked Devices," presentation to IEEE 802, 2005, available online at http://ieee802.org/802_tutorials/05-July/Tutorial%20July%20Nordman.pdf.
[8] "IEEE Std 802.3az: Energy Efficient Ethernet-2010".
[9] H. Barrass, M. Bennett, H. Frazier and B. Nordman, "Energy Efficient Ethernet IEEE 802.3 Call-for-Interest," 2006, available online at http://ieee802.org/3/cfi/1106_1/EEE-CFI.pdf.
[10] U. Hoelzle and L. A. Barroso, "The Datacentre as a Computer: An Introduction to the Design of Warehouse-Scale Machines," Morgan and Claypool Publishers, 2009.
[11] P. Reviriego, J. Hernandez, D. Larrabeiti and J. Maestro, "Performance Evaluation of Energy Efficient Ethernet," IEEE Commun. Lett., vol. 13, no. 9, pp. 1–3, 2009.
[12] P. Reviriego, K. Christensen, J. Rabanillo and J.A. Maestro, "An initial evaluation of energy efficient Ethernet," IEEE Commun. Lett., vol. 15, no. 5, 2011, pp. 578–580.
[13] S. Herrería-Alonso, M. Rodríguez-Pérez, M. Fernández-Veiga, and C. López-García, "How efficient is energy-efficient Ethernet?," in Proceedings of ICUMT, 2011.
[14] M. Ajmone Marsan, A. Fernandez Anta, V. Mancuso, B. Rengarajan, P. Reviriego, and G. Rizzo, "A simple analytical model for energy efficient Ethernet," IEEE Commun. Lett., vol. 15, no. 7, 2011, pp. 773–775.
[15] A. Chatzipapas and V. Mancuso, "Modelling and Real-Trace-Based Evaluation of Static and Dynamic Coalescing for Energy Efficient Ethernet," in Proceedings of ACM e-Energy, 2013.
[16] K. Christensen, P. Reviriego, B. Nordman, M. Bennett, M. Mostowfi and J. A. Maestro, IEEE 802.3az: the road to energy efficient Ethernet," IEEE Commun. Mag., vol. 48, no. 11, pp. 50–56, 2010.

[17] A. Oliva, T. R. Vargas, J. C. Guerri, J. A. Hernandez and P. Reviriego, "Performance analysis of energy efficient Ethernet on video streaming servers," Comput. Networks, vol. 57, no. 3, pp. 599–608, 2013.

[18] D. Dove, "Energy Efficient Ethernet: Switching Perspective," presented at IEEE 802.3 meeting, January, 2008, available online at http://www.ieee802.org/3/az/public/may08/dove_02_05_08.pdf.

[19] W. Diab "LLDP's Use in EEE" presented at IEEE 802.3 meeting, January, 2009, available online at http://www.ieee802.org/3/az/public/jan09/diab_02_0109.pdf.

[20] P. Reviriego, V. Sivaraman, Z. Zhao, J. A. Maestro, A. Vishwanath, A. Sánchez-Macián and C. Russell, "An Energy Consumption Model for Energy Efficient Ethernet Switches," Proc. of the International Conference on High Performance Computing & Simulation (OPTIM Workshop), Madrid (Spain), pp. 98–104, 2012.

[21] NRDC, Small Network Equipment Energy Consumption in U.S. Homes: Connecting devices with less energy, NRDC Issue Paper, 2013.

[22] B. Urban, V. Tiefenbeck, and K. Roth. "Energy Consumption of Consumer Electronics in U.S. Homes in 2010". Final Report by Fraunhofer Center for Sustainable Energy Systems for the Consumer Electronics Association.

[23] J. G. Koomey, "Growth In Data Center Electricity Use 2005 To 2010", Analytics Press, 2011, available online at http://www.analyticspress.com/datacenters.html.

[24] C. Gunaratne, K. Christensen, B. Nordman and S. Suen, "Reducing the energy consumption in Ethernet with Adaptive Link Rate (ALR)," IEEE Trans. Comput., vol. 57, no. 4, pp. 448–461, 2008.

[25] "Next Generation BASE-T IEEE 802.3 Call For Interest," 2012, available online at http://www.ieee802.org/3/NGBASET/public/jul12/CFI_01_0712.pdf.

[26] P. Wu, G. Parnaby and W. Lo "NGBASE-T Requirements Learning from 10GBASE-T," presented at IEEE 802.3 meeting, January 2013, available online at: http://www.ieee802.org/3/NGBASET/public/jan13/WuParnaby_01a_0113_NGBT.pdf.

[27] A. Odlyzko, "Data networks are lightly utilized, and will stay that way," Rev. Netw. Econ., vol. 2, no. 3, pp. 210–37, 2003.

[28] M. Bennett, "Energy-Efficient Ethernet for 100G Backplane and Copper" presentation to IEEE 802.3, November, 2011, available online at: http://www.ieee802.org/3/bj/public/sep11/bennett_01a_0911.pdf.

[29] B. Nordman, "EEE Savings Estimates", prepared for IEEE EEE Study Group, May 2007, available online at: http://www.ieee802.org/3/eee_study/public/may07/nordman_2_0507.pdf.

[30] S. Lanzisera, B. Nordman and R. E. Brown, "Data network equipment energy use and savings potential in buildings", Energy Effic., vol. 5, no. 2, pp. 149–162, (2012)

[31] IEA, Gadgets and Gigawatts "Policies for Energy Efficient Electronics", International Energy Agency, May 2009, available online at http://www.iea.org/w/bookshop/add.aspx?id=361.

[32] "FC-EE - Energy Efficient Fibre Channel" specification draft, November, 2012, available online at: ftp://ftp.t10.org/t11/document.12/12-073v4.pdf.

[33] P. Reviriego, R. Pérez de Aranda and C. Pardo, "Introducing Energy Efficiency in the VDE 0885–763 Standard for High Speed Communication over Plastic Optical Fibers," IEEE Commun. Mag., vol. 51, no 8, pp. 97–102, 2013.

[34] SIEPON Working Group Web Site, http://www.ieee1904.org/1/.

[35] "ENERGY STAR Program Requirements for Small Network Equipment," November, 2012, available online at: http://energystar.gov/products/specs/sites/products/files/ES_SNE_Draft_2%20_V1_Specification_Nov2012.pdf.

[36] M. Gustlin, H. Barrass, M. Bennett, A. Healey, V. Pillai, M. Brown and W. Diab, "Detailed baseline for EEE in 100G" presentation to IEEE 802.3bj, March, 2012, available online at: http://www.ieee802.org/3/bj/public/mar12/barrass_01_0312.pdf.

[37] G. Nicholl and D. Ofelt "Energy Efficient Ethernet (EEE) for 40 Gb/s and 100 Gb/s optical interfaces" presentation to IEEE 802.3bm, November 2012, available online at: http://www.ieee802.org/3/bm/public/nov12/nicholl_01a_1112_optx.pdf.

[38] O. Haran "Applicability of EEE to fiber PHYs" presentation to IEEE 802.az, September 2007, available online at: http://www.ieee802.org/3/eee_study/public/sep07/haran_1_0907.pdf.

[39] M. Bennett "EEE Considerations for RTPGE Reduced Twisted-Pair Gigabit Ethernet Study Group", July, 2012, available online: http://grouper.ieee.org/groups/802/3/RTPGE/public/july12/bennett_01_0712.pdf.

[40] M. Bennett and P. Reviriego "Reach and Energy Efficiency in NGBASE-T" presentation to the NGBASE-T study group, November, 2012, available online at: http://www.ieee802.org/3/NGBASET/public/nov12/bennett_01_1112_ngbt.pdf.

[41] D-Link "Power Saving by Cable Length," available online at: http://www.dlinkgreen.com/energyefficiency.asp.

15

Green Optical Networks: Power Savings versus Network Performance

P. Monti[1], C. Cavdar[1], I. Cerutti[2], J. Chen[1], A. Mohammad[3], L. Velasco[4], P. Wiatr[1] and L. Wosinska[1]

[1]Communication Systems Department, KTH Royal Institute of Technology, Kista, Sweden
[2]Institute of Communication, Information and Perception Technologies, Scuola Superiore Sant'Anna, Pisa, Italy
[3]Electrical Engineering Department, Linköping University, Linköping, Sweden
[4]Department of Computers Architecture, Universitat Politècnica de Catalunya (UPC), Barcelona, Spain

15.1 Introduction

With an ever-increasing demand for bandwidth, connection quality, and end-to-end interactivity, computer networks require more and more sophisticated and power-hungry technologies. This is why every network segment, that is, from the access to the core, has been targeted to find possible ways to reduce the power consumption.

The term core refers to the backbone infrastructure of a network that usually interconnects large metropolitan areas, and may span across nations and/or continents. The term access, on the other hand, refers to the so-called last mile or segment of a network where central offices (COs) and remote nodes (RNs) provide connectivity between the end users and the rest of the network infrastructure. Depending on the reach of the access segment, core and access may or may not be interconnected via a metro infrastructure.

Regardless of the network segment under exam, optical communication plays a central role in reducing the power consumption in communication networks. In the core part, transport

Green Communications: Principles, Concepts and Practice, First Edition.
Edited by Konstantinos Samdanis, Peter Rost, Andreas Maeder, Michela Meo and Christos Verikoukis.

solutions based on wavelength division multiplexing (WDM) technologies are able to significantly lower their overall required power levels because of their ability to limit the number of optical-electrical-optical (O-E-O) conversions per each provisioned connection. Similarly, in the access segment, Passive Optical Networks (PONs) are becoming an attractive alternative to their active counterparts. For this reason, and in order to foster further improvements, energy efficiency in the optical layer has attracted a lot of attention, and a wide range of topics are addressed in the literature.

On the other hand, all these strategies developed to reducing the power drained by the optical layer mainly resort to techniques that turn off unused devices. These techniques can be enabled by the introduction of a sleep mode option in the equipment. However, performing these operations while, at the same time, making sure that other network parameters (e.g., delay, quality of transmission, blocking ratio, reliability) are not affected, requires careful design and/or provisioning operations.

This chapter aims at exploring and evaluating these trade-offs in more detail and to provide a different insight into the green design and provisioning problem in optical networks. More specifically, Section 15.2 presents an overview of the power consumption performance of the main components used in optical access and core networks together with their options in terms of potential power savings. Section 15.3 concentrates on the trade-off between energy efficiency and delay in optical access networks, providing a case study based on WDM optical passive networks. Section 15.4 focuses on green WDM core networks. Three specific trade-offs are analyzed in terms of energy saving versus (i) connection blocking probability (Section 15.4.1), (ii) quality of transmission of the optical signal (Section 15.4.2), and (iii) reliability levels of the provisioned optical connections.

The presented results point out how it is important during the design and provisioning phase not to concentrate on power minimization only. The risk would be to end up with an optical network where the potential savings coming from lower power consumption levels may be nullified by poor performances in terms of other crucial quality of service parameters.

15.2   Device-Specific Energy Characteristics

Access networks are responsible for nearly three quarters of the power consumption of all network equipment [1]. On the other hand, the average utilization of access network components is lower than 15%, allowing some room for energy saving mechanisms. Power optimization can be performed at the device and subdevice levels (see also the techniques described in Section 15.3). The devices responsible for most of the energy consumption are at the end points of each access connection, that is, the optical network unit (ONU) at the user premises and the optical line terminal (OLT) in the CO. There are also some other devices (e.g., splitter, and (possibly) switches) used at intermediate points, but this depends on the specific access architecture under exam. In the case of a PON only optical splitters are used, and they do not require any power for their operations. In the case of an Active Optical (access) Network (AON), or point-to-point (PtP) fiber solutions some switching operation might be included, thus requiring extra power consumption. The list of the typical active optical access network devices together with their energy consumption is presented in Table 15.1.

In core networks, the number of devices is lower compared to the access segment [1]. On the other hand their power consumption value is higher. In addition, it is expected that once end users are able to benefit from connectivity rates in the order of Gbps, the core part of the

Table 15.1 Power consumption values of optical access network equipment

Device	Power consumption [W]	Source
Optical network unit (ONU)	5	[2]
Optical line terminal (OLT)	2/port	[2]
Fiber switch (if needed)	1.5/port	[2]

Table 15.2 Power consumption values of optical core network equipment

Device	Power consumption [W]	Source
OTN 10G transponder	20	[3]
OTN 40G transponder	160	[3]
OTN 100G transponder	360	[3]
Wavelength selective switch (WSS)	60	[3]
Optical demultiplexer	40	[3]
EDFA amplifier	8–16	[3]
Wavelength Cross Connect (WXC)	25	[3]

network will be required to handle a huge amount of traffic with obvious consequences in terms of power consumption. In core networks there are different types of devices for which power consumption can be the subject of optimization. They are listed in Table 15.2 together with their power consumption.

In order to optimize their energy usage access and core network devices can be switched off or put into a low power consumption mode when they are not used. The difference in how and when these options are used lies mainly in the time required to take a device back to a fully operational state. If switching on/off operations can be scheduled ahead of time, then a component can be completely switched off without any fear that it will not be ready when needed. If, on the other hand, a device might become essential at a moment's notice, then it might not be completely switched off, but only some of its components may enter a low power consuming state.

Take, for example, a typical transponder (used not only in core networks but also in OLTs and ONUs). Its components used for receiving are a photodiode, a transimpedance amplifier (TIA), an analog to digital converter (ADC) block, a digital signal processing (DSP) unit, and a deframer. For transmitting, on the other hand, a transponder needs a laser with thermoelectric cooler, a modulator, and a framer. In addition, a transponder normally includes a forward error correction module and some Layer 2 (L2) and/or Layer 3 (L3) electronics. In order to save energy the components responsible for transmission/reception (e.g., the lasers, DSP, ADC, and the modulator) may be switched off or put into a low power mode. However, since the transition from low power mode to the operational mode is not immediate and requires a certain time to be performed, this should be done carefully. On the other hand, it will be easier to activate/deactivate the L2 or L3 electronic. Depending on the components' characteristics different energy saving algorithms can be proposed, but their implementation will trigger a number of performance trade-offs that has to be considered. These aspects are addressed in detail in the rest of the chapter.

15.3   Energy Saving for Optical Access Networks Based on WDM PONs

There are many studies aimed at improving the energy efficiency in access networks and in particular targeting PONs. This is mainly due to the large number of active devices deployed at the user premises. These green approaches can be grouped according to the layer that they target, that is, physical layer (i.e., including hardware) energy-efficient techniques (e.g., component integration or low power circuits), data link layer power optimization strategies (e.g., cyclic sleep) as well as hybrid energy-efficient approaches that work on more than one layer at the same time [4–8]. When focusing on data link layer solutions, the techniques that can be used in PONs include power shedding, dozing, and deep/fast sleep modes, [5, 9]. There are also a number of software-based solutions that can be used to lower the power consumption in PONs, for example, adaptive link rate, traffic shaping, and wavelength shutdown [8].

According to the power shedding concept, the ONU and/or the OLT go into a low power mode by putting only a subset of its components (e.g., Ethernet interface) into sleep mode while keeping both transmitter (Tx) and receiver (Rx) active. The purpose is to not introduce any additional transmission delays (e.g., for wake-up procedures and/or synchronization) in the case of any upcoming upstream and/or downstream traffic. As a result, only a limited amount of power can be saved.

With deep and fast sleep approaches, the Tx and the Rx at the ONU/OLT are in sleep mode when they are not in use (e.g., no upstream/downstream traffic from/to a specific ONU/OLT). This technique can achieve the best energy savings, but a synchronization phase is required to make sure that the OLT/ONU is aware that the transceiver on the other side is active. In addition, the length of the sleep period has a direct impact on packet delay, and as long as the TX is asleep, packets have to be either queued or dropped. Usually, a fast sleep technique is combined with some sleeping policies to optimize the energy savings performance, for example, cyclic sleep [6].

Dozing is a compromise between the advantages of shedding (i.e., little impact on the transmission delay) and the energy saving performance achievable by deep and fast sleep. According to the dozing concept, only the Tx is put into sleep mode while the Rx stays always on (this can be applied at the ONU and at the OLT in the case of WDM-based PONs). One of the main challenges with a dozing approach is to know when to put the Tx into sleep mode and when to wake it up in the presence of incoming traffic. The most straightforward way would be to wake up the Tx as soon as there is traffic to be sent and to put the Tx into sleep mode right after the transmission phase is over. This scheme is referred to as immediate wake-up. It is characterized by a delay that is due to the transition time between sleep and active states. However, this transition also leads to an energy overhead [10], which is paid every time the Tx changes operative state. One way to overcome this drawback is to wait for a certain time (i.e., T_{maxq}), with the intent to collect a number of packets before transmitting them all together in a burst. In this way the number of transitions is minimized, but at the expense of an additional delay. Such a scheme raises the question for "acceptable" delays, especially in the presence of possible delay guarantee requirements for specific services.

The trade-off between energy reduction and delay performance can be addressed by means of schemes that are able to keep the transmitter in a sleep state as long as possible (i.e., to achieve higher energy savings) but are also intelligent enough to wake it up in time to assure the transmission of the collected packets without breaching their maximum delay constraints. This

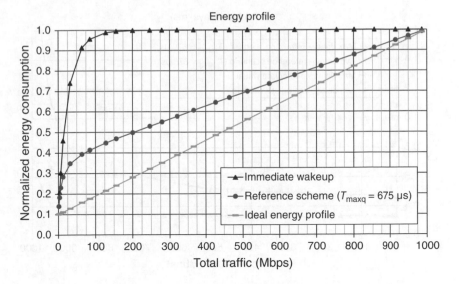

Figure 15.1   Energy profile of Tx

wake-up time calculation takes under consideration the maximum traffic delay constraint, the (nonnegligible) transition time between sleep and active states, the propagation delay (between ONU and OLT), and the transmission delay of all packets waiting for transmission.

The trade-off of energy consumption and packet delay of this scheme is presented in Figures 15.1 and 15.2. More details about the assumptions and the simulation setup are available in Ref. [11]. One conclusion from both figures is that an immediate wake-up scheme has the best delay performance but it is able to offer energy savings only for low link loads, that is, lower than 15% of link occupancy. That is due to the high frequency of transmitter wake-up for nearly each packet. On the other hand, as soon as packets are collected and transmitted in bursts ($T_{maxq} > 0$) significant energy savings can be achieved in medium load conditions. The average and maximum packet delay rises but can be controlled by adapting the value of T_{maxq} to maximum delay constraint for the specific service under exam, as described in Ref. [11].

Additional energy savings can be achieved using the same "burst-formation" intuition but changing the other in which packets are transmitted, on the basis of their priority levels, that is, how stringent their maximum delay constraint is. In fact, low-priority packets, that is, with higher delay tolerance, allow longer sleep times of the transmitter and therefore offer additional energy savings. On the other hand, a high-priority packet can trigger an earlier transmitter wake-up and can be scheduled first. This information can then be incorporated into the packet scheduling operations to maximize the sleep time of the transmitter [11]. Figure 15.3 presents the energy savings that can be observed for traffic composed of two traffic classes: high-priority class and low-priority class with maximum delay restriction of 1 ms and 5 ms, respectively. The scheme with traffic class differentiation can further improve energy savings compared to the previously explained bursting scenario, in particular for traffic with low portion of high-priority packets (Figure 15.4).

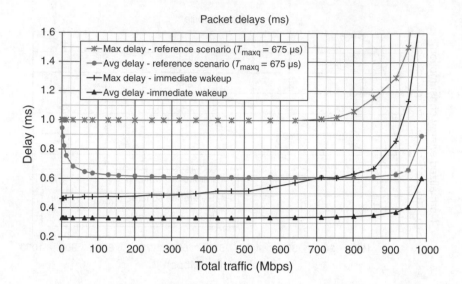

Figure 15.2   Packet delays

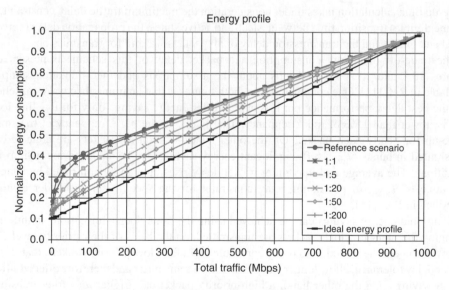

Figure 15.3   Energy profiles when exploiting traffic diversity

15.4   Energy Saving for WDM Core Networks

Transparent optical core networks are a power efficient option, able to reduce the energy consumption of the transport infrastructure. The energy consumption of optical core networks can be further reduced by a proper network design [12–17] and by green routing approaches (as elaborated in Chapter 13) where the (energy) savings come from the minimization of the

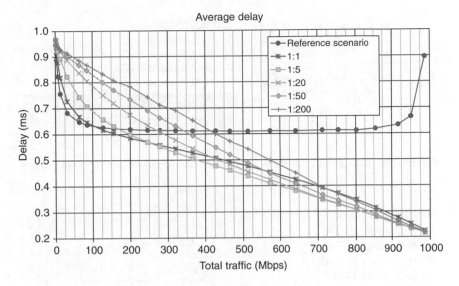

Figure 15.4 Average high-priority packet delays when exploiting traffic diversity

number of active network elements that need to be powered on in order to guarantee the required connectivity. These methods are based on temporarily setting unused network elements in a power saving (*sleep*) mode. However, the minimization of energy consumption may have a negative impact on other network performance metrics, which should be carefully assessed in order to guarantee the practical usefulness of a certain energy saving approach. The next sections look carefully into this problem by addressing a number of trade-offs between energy saving levels and important network metric such as blocking probability, quality of transmission (QoT), and reliability levels.

15.4.1 Energy Saving versus Blocking Probability in Transparent WDM Core Networks

The basic services provided by WDM networks are high speed and all-optical end-to-end channels, also referred to as *lightpaths*. Lightpaths are dynamically created between node pairs to both provide the desired network connectivity and accommodate arriving traffic demands. Each lightpath that needs to be created in a WDM network is assigned both a route and a wavelength – this is, the so-called routing and wavelength assignment (RWA) problem. When traffic demands dynamically enter and depart from the network, the problem is referred to as the *online* RWA problem. One of the online RWA problem objectives is to reserve the minimum number of network resources (wavelengths) for each arriving traffic demand. It is expected that by minimizing the amount of reserved resources per arriving demand, the blocking probability is reduced – where a demand is *blocked* when it cannot be created because of the lack of available wavelengths in the network.

This section explores the trade-off between energy savings and lightpath request blocking. In order to illustrate this trade-off we discuss a simple example, which is shown in Figures 15.5 and 15.6. In this example a given number of lightpaths need to be dynamically provisioned in

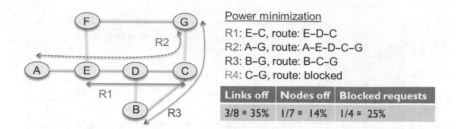

Figure 15.5 Power versus blocking, a trade-off example: power minimization strategy

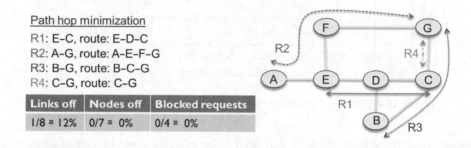

Figure 15.6 Power versus blocking, a trade-off example: path hop minimization strategy

a WDM network with seven nodes and eight bidirectional fiber links. It is assumed to have one fiber link in each direction, with two wavelengths per fiber and equal energy consumption values for all links, that is, for simplicity we consider links with the same length, making the minimum length path problem equivalent to the minimum number of hops. Two provisioning strategies are considered: the first one focuses on finding a solution requiring the minimum amount of power (Figure 15.5), while the second one concentrates on finding the route with the minimum number of hops (Figure 15.6). The provisioning strategy with the objective to minimize the energy consumption will select the path that requires the minimum number of new network elements to be turned on, that is, by forcing the consecutive lightpath requests (R1–R4) to choose links already in use in order to avoid powering on additional network equipment. Hence, in an attempt to reduce power consumption the allocated paths become on average longer, possibly creating bottlenecks in the network, for example, link C-G is soon out of resources, making it impossible to provision connection R4.

On the other hand, the objective of a conventional, that is, not power-aware, RWA algorithm (Figure 15.6), is to balance the network resource utilization in order to minimize the blocking probability.

In the example it is shown that the power minimization strategy is able to save 35% and 14% of the energy consumed by fiber links and network nodes, respectively, at the expense of having to block 25% of the connection requests (Figure 15.5). On the other hand, considering the same traffic, the provisioning strategy based on the minimum number of hops is able to save only 12% of energy consumed by the fiber links, but it will not block any lightpath request (Figure 15.6).

This trade-off can be assessed formally by means of a Weighted Power-Aware Optical Routing (WPA-OR) strategy [18]. WPA-OR is based on a modified version of the k-shortest path algorithm [14], where each fiber link l in the network is assigned a weight (C_l) equal to:

$$C_l = \begin{cases} \alpha \cdot P_{\text{link},l}, & \text{fiber link } l \text{ in use} \\ P_{\text{link},l}, & \text{fiber link } l \text{ not in use} \end{cases},$$

where $P_{\text{link},l}$ represents the power necessary to operate the in-line power amplifier(s) along fiber link l, and α is a weighting factor with values between 0 and 1. Note that $\alpha = 0$ corresponds to a pure power minimization approach, while values of α close to 1 force to provision lightpath requests along shorter routes. Tables 15.3 and 15.4 present a trade-off between the level of energy savings and the blocking probability.

The results are obtained by simulation considering the COST 239 network topology [19], which is a Pan-European test network topology that comprises 11 nodes and 26 bidirectional links. More details about the experimental setup are available at Ref. [18]. For any given value of the load, the average power per request is defined as the ratio between the total network power consumption and the number of provisioned connection requests. The average power saved per request is computed as the difference between the total network power consumption obtained when $\alpha = 1$ and the total network power consumption for any other given value of α. The tables show that considerable power savings (up to 50%) can be achieved, but at the expense of a relevant increase in the network blocking probability.

It is obvious that power efficiency and network blocking probability are two conflicting objectives, because paths obtained by applying a pure power-aware provisioning are typically longer than the shortest possible paths, which leads to network resources fragmentation and an increased blocking probability. This can be avoided by jointly considering power minimization and resource blocking in a single cost function.

15.4.2   Energy Savings versus Quality of Transmission in WDM Core Network Design

Despite the high number of energy-efficient strategies for WDM network design [20], considering static [12] and dynamic [21, 22] provisioning, there is an important aspect of the power minimization problem in transparent WDM networks that is not always properly addressed. The absence of signal regeneration, that is, reamplification, reshaping, and retiming of the optical signal also known as 3R [23] (with its benefits in terms of reduced power consumption), has an impact on the optical signal quality at the receiver because Physical Layer Impairments (PLIs) degrade the signal quality along an optical connection.

PLIs can be divided into linear and nonlinear impairments. Linear impairments do not depend on the signal power and affect each wavelength channel individually. In contrast, nonlinear impairments such as Cross Phase Modulation (XPM) and Four Wave Mixing (FWM) can cause disturbance and interference among channels traversing the same fiber link. Usually, these physical phenomena are accounted during the RWA phase, that is, impairment aware RWA (IA-RWA) strategies [24, 25] aim at minimizing the impact of PLIs on the established connections.

Ignoring the quality of the optical signal while minimizing the power consumption might have detrimental effects on the overall network performance. In fact, energy minimization

Table 15.3 Average power saved per request (%) as a function of the network load and α

	Load [Erlang]												
	15	45	75	105	135	165	195	225	255	285	315	345	375
α = 1.00	0.0%	0.0%	0.0%	0.0%	0.0%	0.0%	0.0%	0.0%	0.0%	0.0%	0.0%	0.0%	0.0%
α = 0.75	9.7%	16.9%	17.9%	16.6%	14.8%	11.3%	9.3%	6.8%	4.8%	2.6%	1.0%	0.3%	0.2%
α = 0.66	13.7%	30.8%	31.0%	26.3%	22.0%	15.6%	12.2%	8.8%	6.5%	4.0%	2.1%	0.9%	0.2%
α = 0.50	21.0%	40.6%	39.3%	32.9%	27.5%	19.5%	15.0%	11.4%	8.6%	5.5%	2.9%	1.3%	0.4%
α = 0.33	27.6%	45.2%	42.5%	36.1%	31.0%	23.0%	18.5%	14.0%	10.3%	7.4%	4.1%	2.5%	1.3%
α = 0.10	38.6%	48.5%	44.0%	39.1%	33.9%	27.4%	22.4%	17.1%	13.5%	10.4%	7.0%	4.1%	2.8%
α = 0.0001	40.2%	49.0%	45.6%	40.0%	35.4%	28.9%	24.0%	18.8%	14.9%	11.9%	8.8%	6.6%	4.3%

Table 15.4 Blocking probability as a function of the network load and α

	Load [Erlang]												
	15	45	75	105	135	165	195	225	255	285	315	345	375
α = 1.00						0.0000	0.0000	0.0007	0.0033	0.0095	0.0207	0.0382	0.0623
α = 0.75						0.0000	0.0002	0.0012	0.0043	0.0103	0.0226	0.0389	0.0624
α = 0.66					0.0000	0.0002	0.0006	0.0019	0.0058	0.0131	0.0251	0.0413	0.0631
α = 0.50				0.0001	0.0002	0.0005	0.0014	0.0037	0.0089	0.0175	0.0277	0.0432	0.0657
α = 0.33			0.0002	0.0006	0.0021	0.0045	0.0086	0.0140	0.0189	0.0269	0.0350	0.0513	0.0724
α = 0.10		0.0006	0.0045	0.0092	0.0178	0.0307	0.0415	0.0518	0.0623	0.0710	0.0773	0.0863	0.1032
α = 0.0001	0.0001	0.0077	0.0173	0.0297	0.0438	0.0519	0.0750	0.0819	0.0930	0.1017	0.1207	0.1292	0.1308

and maximization of the number of connections with their optical signal quality above a certain threshold can be considered as two conflicting objectives. This is mainly because of the way energy minimization provisioning techniques work. Most of the energy efficient RWA approaches tend to concentrate (i.e., "pack") provisioned connection requests on few links to allow as many resources as possible to enter a standby/sleep state. As a consequence, these energy-efficient RWA strategies may result in (i) longer routes on average for established connections, and (ii) higher utilization of the active fiber links, that is, the average number of used wavelength channels per fiber link being higher compared to a conventional (not energy-efficient) RWA approach. Longer paths, on the other hand, translate into worse attenuation levels, and denser fiber links result in higher XPM and cross talk levels. For these reasons, RWA strategies focusing solely on energy efficiency perform insufficiently with respect to signal quality guaranties.

The negative effects of PLI on the optical signal while performing a green WDM design may be limited by introducing regeneration at intermediate nodes inside the network [23, 24]. The idea is to trade higher power consumption values from regeneration operations at selected nodes for a better overall signal quality. Another alternative is to develop a combined impairment and energy-aware RWA approach [26], which already considers the impact of PLIs while solving the RWA problem encountering additional constraints, that is, in addition to the power minimization ones.

The PLI of a connection can be quantified by using the quality factor Q. The Q factor can be computed using the expression in Ref. [25] that includes the effects of amplification noise, the combined effects of optical filtering, XPM, and FWM. The Q factor function in Ref. [25], however, is not linear, and therefore, adding the Q factor to an RWA mathematical formulation entails nonlinear constraints, which need to be added. To avoid nonlinearity, while aiming to provide a model that accurately computes the Q factor in an Integer Linear Programming (ILP) formulation, three main techniques can be used:

- Path precomputation: as linear impairments depend only on the length of the route of the optical connections, the use of arc-path-based ILP formulations, where a set of routes are precomputed, allows for precomputing linear impairments associated to each route.
- Worst case assumption: XPM is the dominant nonlinear effect and its impact is magnitudes larger than FWM [27]. Therefore, FWM can be precomputed assuming a worst-case value (i.e., using a conservative approach) and considered as a fixed penalty for each link.
- Statistical XPM model: the statistical linear model presented in Ref. [28] allows for the estimation of the XPM noise variance, making it usable in an RWA ILP formulation

Compared to existing impairment aware RWA (IA-RWA) and energy-aware (EA-RWA) schemes, it can be shown that a combined energy and impairment aware RWA (EIA-RWA) approach provides energy consumption reduction close to that obtained by EA-RWA, but it still guarantees a sufficient level of the optical signal quality.

Figure 15.7 presents a comparison in terms of three different metrics: (a) power consumption, (b) number of active links, and (c) connections exceeding a predefined Bit Error Rate (BER) threshold, that is, the most common metric used to assess the optical signal quality of a lightpath. Results were obtained by running an ILP solver [29] on a 16-node and 23-link optical topology [30]. Each set of results is generated for three different BER thresholds (i.e., BER-th: 1E-9, 1E-11, and 1E-13) to represent lightpaths that require high, very high, and

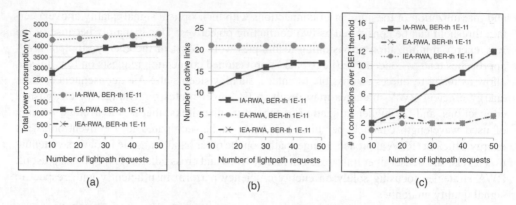

Figure 15.7 Comparison of RWA approaches

extremely high transmission quality, respectively. In terms of Q factor values, the considered BER thresholds translate into Q_{thres} of approximately 6, 6.7, and 7.3. More details on the ILP formulation, the power consumption model as well as on the experimental setup and assumptions are available at [26].

It is clear from Figure 15.7 that both EA-RWA and IEA-RWA achieve the same reduction in total power consumption (ranging from 7% up to 35%) as the IA-RWA approach. In terms of the number of used fiber links (see Figure 15.7(b)), both energy-aware approaches (i.e., EIA-RWA and EA-RWA) use the same amount. IA-RWA, on the other hand, activates every fiber link in the network, in order to minimize the number of requests above the BER threshold. The reason is that IA-RWA tends to choose short routes to minimize the effect of linear impairments, and it encourages the assignment of wavelengths that are spread around in the optical spectrum to avoid nonlinear ones. Finally, Figure 15.7(c) shows how combining energy and impairments objectives in the IEA-RWA approach provides signal quality levels that are very close to the ones provided by the IA-RWA approach while minimizing power consumption. In contrast, EA-RWA performs relatively poorly in terms of the number of lightpaths that satisfy their optical signal quality requirements.

In summary, it is important to have an approach that combines the concept of impairment and energy awareness while designing a WDM network. The reason is that minimizing energy may result in drastic signal degradation as a consequence of reducing the number of active links and thus increasing the average load per link. Conversely, considering only signal quality optimization may result in wasting large amount of energy. Connections are spread among different links to minimize the link load that increases the number of active resources in the network. By contrast, an IEA-RWA approach proves to be able to reduce significantly the power consumption levels (almost reaching the optimum level given by the EA-RWA approach) while providing the required minimum signal quality nearly equivalent to IA-RWA levels.

15.4.3 *Energy Saving versus Resource Utilization in Green and Resilient Core Network Design*

Resiliency to device failures is a fundamental requirement in WDM core networks. It can be achieved by installing or reserving redundant resources that will be used only in the case of a failure to restore the affected connections. Such resources are typically maintained in an active state, independent of the failure pattern, and thus consume power even when they are not utilized.

The energy efficiency of survivable WDM networks can be improved by using various techniques. One solution consists in planning or operating the WDM network in such a way that the power consumption is minimized, in addition to (or instead of) minimizing the resource installation or utilization [14, 31, 32].

Additional improvement can be achieved by enabling a sleep mode state in network equipment [14, 33]. Sleep mode represents a low power, inactive state from which devices can rapidly wake up when necessary. As redundant resources are unused until a failure occurs they can be set in sleep mode, provided that each connection of Service Level Agreement (SLA) can still be satisfied, that is, resource in sleep mode can be back in operation mode within certain time limits. A proper planning [14, 31, 32] and management [34] of the WDM network is required to support sleep mode. Indeed, devices can be put to sleep mode when they support only protection connections. The problem of planning a WDM network supporting sleep mode has been investigated extensively in the literature. Some works look into ways to improve the energy efficiency in all-optical WDM networks with dedicated path protection [14, 32], while others investigate dedicated and shared path protection schemes that can be used in IP-over-WDM networks [31].

Further improvement can be achieved by selecting an energy-efficient protection technique. An example is given by shared path protection that is known to require fewer resources than dedicated path protection, hence leading to energy savings. In the presence of shared protection, additional savings in resource usage and thus in energy can be achieved by enabling differentiated reliability (DiR) levels, which are guaranteed to the provisioned connection requests [35, 36]. If DiR is enabled, the protection path is not always available for each possible link failure scenario, which results in a significant reduction of the number of used network resources (i.e., wavelength) and thus power consumption. A proper selection of the protection resources is fundamental for the effectiveness of the protection technique with the requested reliability level, in order to enable energy savings.

The energy saving techniques for survivable WDM networks discussed above can also be combined to achieve even higher energy efficiency. The impact of three different techniques is assessed [36] using a dynamic WDM network represented by a Pan-European topology with 11 nodes and 26 bidirectional links (i.e., COST 239 [19]) where shared path protection is used. Demands are assumed to arrive randomly at the network nodes and they must be served as they are received. Each demand consists of one lightpath that needs to be provisioned between two nodes with a given level of reliability that needs to be satisfied. A demand is blocked when resources are insufficient for setting up the lightpath with the requested level of reliability. The reliability level is modeled in terms of maximum conditional failure probability (MCFP) that represents the maximum acceptable probability that the connection will not survive when a link failure occurs.

A power minimization (PM) or a resource minimization (RM) strategy is used to solve the RWA problem for the working and protection path of each lightpath connection request. The PM strategy solves the RWA problem giving priority to the power consumption values of the devices and their operational state, that is, active or sleep. The RM strategy, on the other hand, focuses only on the minimization of the number of wavelength resources used to provision each working and protection path. For each request, the working and protection paths are selected

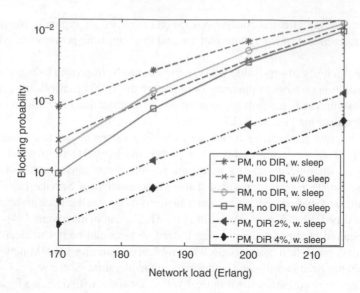

Figure 15.8 Blocking probability versus offered network load when minimizing power consumption and resource utilization

among a list of precomputed paths, according to the PM or the RM strategy, while ensuring the required level of reliability, that is, MCFP. Details of the PM and RM routing strategies and the power consumption model as well as of the experimental setup and assumptions are available at [36].

Figure 15.8 shows the blocking probability versus network load for PM and RM routing strategies. PM blocking probability performance is not as good as RM especially at low loads when the impact of blocking probability performance is less detrimental. The impact of the sleep mode and DiR are also assessed. Two insights can be gained. First, the blocking probability slightly increased when enabling sleep mode. The reason is that, for the working lightpaths, in the presence of sleep mode the cost function forces the selection of longer paths. In turn, this strategy allows setting as much as possible resources in sleep mode. This approach leads to a higher resource utilization and consequently higher blocking. On the other hand, an improvement of more than one order of magnitude in blocking probability can be achieved when a lower reliability level is requested. However, if the reliability level is further decreased (i.e., passing from 2% to 4% MCFP), only a marginal improvement in blocking probability can be achieved.

The average power savings of the different energy-efficient techniques are depicted in Figure 15.9. The PM strategy enables savings of up to 10% compared to the RM strategy, while additional savings up to 10% can be gained at low loads because of the sleep mode. However, these benefits decrease at high loads. More significant power savings, almost independent of the load, are gained by exploiting DiR. The maximum of about 25% is achieved by jointly optimizing power consumption and resource utilization while enabling DiR and sleep mode.

In summary, energy-aware strategies help to save power at different loads but higher blocking probability especially at low loads. This trade-off between network performance

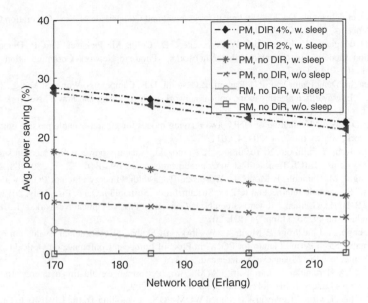

Figure 15.9   Average power saving versus offered network load

and power savings exists also when introducing a sleep mode, which is energy efficient at low loads but increases the blocking probability. On the contrary, DiR may effectively provide energy efficiency, while introducing significantly less blocking probability, independent of the load. The combined effects of the three different energy-efficient techniques can lead to energy savings of up to 25% while improving the blocking probability compared to a conventional protection techniques.

15.5   Summary

Although it is obvious that optical networks are a promising alternative to reduce the power consumption of telecommunication networks, energy can be further saved by proper power-aware network design and/or connection provisioning approaches. While investigating the various aspects of green optical networks this chapter showed how power minimization and some important network performance parameters (i.e., delay, connection blocking probability, lightpath quality of transmission, and connection reliability levels) are in conflict. The reason lies either in the length of the provisioned paths originating from a pure power-aware provisioning approach or in the devices transition times required to go from a low power to a fully functional operational state. The presented results highlight this crucial aspect of the green optical networking optimization problem and show how it can be mitigated by carefully balancing the importance of the objective functions to be optimized.

References

[1] R. Bolla, R. Bruschi, F. Davoli, F. Cucchietti, "Energy Efficiency in the Future Internet: A Survey of Existing Approaches and Trends in Energy-Aware Fixed Network Infrastructures," IEEE Communications Surveys & Tutorials, vol.13, no.2, pp.223–244, 2011.

[2] OASE project deliverable D4.2.2, "Technical Assessment and Comparison of Next-Generation Optical Access System Concepts, March 2013".

[3] W. Van Heddeghem, F. Idzikowski, W. Vereecken, D. Colle, M. Pickavet, and P. Demeester, "Power consumption modeling in optical multilayer networks," Photonic Network Communications (2012), DOI: 10.1007/s11107-011-0370-7.

[4] L. Valcarenghi, Dung Pham Van; P.G. Raponi, P. Castoldi, D.R. Campelo, S. Wong, S. Yen, L.G. Kazovsky, S. Yamashita, "Energy efficiency in passive optical networks: where, when, and how?," Network, IEEE, vol. 26, no. 6, pp. 61–68, 2012.

[5] B. Skubic, D. Hood, "Evaluation of ONU power saving modes for gigabit-capable passive optical networks," IEEE Network, vol. 25, no. 2, pp. 20–24, 2011.

[6] R. Kubo, J.I. Kani, Y. Fujimoto, N. Yoshimoto, K. Kumozaki, "Adaptive power saving mechanism for 10 Gigabit class PON systems", IEICE Transactions on Communications, vol. E93-B, no. 2, pp. 280–288, 2010.

[7] L. Valcarenghi, M. Chincoli, P. Monti, L. Wosinska, P. Castoldi, "Energy efficient PONs with service delay guarantees," Sustainable Internet and ICT for Sustainability (SustainIT), Oct 4, Pisa, Italy, 2012.

[8] R. Dhaini, P.-H. Ho, G. Shen, "Toward Green Next-Generation Passive Optical Networks," IEEE Communication Magazine, vol. 49, no. 11, pp. 94–101, 2011.

[9] M. Fiammengo, A. Lindström, P. Monti, L. Wosinska, B. Skubic, "Experimental Evaluation of Cyclic Sleep with Adaptable Sleep Period Length for PON," in Proc. of European Conference on Optical Communications (ECOC), September 18–22, Geneva, Switzerland, 2011.

[10] P. Reviriego, J.A. Hernandez, D. Larrabeiti; J.A. Maestro, "Performance evaluation of energy efficient Ethernet," Communications Letters, IEEE, vol. 13, no. 9, pp. 697–699, 2009.

[11] P. Wiatr, J. Chen, P. Monti, L. Wosinska, "Green WDM-PONs: Exploiting Traffic Diversity to Guarantee Packet Delay Limitation," in Proc. of IEEE Optical Network Design and Modeling (ONDM), April 16–19, Brest, France, 2013.

[12] Y. Wu, L. Chiaraviglio, M. Mellia, and F. Neri, "Power aware routing and wavelength assignment in optical networks," in Proc. of IEEE European Conference on Optical Communications (ECOC), September 20–24, Vienna, Austria, 2009.

[13] A. Coiro, M. Listanti, A. Valenti, F. Matera, "Reducing Power Consumption in Wavelength Routed Networks by Selective Switch Off of Optical Links," Selected Topics in Quantum Electronics, IEEE Journal, vol. 17, no. 2, pp. 428–436, 2011.

[14] A. Muhammad, P. Monti, I. Cerutti, L. Wosinska, P. Castoldi, and A. Tzanakaki, "Energy-efficient WDM network planning with dedicated protection resources in sleep mode," in Proc. of GLOBECOM, December 6–10, Miami, USA, 2010.

[15] C. Cavdar, F. Buzluca, and L. Wosinska, "Energy-Efficient Design of Survivable WDM Networks with Shared Backup," in Proc. of GLOBECOM, December 6–10, Miami, USA, 2010.

[16] R. Tucker, "Green optical communications - Part I: Energy limitations in transport," IEEE J. Selected Topics in Quantum Electronics, vol. 17, no. 2, pp. 245–260, 2011.

[17] R. Tucker, "Green optical communications - Part II: Energy limitations in networks," IEEE J. Selected Topics in Quantum Electronics, vol. 17, no. 2, pp. 261–274, 2011.

[18] P. Wiatr, P. Monti, L. Wosinska, "Power savings versus network performance in dynamically provisioned WDM networks," Communications Magazine, IEEE, vol. 50, no. 5, pp. 48–55, 2012.

[19] P. Batchelor et al., "Study on the Implementation of Optical Transparent Transport Networks in the European Environment-Results of the Research Project COST 239," Photonic Network Communications, vol. 2, no. 1, pp. 15–32, 2000.

[20] A. Ahmad, A. Bianco, E. Bonetto, D. Cuda, G. G. Castillo, and F. Neri, "Power-aware logical topology design heuristics in wavelength-routing networks," in Proc. of IEEE Optical Network Design and Modeling (ONDM), February 8–10, Bologna, Italy, 2011.

[21] P. Wiatr, P. Monti, and L. Wosinska, "Green lightpath provisioning in transparent WDM networks: pros and cons," in Proc. ANTS, December 16–18, Mumbai (Bombay), India, 2010.

[22] C. Cavdar, "Energy-efficient connection provisioning in WDM optical networks," in Proc. OFC, March 6-10, Los Angeles, USA, 2011.

[23] M. S. Savasini, P. Monti, M. Tacca, A. Fumagalli, H. Waldman, "Regenerator Placement with Guaranteed Connectivity in Optical Networks," in Proc. of International Conference on Optical Networking Design and Modeling (ONDM), May 29–31, Athens, Greece, 2007.

[24] S. Azodolmolky, M. Klinkowskiv, E. Marin, D. Careglio, J. S. Pareta, and J. Tomkos, "A survey on physical layer impairments aware routing and wavelength assignment algorithms in optical networks," Computer Networks Journal, vol. 7, no. 53, pp. 926–944, 2009.

[25] A. Jirattigalachote, P. Monti, L. Wosinska, K. Katrinis, A. Tzanakaki, "ICBR-Diff: an Impairment Constraint Based Routing Strategy with Quality of Signal Differentiation," Journal of Networks, special Issue on All-Optically Routed Networks, 2010, vol. 5, no. 11, pp. 1279–1289, 2010.

[26] C. Cavdar, M. Ruiz, P. Monti, L. Velasco, and L. Wosinska, "Design of Green Optical Networks with Signal Quality Guarantee," IEEE International Conference on Communications (ICC), June 10–15, Ottawa, Canada, 2012.

[27] L. Velasco, A. Jirattigalachote, M. Ruiz, P. Monti, L. Wosinska, G. Junyent, "Statistical Approach for Fast Impairment-Aware Provisioning in Dynamic All-Optical Networks," IEEE/OSA Journal of Optical Communication and Networking (JOCN), vol. 4, no. 2, pp. 130–141, 2012.

[28] M. Ruiz, L. Velasco, P. Monti, and L. Wosinska, "A Linearized Statistical XPM Model for Accurate Q-factor Computation," IEEE Communications Letters, vol. 16, pp. 1324–1327, 2012.

[29] IBM ILOG Optimization Products, Licence provided by Academic Initiative, https://www-304.ibm.com /support/docview.wss?uid=swg21419058.

[30] S. De Maesschalck, D. Colle, I. Lievens, M. Pickavet, and P. Demeester, "Pan-European optical transport networks: an availability-based comparison," Photonic Network Communications, vol. 5, no. 20, pp. 203–225, 2003.

[31] F. Musumeci, M. Tornatore, J. Lopez Vizcaino, Y. Ye, and A. Pattavina, "Energy efficiency of protected IP-over-WDM networks with sleep-mode devices," J. High Speed Networks, vol. 19, no. 1, pp. 19–32, 2013.

[32] P. Monti, A. Muhammad, I. Cerutti, C. Cavdar, L. Wosinska, P. Castoldi, A. Tzanakaki, "Energy-Efficient Lightpath Provisioning in a Static WDM Network with Dedicated Path Protection," in Proc. of IEEE International Conference on Transparent Optical Networks (ICTON), June 27–30, Stockholm, Sweden, 2011.

[33] A. Morea, S. Spadaro, O. Rival, J. Perell'o, F. Agraz, and D. Verchere, "Power management of optoelectronic interfaces for dynamic optical networks," in Proc. of IEEE European Conference on Optical Communications (ECOC), September 18–22, Geneva, Switzerland, 2011.

[34] A. Jirattigalachote, C. Cavdar, P. Monti, L. Wosinska, and A. Tzanakaki, "Dynamic provisioning strategies for energy efficient WDM networks with dedicated path protection," Optical Switching and Networking, vol. 8, no. 3, pp. 201–213, 2011.

[35] J. L. Vizcaino, Y. Ye, V. Lopez, F. Jimenez, R. Duque, F. Musumeci, A. Pattavina, and P. Krummrich, "Differentiated quality of protection to improve energy efficiency of survivable optical transport networks," in Proc. of OFC, March 17–21, Anaheim, USA, 2013.

[36] A. Muhammad, P. Monti, I. Cerutti, L. Wosinska, P. Castoldi, "Reliability Differentiation in Energy Efficient Optical Networks with Shared Path Protection," in Proc. of IEEE Online Conference on Green Communication (Greencomm), October 29–31, 2013.

16

Energy-Efficient Networking in Modern Data Centers

Dominique Dudkowski and Peer Hasselmeyer
NEC Europe Ltd., NEC Laboratories Europe, Heidelberg, Germany

16.1 Introduction

Achieving energy-efficient operation of information and communication (ICT) technology has become one of society's primary objectives, being actively investigated in research and development. It is estimated that 8–10% of global carbon dioxide (CO_2) emissions are because of the ICT industry (http://www.vertatique.com/ict-10-global-energy-consumption, http://europa.eu/rapid/press-release_IP-13-231_en.htm).

Data centers play an essential role in today's ICT infrastructure by serving as the backbone for many kinds of electronic services. They make up a significant part of ICT's total energy consumption, accounting for 23% of ICT's global CO_2 emissions (http://www.gartner .com/it/page.jsp?id=530912). It is assumed that in the United States all data centers combined consumed between 1.7% and 2.2% of the total US electricity consumption in 2010, while worldwide data center energy consumption was around 1.3% [1].

This amount is certainly significant, and data center operators and researchers are working hard to reduce data center energy consumption. Approaches toward increasing data center energy efficiency mainly target computing, storage, the network, and cooling infrastructure individually or in a joint way, as illustrated in Figure 16.1. Computing nodes, including high-end, mid-range, and volume servers, together may consume as much as 40% of a data center's total energy [2]. Accordingly, computing has received the most attention by researchers and developers. In contrast, the typical energy consumption of a data center network may range from 5% [2] to 12% of overall data center consumption when servers are fully loaded [3].

Due to the small fraction of the overall energy consumption that is taken up by the network, it is necessary to motivate the need for energy-efficient solutions in the network. The most relevant arguments are:

Green Communications: Principles, Concepts and Practice, First Edition.
Edited by Konstantinos Samdanis, Peter Rost, Andreas Maeder, Michela Meo and Christos Verikoukis.
© 2015 John Wiley & Sons, Ltd. Published 2015 by John Wiley & Sons, Ltd.

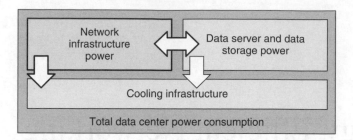

Figure 16.1   Energy efficiency considerations in the data center

16.1.1   Energy-Proportional Computing

Research on the computing side of the data center has been focused primarily on making individual servers and complete server farms *energy proportional* [4, 5], meaning that their energy consumption depends linearly on their workload. The same concept can be applied to the network, meaning that the energy consumption of network links, network elements, and the network as a whole depend linearly on their utilization (in terms of fraction of the maximum bandwidth provided).

Given that the server side is managed energy proportionally, the network's relative share of overall energy consumption increases with a decrease in computing load on the data center. Abts et al. [3] illustrate this under the premise of energy proportionality implemented on the computing side, but not on the network side. While 12% of the DC energy consumption can be attributed to the network when the servers are fully loaded, the network's share increases to 50% when the data center (both servers and network) is utilized only 15%.

16.1.2   Boost in Link Bandwidth

In data center networks, link bandwidth is constantly increasing. Common link rates are currently 1–40 Gbps, with different bandwidths typically used in different tiers of a data center network, for example, 10 Gbps links in the core and aggregation tiers, and 1 Gbps links in the access tier [6]. Reviriego et al. [7] assume that 1 Gbps Ethernet links consume 1 W while 10 Gbps Ethernet links reach 5 W. Although efficiency (in bps/W) is increasing with every generation of networking equipment and absolute power requirements are decreasing for the same bandwidth, with growing bandwidth, for example, from 10 Gbps to 100 Gbps, the power consumption still increases fivefold [8]. Because a data center network is composed of a large number of elements, changing link bandwidth results in a large increase in power consumption and therefore deserves consideration.

16.1.3   Impact on Cooling Infrastructure

Besides the direct benefits from saving energy in the network, less energy consumed by network elements decreases the requirements on cooling (Figure 16.1). While this fact is recognized on the server side and, for example, considered by Ahmad and Vijaykumar [5], it also holds for networking equipment and can make a big difference. Only when all servers *and* switches in the same rack are deactivated can the cooling auxiliaries be shut down in that segment too, reducing energy consumption accordingly.

16.1.4   Impact on Power Distribution Infrastructure

Similar to the cooling infrastructure, a second indirect impact is on the power distribution auxiliary infrastructure. For example, Pelley et al. [9] note that it is very unlikely that all servers in a network peak at the same time. A time series of 5000 servers is also given by Barroso and Hölzle [4], noting that peak power is achieved only in very few cases. Furthermore, according to Pelley et al., statistical variation of usage is high. These reasons have led to power distribution infrastructures being highly oversized due to conservative measures. While this does not primarily mean OPEX-side problems, this time it is capital expenditures on power distribution equipment that makes up a significant part of overall data center CAPEX (e.g., $10–$100 millions, according to Pelley et al. [9]).

The same reasoning applies to data center networks, as, intuitively, network traffic also correlates with computing load. Applying energy control on the network side therefore leads to similar effects in reducing power distribution needs, translating into additional CAPEX savings.

While these are strong arguments for considering energy saving in the network, current practices in operational data centers are still limited to solutions on the computing side, while the complete network remains operating 24/7. One reason for this disparity is that enabling technologies are readily available on the computing side, in particular, virtualization and virtual machine migration enabling freeing and shutting down physical machines. On the network side, however, enabling technologies for flexible traffic management are not yet available in production data center networks, although appearing in recent research, for example, Open-Flow (http://www.openflow.org).

16.2   Energy Efficiency in Data Center Networks

Work on energy-efficient networks is amply available in the literature, some of which is applicable to any type of fixed network, and some being specific to data centers. The main consensus is that energy savings should never occur at the expense of performance. The main techniques are described in this section.

16.2.1   Dynamic Link Rate Adaptation

With the increase in link bandwidth, it makes sense to allow links to run at different speeds, as the maximum link bandwidth is usually not always needed. Such scaling is supported by current technology, such as InfiniBand, which can operate at single, double, and quad data rates ranging from 2.5 to 40 Gbps [3]. Such variation in link bandwidth is sometimes classified as *performance state*, in contrast to power state, such as sleep mode [2].

If energy proportionality is given, link rate adaptation influences energy consumption according to link speed. As a network is often loaded only lightly, link rate adaptation bears large potential for energy savings. Abts et al. [3], who apply link rate adaptation in their approach, indicate a strong benefit of scaling link bandwidth: 73% of the time they are able to operate a link at only 2.5 Gbps, while it runs at 40 Gbps only 5% of the time. Gunaratne et al. [10] similarly report that a link can be run at a lower data rate than its maximum for more than 80% of the time. Even better energy proportionality can be achieved by using traffic bursts with long intermittent sleep phases [7].

16.2.2   Link and Switch Sleep Modes

Changing the operational mode (also called the *power state*, in contrast to performance state [2]) of network elements from active to a lower power state such as "idle" or "off" is another promising approach. Controlling the state of network elements is possible at the level of individual links and/or complete switches, a choice that depends on the capabilities of the network element. While link rate adaptation is suitable for elements that operate near energy proportionality, network elements are often not energy proportional and according to Heller et al. [11] consume just 8% additional energy when fully loaded compared to the idle state. In that case, turning links and/or switches on and off makes more sense, which Heller et al. consider in the proposed ElasticTree.

In order to shut down and start up network elements and individual links the data center needs to offer appropriate capabilities under programmatic control. As switches at this time do not generally support a suitable shutdown function, a viable alternative for controlling the power states of switches is via remotely switchable power strips. In case native shutdown functions are supported, reactivating a network element can be implemented via Wake-on-LAN or similar techniques. Controlling individual links of a network switch, on the other side, may be commonly achieved via SNMP.

16.2.3   Network Topology

The topology is probably the most critical parameter of a data center's network because it determines various performance and operational limits, such as latency, redundancy, and maximum aggregate end-to-end bandwidth between computing nodes. Besides performance, different network topologies have different characteristics regarding the ability to apply energy saving strategies. Figure 16.2 provides an overview of data center network topologies from the literature that are considered valuable with respect to energy efficiency.

The common data center network topology as shown in Figure 16.2.a has large oversubscription ratios (i.e., the aggregate bandwidth between layers decreases toward the top of the tree) and is therefore inadequate for cloud-style traffic loads that exhibit unpredictable traffic patterns with a large fraction of traffic crossing rack boundaries. Fat trees, and in particular their incarnation as folded-Clos networks (Figure 16.2b), have been proposed as alternative topologies featuring full bisection bandwidth. Compared to large switches commonly used in hierarchical networks, fat tree networks built from multiple smaller off-the-shelf switches can save as much as 56.6% of energy according to Al Fares et al. [6] for comparable bandwidth and performance, even if all switches are on.

Because the fat tree provides much more flexibility in deactivating individual links and whole switches while connectivity can still be maintained, Heller et al. [11] propose the Elastic Tree, which is backed by a subset of the fat tree (e.g., configuration shown in Figure 16.2c). The authors propose to turn links and switches on and off, constantly adjusting the set of active

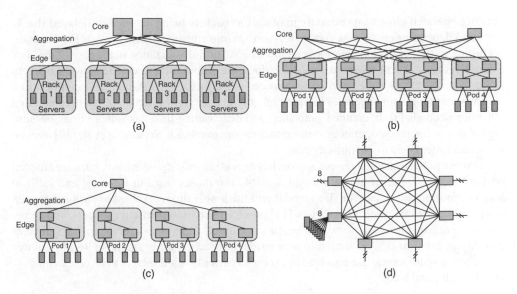

Figure 16.2 Network topologies considered in the context of energy efficiency. (a) Common date center network topology [6, 10]. (b) Fat tree topology, based on [6, 10]. (c) Elastic tree (fat tree subset), based on [10]. (d) Flattened butterfly topology for 64 nodes based on [3]

network elements according to network performance and fault tolerance requirements, leading to power savings of 25–40% compared to an always-on fat tree. If used in a large network, the approach exhibits energy-proportional behavior on the network level despite the individual elements not being energy proportional.

The butterfly topology proposed by Abts et al. [3] and shown in Figure 16.2d is claimed to be inherently even more energy-efficient than fat tree networks.

16.2.4 Combination of Approaches

Energy saving techniques for data center networks can be combined with one another relatively easily. The conceptual difference of performance states and power states according to Ref. [2] is exploited in an illustrative way by Nedevschi et al. [12]. The authors apply both sleeping and dynamic link rate adaptation by putting network elements to sleep while idle and by adapting network link rates according to traffic volume. The authors' approach is particularly interesting because it exhibits high network-wide energy proportionality on the network as a whole.

Another important aspect is how to combine network-centric energy saving techniques with computing and cooling management in the data center. Ahmad and Vijaykumar [5] explore the combination of computing and cooling in energy management (also illustrated in Figure 16.1), noting that only a joint approach can optimize the overall energy consumption of server and cooling power, instead of, for example, creating hotspots that lead to increased cooling cost.

16.2.5 Network Performance

Last but not least, maintaining data center performance is vital despite the attempts to save energy using any of the aforementioned techniques. The general observation [3, 11] is that a network operated close to its capacity may lead to packets being dropped or delayed (back-pressure) depending on the flow control mechanism implemented in the network. For instance, according to Ref. [3], for values above as low as 75% of link utilization in the case of the considered dynamic link rate adaptation, latency increases substantially, and the network is likely to saturate. An additional constraint to consider is the fact that changing the speed of a link involves reconfiguration latencies that are typically in the range between 100 nanoseconds and 100 microseconds [3]. In contrast, activating and deactivating links takes more time, because higher layers (memory, operating system) incur more overhead. Nedevschi et al. [12] assume transition times of up to 10 milliseconds.

For these reasons state-of-the-art approaches provide heuristics, where link rates are adapted on the basis of expected utilization with suitable thresholds (e.g., in Refs. [3] and [10]) to avoid saturation. In the case of sleep modes and the longer times involved to change between sleep and operational state, Heller et al. [11] provide heuristics, predictive models, and safety margins, such that links and network elements are turned on soon enough to provide sufficient capacity at any time. This is similar to server side approaches, for example, in Ref. [5], where a number of spare servers are provided in excess to currently used servers to be able to quickly handle additional load.

16.3   A Joint Energy Management Solution

Most work on energy efficiency has so far focused on particular aspects of single domains, in particular, servers or networks. It is nevertheless apparent that energy efficiency can be improved by looking at multiple domains together than at a single one in isolation. We therefore developed an approach to network energy management that performs traffic placement in the data center network using knowledge from the server and the network domains to achieve higher energy savings than would be possible by looking at the network domain alone. The main idea of our approach is to aggregate traffic in certain parts of the data center in order to increase the chances for shutting down switch ports and complete switches in the remaining parts of the data center, similar to the Elastic Tree [11]. The decisions for placing traffic are guided by utilization metrics from both the network and the server side.

Our approach rests on the notion of traffic flows. A flow is a sequence of related packets. Typical examples of flows are individual TCP connections, all traffic to a particular server, and traffic related to a particular service, for example, Web traffic. The granularity of flows can vary widely and is influenced by the traffic patterns exhibited by the ensemble of applications deployed in the data center. Our approach works with any flow definition, although its effectiveness may be influenced by traffic patterns.

All traffic flows need to be routed inside the data center network. Our approach is inspired by OpenFlow, but can also be realized with other technologies, such as SNMP and MPLS. OpenFlow is a protocol that makes routing control in the form of forwarding tables accessible to external controlling entities. Forwarding tables contain entries that specify which actions to perform to packets matching a given set of criteria. Criteria include information gathered from the Ethernet, IP, and TCP/UDP headers of data packets. Actions include rewriting certain fields of packet headers as well as forwarding instructions.

We assume that controllers decide on the routes that traffic must follow. The routes are established on demand when a new flow is encountered the first time. Once a route is determined, the flow is enforced by setting appropriate flow entries in OpenFlow-enabled switches that are on the route. Contrary to some existing approaches, like Elastic Tree, our approach does not actively reroute existing flows in the data center network. It rather restricts itself to influencing the placing of new flows on their creation, relying on a gradual accumulation of traffic in certain parts of the data center over time. As a consequence, and depending on network load, links and switches can eventually be freed from any traffic and shutdown. This approach is computationally inexpensive as computational effort for calculating energy-efficient routes is needed only at flow creation time.

Because our approach works on both switch and link level, it provides good energy proportionality, which our performance evaluation shows. Although we consider power states only according to Ref. [2], energy saving techniques that relate to performance states, such as dynamic link rate adaptation [3, 10], can be integrated easily. Moreover, our approach does not depend on a particular network topology, and besides the fat tree that we assume in this chapter, any other topology is also supported.

16.3.1 Description of Approach

The idea of our approach is to anticipate which network elements or individual ports are likely to become inactive in the near future. By inactive, we mean that a network element or port does not process any flow, that is, it does not forward any packets. Intuitively, the smaller the amount of flows processed by a network element, the easier it becomes to turn off the switch by either waiting for the remaining flows to time out or to be actively rerouted. In this chapter we consider timeouts only.

Figure 16.3 illustrates the overall approach in a small fat tree network that is composed of commodity network switches supporting OpenFlow. The switches are controlled by the management system that is responsible (among other things) for traffic assignment. The management system is enhanced by energy control functions as described here.

In order to characterize a network element in terms of its current utilization, a scalar utilization metric is calculated for each network element. The metric expresses the approximate utilization of the switch, for example, in terms of the number of flows the switch is processing. The utilization metric of a switch is recalculated whenever the load changes, for example, when a flow is created or removed. The management system retains a record of the utilization metrics of all the switches. The metrics are used later on to find optimal traffic paths.

As an example, switch S_1 in Figure 16.3 reports its flow information to the energy management module in the management system in step 1. The energy management module then recalculates the metric for switch S_1 in step 2 and stores the information for later use.

The main process is triggered by the arrival of an unknown flow. In Figure 16.3, switch S_7 receives a traffic flow for which it does not have routing information. In step 3, it hands the request over to the management system's interceptor.

In a typical data center environment, the request is then handed to the load balancer (step 4), which selects possible target hosts (virtual machines) for the request. For instance, the request may originate from a Web server application, which requests one instance of a database application, of which multiple instances are running on different hosts in the data center. In current setups, the load balancer outputs exactly one instance. We assume that the load balancer function is changed slightly to output a small set of potential targets in order to provide host diversity for increased flexibility in path selection.

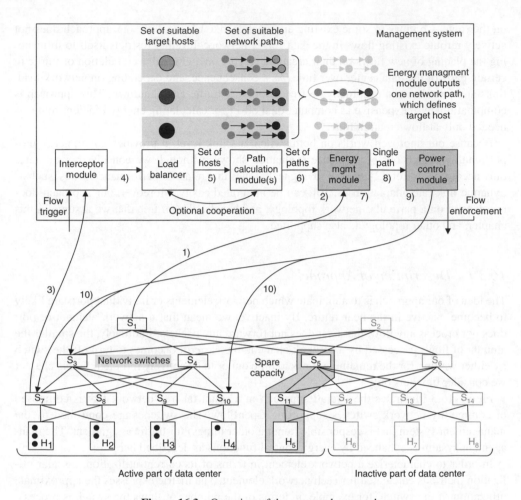

Figure 16.3 Overview of the proposed approach

In step 5, the set of hosts determined by the load balancer is handed to the path calculation module that computes paths under consideration of performance constraints such as bandwidth, latency. For each target host determined by the load balancer, this module calculates a set of suitable network paths, for example, all paths that have a minimum amount of bandwidth available to serve the request. This step results in multiple and diverse paths that are input to the energy management module in step 6.

Recall that the energy management module possesses, for any network element and port, a metric that characterizes the element and port in terms of utilization. The energy management module uses these values to select the path from the previously calculated ones, which it considers most suitable in terms of energy efficiency. Note that all paths at this point satisfy performance requirements, so the energy module can freely decide which one is the most suitable from its perspective.

Assuming network elements as the unit of granularity, the path selection algorithm, executed as step 7 in Figure 16.3, works as follows:

1. Create set S of switches containing those switches that are part of at least one input path.
2. Unmark all switches in all input paths.
3. Select and remove the switch with the highest utilization metric from S.
4. Mark the selected switch in all paths containing it.
5. If all switches in a path are marked, output that path, otherwise, go to 3.

This approach ensures that switches with the highest utilization are always preferred for placing additional flows, and the least utilized switches remain with no additional flows to increase their chances of being freed from traffic eventually.

The path determined by the aforementioned algorithm is then forwarded to the control module (step 8 in Figure 16.3), which enforces the path after ensuring that all network switches are active (step 9). Although this is usually the case, the selected path may contain network elements that need to be started up first.

In step 10, the route is finally set up by appropriately configuring all switches involved in the path. In the example in Figure 16.3, new traffic forwarding entries are created in switches S_7, S_3, and S_{10}.

The main feature of this approach is that the path selection algorithm looks at the utilization of both the network and the computing domains. By having the load balancer (computing domain) propose multiple potential destination hosts, the set of candidate paths (network domain) is increased. With it, the chances of aggregating traffic are increased and so are the chances of relieving switches from traffic and shutting them down. The joint consideration of the computing and network domains therefore results in better decisions and increased energy savings than could be achieved by looking at the individual domains separately.

16.4 Performance Evaluation

The key effect of the proposed method is that network flows are consolidated on as few network elements as possible while observing all performance constraints. The efficiency of the approach has been evaluated by simulation. We developed an event-based simulator that is designed to track energy consumption of a data center's constituting components. The simulator operates on descriptions of the physical data center model and models of the applications and their traffic. The physical model is described in terms of network and server resources as well as their interconnections. The application models describe behavior in terms of how much load applications put on servers and how much traffic they generate in response to incoming requests. Traffic is modeled with packet granularity. Load is put on the data center by clients outside the data center, which send requests to particular applications.

Multiple instances of applications can be deployed on multiple servers in order to cope with large numbers of requests. Load balancers distribute incoming requests among all running instances of an application according to a defined scheme (we used random distribution in our simulations). Deployments of applications can be scaled dynamically according to the size of the work load they receive. If the currently active application instances are close to their saturation points, an auto scaler component automatically deploys additional application instances. Similarly, if load decreases, unnecessary instances are undeployed from servers and removed from the set of application instances. In our experiments, we assume that a maximum of two applications is deployed per server.

For our simulations we set up a fully populated 24-ary fat tree, which contains 3456 servers and 720 24-port switches. All switches exhibit OpenFlow behavior. In particular, they forward packets according to a flow table that can be influenced by an OpenFlow controller. Connection to the Internet, and therefore to the clients of deployed applications, is provided via routers connected to the core switches of the network.

We modeled three kinds of applications: Web serving (WEB), video streaming (STREAM-ING), and computing (COMPUTING). The three applications differ in the amount of processing time they need from servers and how much traffic they produce. For each incoming request, the WEB application puts a short peak of load on the server. It then responds with a small burst of data that mimics the traffic produced by serving a medium-sized Web page (64 kB). In the STREAMING application, each request is followed by a 1-minute stream of data bursts (20 chunks of 1 MB each). The COMPUTING application puts a significant amount of load on a server for 5 seconds before it responds with a packet similar in size to the WEB application. We simulated and analyzed each of the applications in isolation, meaning that only that particular application is deployed and used. In addition, a combined setup (HYBRID) was examined in which all three applications are active in parallel and are requested at a ratio of 2:1:1 with the WEB application receiving twice as many requests as the other two.

The number of requests sent to applications follows a normal distribution, aimed at resembling the traffic load distribution of a 24-hour period. In order to reduce simulation time, we map the 24-hour distribution to 10 minutes of simulated time. During that time, up to 10^6 requests are sent to applications (with 28 distinct request numbers evaluated). For the largest simulation, this equates to a maximum of about 5000 requests per second.

Energy consumption of server and network equipment is described by energy models that detail a component's energy consumption in relation to its load. In addition, we assume that components can be switched to a low power sleep mode in which significantly less energy is consumed than when the equipment is on but idle. Moving into and out of sleep mode takes some time and potentially increases energy consumption during the state migration. Although each piece of equipment can have its own specific energy model, all servers and all switches have the same energy models in our simulations. The energy model of the switches is shown in Figure 16.4. The switch's energy consumption is independent of the traffic that crosses the switch, but starting or stopping it requires time and additional power. Energy consumption of servers was modeled as well but is not presented here.

```
begin switch-energy-model
  idle-power 100
  port-power 2
  sleep-power 2
  start-up-power 120
  shut-down-power 120
  start-up-duration-millis 10000
  shut-down-duration-millis 1000
end switch-energy-model
```

Figure 16.4   Switch energy model

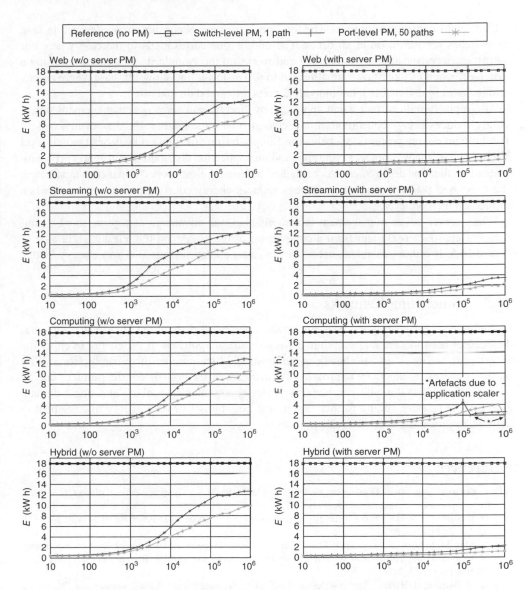

Figure 16.5　Energy consumption E of network equipment in kW h for various scenarios

Figure 16.5 shows the simulation results of the network's energy consumption for each application and the hybrid scenario, aggregated over 10 minutes of simulated time. For each setup, the chart on the left shows the energy consumption when all servers are running continuously and applications are deployed randomly across the data center. Accordingly, the network must ensure connectivity to large parts of the data center. The charts on the right side show the network's energy consumption when server power management is enabled, meaning that servers are shutdown if not needed. Application deployment in this case happens "from left to right" (vs randomly) as servers are gradually activated with increasing load.

Each chart contains three lines. The top line is the energy consumption reference in case all switches are turned on at all times. The middle line corresponds to the case where our energy management strategy is enabled and works on the granularity of switches, turning a switch off when no more flows are allocated to it. The lower line shows the case where energy management is also enabled, but individual ports are turned on and off.

All graphs have in common that applying any kind of network power management significantly reduces energy consumption in accordance with the load on the data center, closely matching an energy-proportional behavior. Simply turning off unused switches (switch-level PM) can yield 25–30% power savings on random application distribution and nearly 90% with ordered application deployment. As explained before, application performance is unaffected. Our proposed power management strategy reduces energy consumption even further when applied on the port level.

When server power management is also enabled, the advantage of a joint energy management approach is evident from a comparison of the graphs on the left and right side of Figure 16.5. It is then possible to achieve even larger energy savings also under heavy load.

16.5   Concluding Remarks

Reducing energy consumption is an important topic for all parts of the IT infrastructure, and data center networks are no exception. Saving significant portions of the energy in either the server or the network domain has been demonstrated by a number of research efforts. With our joint energy management approach we illustrate how the holistic consideration of energy management across domains can help save more energy than by looking at a single domain alone. More research should further explore the possibilities of joint energy management to help make green data centers a reality.

References

[1] J. Koomey, "Growth in Data center electricity use 2005 to 2010." Analytics Press, Oakland, CA, 2011. http://www.analyticspress.com/datacenters.html.
[2] D. J. Brown, C. Reams, "Toward energy-efficient computing," Communications of the ACM, vol. 53, no. 3, pp. 50–58, March 2010.
[3] D. Abts, M. R. Marty, P. M. Wells, P. Klausler, H. Liu, Energy Proportional Datacenter Networks, Proceedings of the 37th Annual International Symposium on Computer Architecture (ISCA'10), pp. 338–347, Saint-Malo, France, June 19-23, 2010.
[4] L. A. Barroso, U. Hölzle, "The case for energy-proportional computing," IEEE Computer, vol. 40, no. 12, pp. 33–37, December 2007.
[5] F. Ahmad, T. N. Vijaykumar, "Joint optimization of idle and cooling power in data centers while maintaining response time," ACM SIGARCH Computer Architecture News, vol. 38, no. 1, pp. 243–256, March 2010.
[6] M. Al-Fares, A. Loukissas, A. Vahdat, "A scalable, commodity data center network architecture," ACM SIGCOMM Computer Communication Review, vol. 38, no. 4, pp. 63–74, October 2008.
[7] P. Reviriego, J. A. Maestro, J. A. Hernández, D. Larrabeiti, "Burst transmission for energy-efficient Ethernet," IEEE Internet Computing, vol. 14, no. 4, pp. 50–57, 2010.
[8] C. F. Lam, H. Liu, B. Koley, X. Zhao, V. Kamalov, V. Gill, "Fiber optic communication technologies: What's needed for datacenter network operations," IEEE Communications Magazine, vol. 48, no. 7, pp. 32–39, 2010.
[9] S. Pelley, D. Meisner, P. Zandevakili, T. Wenisch, J. Underwood, "Power routing: dynamic power provisioning in the data center," ACM SIGARCH Computer Architecture News, vol. 38, no. 1, pp. 231–242, March 2010.
[10] C. Gunaratne, K. Christensen, B. Nordman, S. Suen, "Reducing the energy consumption of Ethernet with adaptive link rate (ALR)," IEEE Transactions on Computers, vol. 57, no. 4, pp. 448–461, 2008.

[11] B. Heller, S. Seetharaman, P. Mahadevan, Y. Yiakoumis, P. Sharma, S. Banerjee, N. McKeown, ElasticTree: Saving Energy in Data Center Networks, Proceedings of the 7th ACM/USENIX Symposium on Networked Systems Design and Implementation (NSDI'10), San Jose, California, USA, April 28-30, 2010.

[12] S. Nedevschi, L. Popa, G. Iannaccone, S. Ratnasamy, D. Wetherall, Reducing Network Energy Consumption via Sleeping and Rate-Adaptation, Proceedings of the 5th USENIX Symposium on Networked Systems Design and Implementation (NSDI'08), pp. 323–336, San Francisco, California, USA, April 16-18, 2008.

17

SDN-Enabled Energy-Efficient Network Management

Michael Jarschel[1,2], Tobias Hoßfeld[1,3], Franco Davoli[4,5], Raffaele Bolla[4,5], Roberto Bruschi[5] and Alessandro Carrega[4,5]

[1] University of Würzburg, Communication Networks, Würzburg, Germany
[2] Nokia Networks, Munich, Germany
[3] University of Duisburg-Essen, Modeling of Adaptive Systems, Essen, Germany
[4] Department of Electrical, Electronic and Telecommunications Engineering, and Naval Architecture (DITEN), University of Genoa, Genoa, Italy
[5] National Inter-university Consortium for Telecommunications (CNIT), University of Genoa Research Unit, Genoa, Italy

17.1 Introduction

The continuing growth of network content, applications, and services have ultimately resulted in increasing requirements for data storage capacity and data transfer speed, which raise significantly the energy consumption. In order to cope with the limited resources of user devices, more and more applications and services follow the "cloud" paradigm in order to move most of the computational and storage weight to large-scale powerful datacenters [1, 2]. In 2012 alone, such datacenters required 30 billion watts of electricity worldwide [3], which demonstrates the unsustainability of current network infrastructures to satisfy an ever-growing demand for mass ICT services considering the environmental/economic concern.

Indeed, since the Future Internet is currently shaped, it makes sense to investigate new concepts and analyze key aspects, which have the potential to impact the evolving network design criteria. One of these aspects is energy efficiency.

Achieving energy-efficient network operation requires a more flexible network control and management, since resource allocation becomes more dynamic anticipating a higher degree of adaptability for new applications and network services.

Green Communications: Principles, Concepts and Practice, First Edition.
Edited by Konstantinos Samdanis, Peter Rost, Andreas Maeder, Michela Meo and Christos Verikoukis.
© 2015 John Wiley & Sons, Ltd. Published 2015 by John Wiley & Sons, Ltd.

In this context, software defined networking (SDN) [4] and network functions virtualization (NFV) [5] are viable solutions to boost network capacity swiftly and flexibly ensuring smooth connectivity across increasingly complex networks and clouds for global users, who are often on the move. Not only that, but as networks are used for an increasing number of tasks of varying complexity, all of this needs to be done in a more efficient and cost-effective way.

The development of SDN-enabled applications that allow more efficient energy consumption requires the presence of specific functionalities inside the network devices. We refer to these functionalities as "Power Management Primitives" (PMPs). The remainder of this chapter is structured as follows. First we introduce the PMPs as well as global network primitives. Then, we define an SDN-based network architecture for energy-efficient networking and formulate the notion of green abstraction layer (GAL) [6] for the power management of an individual device. Finally, we draw our conclusions.

17.2   Background: Concepts for Network Operation

Two promising concepts to enable energy-efficient networking in the future are SDN and NFV. We briefly introduce both in this section.

17.2.1   Software Defined Networking

The key principle of SDN is the separation of the network elements' control plane to a central external entity. There are several advantages in this concept. The external control plane can be a software program run on commodity hardware. This removes the control plane from a monolithic device and enables development and adaptation of the control plane without the need for new hardware. On the other hand, the data plane has now become interchangeable and can consist of hardware built from standard components. Essentially, both control and data plane hardware are now commodity, and products of different vendors can be combined in a "mix-and-match" approach.

This is possible only because of the open interface, which SDN requires, between control and data plane. This interface is often called the "Southbound-API." The most popular realization of this interface is the OpenFlow protocol [7]. It provides a set of standard messages that allow an external control plane to operate its connected network elements via a management network. On top of the controller, the network functionalities are run as network control modules (NCMs). These modules are freely programmable and can be changed and combined according to the requirements of the network they operate. In addition, these control applications can communicate with the control plane of applications running on top or in conjunction with the network, for example, a cloud orchestration software, to optimize the network according to the applications' requirements. This interface is called the "Northbound-API." Figure 17.1 illustrates the components of the SDN architecture, as well as an OpenFlow-based realization. The OpenFlow switch (=data plane) is connected to the controller (=network control plane) via the OpenFlow protocol. In turn, the controller exchanges information with the network component (=application control plane) of a cloud orchestration software, that is, OpenStack, via a special purpose module. By introducing an abstraction or virtualization layer between controller and switch, it is also possible to change the modules entirely during the operation of the network. This provides the required flexibility for energy-efficient networking.

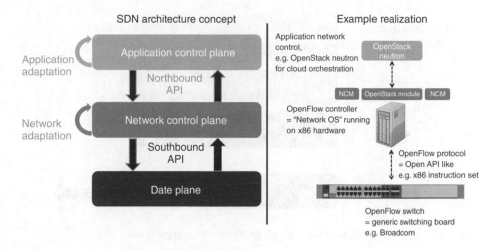

Figure 17.1	SDN concept sketch

## 17.2.2	Network Functions Virtualization

The concept of NFV is the next logical step in the commoditization process of networking. The central idea is to run network functions, for example, deep packet inspection or intrusion detection, in virtual machines on commodity hardware. Currently, these functions are realized on special purpose hardware, that is, middle boxes. With virtualized network functions the advantages of cloud computing can be leveraged. Network functions can be dynamically instantiated and removed, scaled flexibly, and relocated according to the current demands on the respective function. This way, resources that are not used can be shut down and energy can be saved. There are two key criteria for this concept to be deployed successfully. First, a swift I/O performance between the physical network interfaces of the hardware and the software user-plane in the virtual functions is required to enable sufficiently fast processing. Second, a well-integrated network management and cloud orchestration system is necessary to benefit from the advantages of dynamic resource allocation and to ensure a smooth operation of the NFV-enabled networks. Here, NFV can benefit from being deployed in conjunction with SDN. However, SDN is not a requirement. Figure 17.2 shows an example of such a joint SDN/NFV deployment enabling the flexible composition of multiple network functions.

In this example, an SDN switch is used to selectively redirect a portion of the production traffic to a server running virtualized network functions. This way the server and functions do not need to cope with all production traffic, but only the relevant flows. The virtual switch running inside the server's hypervisor is SDN-enabled and can dynamically redirect traffic flows transparently to an individual network function or to a chain of network functions. This enables a very flexible operation and network management, as functions can be plugged in and out of the service chain at runtime.

## 17.3	Energy-Efficient Network Management Practices

There is a variety of ways to improve the energy footprint of a network. The focus of this section concentrates on power management as well as on the energy efficiency of devices and

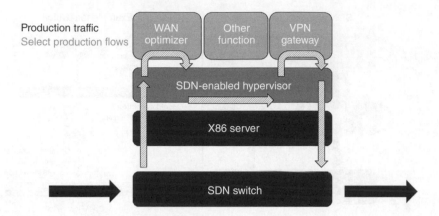

Figure 17.2   SDN/NFV functional composition

network primitives. We define a primitive as an inherent function or feature of a device or network that does not require any additional components to work.

17.3.1   Power Management Primitives

The vast majority of currently deployed network links and devices are designed to operate (and, consequently, to consume power) constantly at their maximum capacity, irrespective of the traffic load and even though their average utilization lies far below the maximum [8–10].

These observations have suggested a profitable power conservation potential for adapting the network offered resources to the actual traffic profiles [11, 12]. Similarly to general purpose computing systems, this ability can be realized by including PMPs into the hardware platforms of networking devices, where energy absorption physically takes place.

PMPs allow a highly dynamic adaptation of the energy consumption in networking devices or some of their components, by putting them into standby states when they are not used, or by decreasing their maximum performance in the presence of low incoming traffic volumes. The best performance is provided when the device operates under no power limitation, while the maximum power saving is obviously obtained when the equipment is completely turned off. There is a whole range of intermediate possibilities between these two extremes.

In principle, the main PMPs can be classified into two categories:

1. Dynamic adaptation (DA)
2. Sleeping/standby

The DA of network/device resources is designed to modulate (i.e., dynamically adapt) capacities of packet processing engines and of network interfaces, to meet actual traffic loads and quality of service (QoS) requirements. This can be performed by using two power-aware capabilities, namely, performance scaling (PS) and idle logic (IL), which both allow the dynamic trade-off between packet service performance and power consumption. PS adapts the processing speed (by changing operating frequency, possibly together with voltage, or throttling the clock), whereas IL exploits idle times (when no processing is required) to put the processors into low power states and to resume the processing at the chosen speed when new packets arrive (thus, operating at the packet timescale).

Sleeping/Standby approaches are used to smartly and selectively drive unused network /device portions to low standby modes (or to switch them off altogether) and to wake them up only when needed (thus, operating at a much longer timescale with respect to packet processing/transmission times). However, since today's networks, related services, and applications are designed to be continuously and always available, standby modes have to be explicitly supported by using special proxying techniques that are able to maintain the "network presence" of sleeping nodes/components and to reactivate the sleeping device when needed (again, with a sensibly longer timescale than IL-related reactivation).

The PMP-enabled devices need control loops to dynamically tune hardware capabilities to provide the required QoS level for incoming traffic with minimal power consumption. It is worth noting that the PMPs have features locally available in network nodes, and their efficiency may heavily depend on the specific implementation and low-level details of a device's hardware platform; the latter may be quite heterogeneous (in terms of hardware components or firmware), even when considering equipment of the same market segment or vendor.

Owing to these considerations, it is necessary to provide each network device with its own independent control loop, namely, to provide local control policies (LCPs). These control loops may dynamically orchestrate the configuration of internal components (e.g., line-cards, link interfaces, network processors) to meet the desired QoS with the minimum power consumption. However, when each device independently performs energy optimizations, the resulting overall network power conservation is not as high as in the case of cooperation among the participant network nodes. Along this direction, a number of approaches [13–15, 16] have been recently proposed in order to extend current routing and traffic engineering policies beyond conventional network QoS metrics and to also explicitly consider the energy consumption of the entire network.

Figure 17.3 shows the necessary steps in order to optimize the power consumption of the network device and its hardware (HW) components. Internally, a network device can be provided with several LCPs, each of which is designed for a specific goal (e.g., controlling a fan, scaling an operating frequency, to meet given requirements). These LCPs set the energy configuration of the HW components, by using the relative PMPs (a) that, in turn, act directly on the hardware components (b). The critical points of this scenario concern:

1. how the LCPs know the PMPs enabled in the specific devices;
2. how the PMPs can set the HW components considering heterogeneous equipment.

In addition to these problems, it is necessary to take into account that LCPs alone are not sufficient to optimize the energy consumption, while at the same time ensuring a given level of performance. Hence, it is necessary to introduce centralized control policies that consider the network as a whole. We refer to these policies as "Network-wide Control Policies" (NCPs).

Despite their high potential effect compared to LCPs, NCPs applied along suffer from certain drawbacks. First, NCPs can exhibit much higher feedback/convergence delays. Secondly, routing and traffic engineering frameworks generally may not have the ability of distinguishing how logical network entities can be mapped to physical resources, which directly cause energy absorption. Finally, NCPs often represent a network device simply as a node in a graph, whose arcs are the virtual/physical network links.

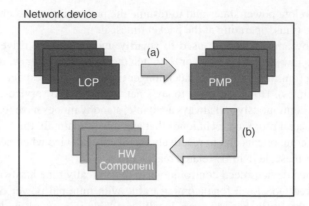

Figure 17.3   LCP–PMP interactions (from LCP control action (a) to its implementation on the physical component (b))

Such a simplistic representation does not allow retaining the knowledge of some hardware peculiarities (e.g., the possible different operating and idle states) that may be very important for reducing energy consumption. Such drawbacks suggest that jointly adopting LCPs and NCPs may optimize the device energy consumption, by following, for example, a hierarchical network management architecture.

Figure 17.4 shows a network example with 5 network devices (ND) connected to each other. Each ND supports LCPs and PMP functionalities, while one ND provides NCPs in order to control the other devices, by considering network-wide constraints. Such a case requires coordinated and well-defined interactions between the NCPs and the other NDs, where NCPs should not experience difficulties in obtaining and setting the energy-aware configuration of each ND.

On the basis of considerations, it is concluded that there is still a significant gap between hardware power management and LCPs/NCPs, as well as certain open issues regarding the adoption of control loops and their effective management, by considering the relationships among multiple local and/or network-wide control loops. Besides devising new energy-aware LCPs and NCPs, an almost necessary condition for their effective development and adoption is the representation of management and control actions, as well as device/network status information, in some standard abstract form, independently of the details of the specific manufacturers' implementations.

So, in order to better exploit the energy-aware functionalities to optimize the network devices considering the energy consumption, it is necessary to define an interface that provides a way to expose green networking capabilities of devices toward the network control plane. We refer to this interface as "Green Abstraction Layer."

17.3.2   Network Primitives

Apart from actively influencing and optimizing the power consumption of intermediate devices and middle boxes locally, the design and operation of the network as a whole can be adapted

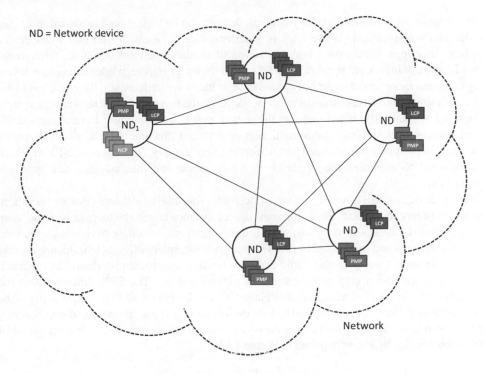

ND = Network device

Network

Figure 17.4 LCP–NCP interactions

to be as resource- and, therefore, power-efficient as possible. Hence, we identify two classes of power consumption optimization methods:

1. Local, hardware-specific, and device-based
2. Global, parameter-based, and network-wide

The former approach was discussed in the previous section. In this section we discuss how to make energy-efficient operation and resource management intrinsic to the network, that is, a network primitive.

To enable such a network primitive, the network needs to be able to instantiate and operate network devices in a highly dynamic way [17]. This is necessary because the demands on the network infrastructure change frequently. Technologies that meet the criteria to create such a network are virtualization and the related NFV, as well as SDN. Together they form the foundation for a network substrate based on commodity hardware. At first glance, the advantage of commodity hardware opposed to special purpose devices in terms of energy efficiency may not be obvious. The designers of special purpose hardware can build the equipment for one specific purpose and thus optimize the power usage of each device toward that. However, during the normal operation of a network, not all devices are needed all the time or only at a relatively low level of utilization. This means that even though the devices themselves may be energy efficient, their operation is not.

On commodity hardware, arbitrary network devices can be instantiated as virtual machines on-demand when needed. While there is a performance overhead for an individual device instance, this approach allows a high utilization of available network resources through the consolidation of individual network functions and devices. Therefore, while each virtual device may consume more power under heavy load than its hardware counterpart, the reduction in the amount of underutilized resources in the network yields the benefit that less network equipment is running permanently, which reduces the overall power consumption. Leveraging today's cloud management systems, the instantiation of a virtual machine with a specific network function can be highly automatized and performed very quickly. In the area of NFV, research is conducted to minimize the overhead of virtualization and thus advance this approach further [18].

Virtual devices alone, however, are not sufficient for this kind of efficient network operation. Decisions to reroute traffic to and from a new device instance have to be on pace with the setup of the device. In the traditional distributed routing approach, no single device has a view on the entire network and the flow of traffic. This makes it extremely difficult to influence routing decisions in such a way that the traffic flow can be directed efficiently. Here, SDN offers a way to bring the network up to the speed of the cloud systems. The SDN concept allows the separation of a network device's control plane that can be relocated to a central entity – the SDN controller. Here, monitoring and control information can be aggregated and subsequently be used to make network-wide routing decisions that ensure an energy-efficient operation of the network using NCPs, as depicted in Figure 17.5.

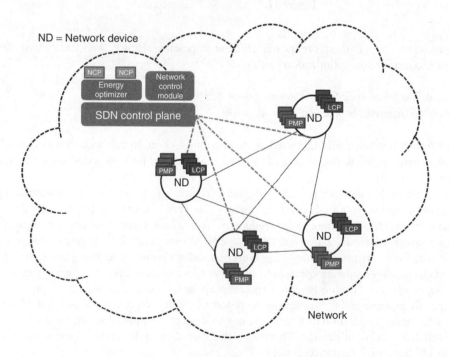

Figure 17.5   SDN-based NCPs

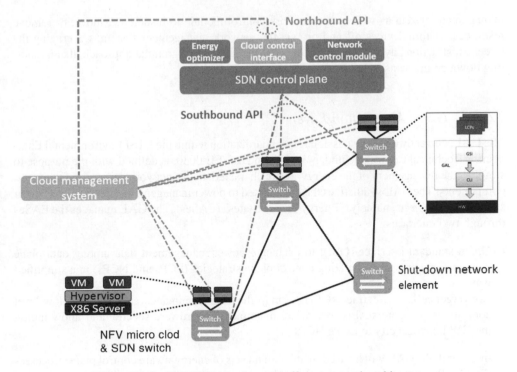

Figure 17.6   SDN/NFV-based energy-efficient network architecture

17.4   Energy-Efficient Network Management Enablers

17.4.1   SDN/NFV-based Energy-Efficient Network Architecture

Figure 17.6 shows a simple example of SDN/NFV-based network architecture, wherein a network element consists of an SDN-enabled switch and an adjacent NFV micro cloud. The SDN switch is responsible for forwarding traffic as dictated by the rules from the SDN control plane. The micro cloud serves to rapidly instantiate network functions as virtual machines on-demand, a process controlled by the cloud management system. The SDN control plane is connected to the cloud management system via a cloud control interface module that represents a realization of SDN's "Northbound-API." Both control planes receive monitoring information from an external network management system not depicted in Figure 17.6 for the purpose of simplicity.

This system is also responsible for the initial configuration of the network elements. The energy optimizer module running on top of the SDN controller uses the monitoring information to calculate the traffic in the network, while it receives load information for the deployed network functions, for example, caches, from the cloud management system via the cloud control interface. Using a scoring system based on, for example, service level agreements (SLAs), the energy optimizer can calculate a network arrangement where the placement of functionality preserves a good trade-off between the maintenance of service quality and energy conservation. Once such a placement is determined, the energy optimizer can then notify the cloud management system as well as the network control module to instantiate, move, or remove resources, or turn devices off and on, via the network management system. Leveraging the Green Abstraction Layer described in the next section, a more granular approach, that is, shutting down components of devices like ports, is also possible.

17.4.2   Green Abstraction Layer

The GAL, currently under discussion for standardization within the ETSI Environmental Engineering Technical Committee under Work Item DES/EE-0030, is defined with the purpose to simplify the management of the energy-aware characteristics of network devices and their subcomponents. The GAL synthetizes the data related to power management settings into a sort of standard data objects, namely, "Energy-Aware States" (EASes). The GAL manages the EASes through two interfaces:

- Green standard interface (GSI): to exchange power management data among data plane elements and processes realizing control plane strategies (LCPs and NCPs) in a simplified way;
- Convergence layer interface (CLI): to map the GAL commands and data into low-level configuration registers/APIs, which are manufacturer/hardware specific and allow hiding the HW heterogeneity to LCPs/NCPs.

In general, the GAL will be used by three main sets of energy-aware control plane processes (LCPs, NCPs, monitoring and operation, administration & management – OAM).

The energy-aware LCPs aim to optimize the configuration at the device level, in order to achieve the desired trade-off between energy consumption and network performance, according to the incoming traffic load. To this purpose, such processes need to know in detail the internal architecture of the device (or parts thereof), the number, the typology, and the capability of energy-aware elements, as well as to have access to network performance indexes.

Instead, the energy-aware NCPs aim to autonomously control and optimize the network behavior considering a set of devices. Typical examples of this kind of processes are traffic engineering, routing, and signaling algorithms/protocols (e.g., OSPF-TE/RSVP-TE) with "green" extensions.

Finally, the energy-aware OAM processes are used by the operator to control and optimize the behavior of a network (as in network management systems with "green" capabilities).

The adoption of local, network-wide, and OAM policies is not necessarily exclusive, but they are conceived to work with different goals and in different contexts in a complementary way. Notwithstanding these differences, the behavior of local policies cannot be completely independent of network-wide and OAM control frameworks. A simple example of such dependency is given by the fact that the optimal local configuration relies on the traffic load incoming from device links, which, in turn, is influenced by network-wide and OAM policies.

17.4.3   GAL Main Design

The goal of the GAL is to extend and reengineer the ACPI standard [19] for general purpose computing systems and adapt it to architectures, functionalities, and paradigms of network devices, and especially of their data plane components.

It is worth noting that the GAL is a device internal interface, so it does not behave as a network signaling protocol. Control plane processes implement signaling protocols to make network nodes converge to a certain configuration. This is not a direct goal of the GAL, but these control plane processes need to be interfaced with the GAL for acquiring/setting power-related parameters. For example, considering the SDN area, such signaling protocols can be based on the OpenFlow specification.

In other words, the GAL is conceived to make control processes acquiring information on the green capabilities available at the data plane, configuring them, and carrying measurements on the energy consumption.

Thus, unlike the ACPI standard, the specification of the GAL must consider the presence and the main features of multiple and heterogeneous internal components (e.g., link interfaces, multiple chips for packet processing, fans) with energy adaptation and monitoring capabilities, and hide the complexity to the high-level control policies. In addition, it can simplify the management of complex network devices composed of several HW components, by providing methodologies for aggregating all the information from single internal components (e.g., single component, link level, line-card, chassis, and node levels). Obviously, the GAL has to characterize the effect of putting components in a certain energy configuration in terms of power consumption and network performance (e.g., minimum and maximum energy consumption, maximum throughput), by exposing this information to LCPs and NCPs.

The energy-aware configuration is represented by a structure called "Energy Aware State" – EAS. The interaction between NCPs and LCPs takes place via SDN communication by using the primitives provided by the GAL framework.

Figure 17.7 depicts an example, which illustrates how the GAL framework is implemented inside a network device wherein several LCPs are installed, too. These LCPs set and get the

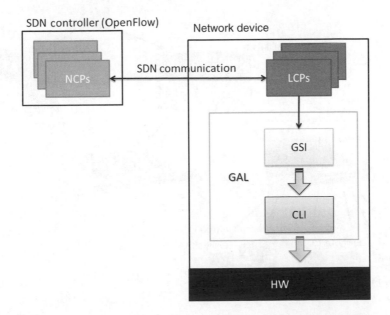

Figure 17.7   GAL–NCP–LCP communications in the SDN framework

energy-aware configuration by means of the EASes and by using the GSI. Inside the GAL framework, each GSI request is translated by the CLI in a specific command for the underlying HW components. The NCPs are installed in a remote device as modules of an SDN controller (in this specific example, we consider OpenFlow). Each interaction between the NCPs and the LCPs is performed according to the OpenFlow specification.

17.4.4 GAL Hierarchical Structure

In this section, we describe in more detail the hierarchical structure provided by the GAL framework.

The GAL architecture is designed as a modular and easily extendable software framework, providing interface capabilities toward heterogeneous HW, as well as multiple hierarchical interfaces toward control processes, in order to set energy configurations at various detailed levels of the internal architecture.

Figure 17.8 shows the case of a multichassis network device including different HW components with energy-aware capabilities, which may be managed by control plane processes. Such device can be represented at various levels: single HW component and/or physical link levels, line-card, chassis, and node levels.

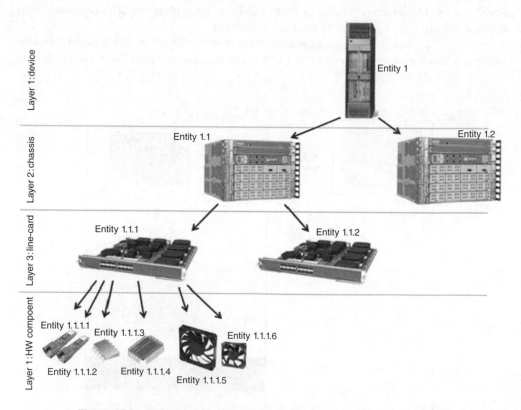

Figure 17.8 The hierarchical architecture of a multichassis network device

The GAL allows the interaction between the power management usually realized inside HW components and network control processes (e.g., routing and traffic engineering signaling protocols) run at the device level. When new device configurations are determined by such control processes, the GAL translates them into specific settings of components at underlying levels. For example, if the network-wide control decides to scale the speed of a link, it has to pass this command to the chassis and then to the line-card hosting the link. The line-card can scale the speed of the physical link, as well as the speed of the related HW (which can be composed of multiple energy-aware components). Thus, a single configuration decided at the device level may impact on multiple underlying components.

Given the fact that power management primitives (PMPs) reside at the lowest levels of the hierarchy, they shall require specific LCPs to directly manage them to achieve the desired operating behavior. For this reason, the GAL is used for interfacing such lowest level energy-aware components with some control plane processes implementing HW-specific optimization strategies.

Then, at intermediated levels (e.g., line-card or chassis), new LCPs (one for each entity at that level) are needed to orchestrate the settings and the operating behaviors of underlying energy-aware components, and to expose a synthetic and aggregate set of operating characteristics and available configurations (of the line-card, or of the chassis) to higher levels.

This process terminates at the device level, where the highest LCP orchestrates the high-level configuration of the device and needs to expose a simplified view of it to network signaling protocols (network-wide and OAM control applications).

The approach pursued here consists of a chain of LCPs, which control from single energy-aware HW components to the entire device. LCPs at different levels need to interoperate, and the highest level LCP needs to interact with processes realizing network signaling protocols.

The result of such hierarchical approach consists of a tree, where root nodes are LCPs and control applications, and leaf nodes correspond to HW elements. The interface among the tree nodes is realized by means of the GAL. Leaf nodes may reside at different levels of the hierarchy for two main reasons:

- Some HW energy-aware component may be accessible at higher hierarchical levels (e.g., fans in a chassis);
- Some manufacturers shall prefer not to expose the internal organization and the subcomponents of a "composite" part of the device (e.g., line-card, chassis).

In the latter case, the "composite" part of the device will be treated as a single HW energy-aware element.

17.5 Conclusions

This chapter focuses on the problem of energy consumption, considering the current network infrastructures, which are not prepared to satisfy an ever-growing demand for mass ICT services with the scalability, flexibility, and effectiveness needed to cope with future technology and traffic trends, as well as environmental and economic challenges. Therefore, it is desired to extend the network design criteria including new different aspects, such as energy efficiency.

The design of energy-efficient networks requires a more flexible approach, making necessary the dynamic allocation of resources that the traditional networks are not able to realize.

The SDN/NFV paradigms can be a viable solution to improve the network functionalities and capacity, by allowing a flexible and dynamic resource allocation. The development of SDN-enabled applications that allow more efficient-energy consumption requires the presence of specific functionalities inside the network devices necessary to manage the energy/performance trade-off.

In this chapter, we introduced the notion of PMPs and also described global network primitives. There is a variety of ways to improve the energy footprint of a network. We focus on providing power management and network primitives, which are inherent functions or features of a device or network that does not require any additional components to operate.

Furthermore, we analyzed an SDN-based network architecture for providing energy-efficient networking, by formulating the notion of a GAL for the power management of individual devices. The architectural considerations and interfaces enabling the GAL have been elaborated, which enable to export data plane energy-aware capabilities from network devices toward the Control Plane. Specifically, the GAL synthetizes the data related to power management settings into a sort of standard data objects, namely, "Energy-Aware States" (EASes).

We have outlined the definition of two interfaces, the Green Standard Interface, which is the "external" interface used to interact with clients and applications and an internal interface (convergence layer).

Finally, we described the hierarchical architecture of the GAL that, with a functionally complete set of primitives, can be very useful and suitable for the management of resources in energy-aware SDN platforms.

References

[1] K. W. Cameron, "Energy oddities, part 2: why green computing is odd," IEEE Comput., vol. 46, no. 3, pp. 90–94, 2013.

[2] P. M. Corcoran, "Cloud computing and consumer electronics: a perfect match or a hidden storm?," IEEE Consumer Electron. Mag., vol. 1, no. 2, pp. 14-19, 2012.

[3] J. Glanz, "The cloud factories – power, pollution and the Internet," The New York Times, Sept. 22, 2012, http://www.nytimes.com/2012/09/23/technology/data-centers-waste-vast-amounts-of-energy-belying-industry-image.html?_r=0.

[4] White paper on "Software Defined Networking", https://www.opennetworking.org/sdn-resources/sdn-library/whitepapers.

[5] White paper on "Network Functions Virtualisation", http://portal.etsi.org/NFV/NFV_White_Paper.pdf.

[6] R. Bolla, R. Bruschi, F. Davoli, L. Di Gregorio, P. Donadio, L. Fialho, M. Collier, A. Lombardo, D. Reforgiato, and T. Szemethy, "The green abstraction layer: a standard power-management interface for next-generation network devices," IEEE Internet Comput., vol. 17, no. 2, pp. 82–86, 2013.

[7] The OpenFlow Specification, URL: http://www.openflow.org.

[8] R. Bolla. R. Bruschi, A. Carrega, F. Davoli, D. Suino, C. Vassilakis, and A. Zafeiropoulos, "Cutting the energy bills of Internet service providers and telecoms trough power management: an impact analysis," Comput. Netw., vol. 56, no. 10, pp. 2320–2342, 2012.

[9] The Climate Group and Global e-Sustainability Initiative, "SMART 2020: enabling the low carbon economy in the information age," http://www.theclimategroup.org/, pp. 1–87, 2008.

[10] S. Nedevschi, L. Popa, G. Iannaccone, S. Ratnasamy, and D. Wetherall, "Reducing network energy consumption via sleeping and rate-adaptation," Proceedings of the 5th USENIX Symposium on Networked Systems Design and Implementation, pp. 323–336, 2008.

[11] R. Bolla, F. Davoli, R. Bruschi, K. Christensen, F. Cucchielti, and S. Singh, "The potential impact of green technologies in next-generation wireline networks: is there room for energy saving optimization?," IEEE Commun. Mag., vol. 49, no. 8, pp. 80–86, 2011.

[12] R. Bolla, R. Bruschi, F. Davoli, and F. Cucchietti, "Energy efficiency in the future Internet: a survey of existing approaches and trends in energy-aware fixed network infrastructures," IEEE Commun. Surveys Tutorials, vol. 13, no. 2, pp. 223–244, 2011.

[13] L. Chiaraviglio, M. Mellia, and F. Neri, "Minimizing ISP network energy cost: formulation and solutions," IEEE/ACM Trans. Netw., vol. 20, no. 2, pp. 463–476, 2011.

[14] A. Cianfrani, V. Eramo, M. Listanti, and M. Polverini, "An OSPF enhancement for energy saving in IP networks," Proceedings of the 2011 IEEE Conference on Computer Communications, pp. 325–330, 2011.

[15] J. C. C. Restrepo, C. G. Gruber, and C. M. Machuca, "Energy profile aware routing," Proceedings of the 2009 IEEE International Conference on Communications, pp. 1–5, 2009.

[16] R. Bolla, R. Bruschi, A. Cianfrani, and M. Listanti, "Enabling backbone networks to sleep," IEEE Netw. Mag., vol. 25, no. 2, pp. 26–31, 2011.

[17] B. Heller, S. Seetharaman, P. Mahadevan, Y. Yiakoumis, P. Sharma, S. Banerjee, and N. McKeown, "ElasticTree: saving energy in data center networks," In NSDI'10 – Proc. 7th USENIX conference on Networked Systems Design and Implementation, San Jose, CA, 2010, vol. 3, pp. 1–16; https://www.usenix.org/legacy/event/nsdi10/tech/full_papers/heller.pdf.

[18] S. Niccolini, "Free Your Middlebox Functions down to the Data Plane with Tiny, Fast Network VMs," Invited talk at European Workshop on Software Defined Networks, Darmstadt, Germany, 2012.

[19] Advanced Configuration & Power Interface (ACPI), URL: http://www.acpi.info.

18

Energy-Efficient Protocol Design

Giuseppe Anastasi[1], Simone Brienza[1], Giuseppe Lo Re[2] and Marco Ortolani[2]
[1]*Department of Information Engineering, University of Pisa, Pisa, Italy*
[2]*DICGIM, University of Palermo, Palermo, Italy*

18.1 Introduction

In developed countries, the total energy consumed by the Internet accounts for approximately 2–3% of the overall worldwide energy consumption [1, 2]. Although this percentage is not so high, its absolute value is very remarkable and has followed an increasing trend over the years [3]. More important, it has been estimated that a large fraction of the overall energy consumed by the Internet is wasted due to an inefficient utilization of infrastructure and user equipment [4]. Hence, significant energy savings could be achieved through appropriate power management strategies. This has stimulated the interest and efforts of the research community. So far, most of the research projects and activities have been driven by telcos and Internet Service Providers (ISPs) and, thus, they have been aimed at reducing the energy consumption mainly in the Internet core (i.e., at routers) and at data centers [5]. Less attention has been devoted to reducing the energy consumption of *edge devices* at user premises (i.e., PCs, printers, IP phones, displays).

This chapter focuses on solutions for optimizing the energy consumption of PCs and other user equipment connected to the Internet. These edge devices account for the major fraction of the overall Internet-related energy consumption [6]. Even if the power consumed by a single edge device is limited (e.g., about 100 W for desktop PCs and about 20 W for notebooks) – because of their large number and utilization time – the total consumed energy can be huge. Moreover, edge devices are typically used with little or no attention to the energy problem. For example, many PCs are left on, even overnight and during the weekend, because of laziness, carelessness, or to maintain network connectivity (e.g., for peer-to-peer file sharing). In addition, users often do not use energy saving policies (e.g., automatic hibernation after a certain period of inactivity). Other edge devices, such as printers, IP phones, and displays, are typically kept *always on*, especially in offices and public buildings. Every year, the estimated

Green Communications: Principles, Concepts and Practice, First Edition.
Edited by Konstantinos Samdanis, Peter Rost, Andreas Maeder, Michela Meo and Christos Verikoukis.
© 2015 John Wiley & Sons, Ltd. Published 2015 by John Wiley & Sons, Ltd.

overall energy consumption due to edge devices in United States is in the order of tens of TWh, causing an expense of billions of dollars. The need for specific solutions to the problem is, thus, quite apparent.

In the next sections, we consider the main approaches to power management of PCs and other user equipment connected to the Internet. Specifically, we introduce a general taxonomy to classify the proposed solutions. Then, according to the introduced taxonomy, we survey the main proposals presented in the literature. Obviously, most of the proposed solutions refer to PCs. However, some of them could be extended to other edge devices as well.

18.2   General Approaches to Power Management of Edge Devices

In order to identify possible approaches to energy efficiency in edge devices, it is necessary to determine preliminarily the cause of energy waste. One of the fundamental causes is the fact that many users leave their PC always on (especially in their workplace), due to laziness and/or carelessness. This clearly emerges, for example, in the PC energy report [7] about the energy consumption of PCs used at work, issued by the UK National Energy Foundation. This report highlights that about 21% of PCs used at work are almost never turned off (during nights and weekends), thus resulting in a waste of energy equal to approximately 1.5 TWh per year (corresponding to 700,000 tons of CO_2). In order to reduce this energy waste due to laziness and carelessness, PCs and other edge devices could be forcedly turned off at a certain time, employing common solutions, such as Nightwatchman [7], which is already used in many environments.

There are cases, however, in which PCs are deliberately left on for the execution of certain network activities, such as remote connection or P2P file sharing. Since the PC is used for the execution of that particular application only for a limited time interval, most of the energy consumed to maintain connectivity could be saved by introducing appropriate mechanisms for power management [8]. However, to be effective these mechanisms should save energy without introducing a significant degradation in performance. Some studies related to network traffic [9, 10] have shown that PCs (and other edge devices) experience long idle periods, during which they might be turned off or placed in *sleep* state, thus resulting in significant energy savings. Specifically, we need mechanisms that can allow a PC to sleep during idle periods and to resume promptly whenever an external packet is received or the user wants to use her/his PC, so as to minimize the impact on the system responsiveness.

Figure 18.1 shows the possible approaches aimed at reducing the energy consumption of edge devices (with particular reference to PCs) by eliminating wastes during idle time. The approach based on *on-demand wake-up* consists in putting the PC in *sleep* state during periods of inactivity and waking it up, later, by means of a special message called *Magic Packet* [11]. Conversely, *proxying* can be used to allow the PC to be in *sleep* state during periods of inactivity and to interact, at the same time, with any remote *host* through the network. This technique allows to delegate to an entity, called *proxy*, the management of interactions with the network. The two mechanisms can also be combined. In this case, when the *proxy* is not able to handle the request received through the network, it wakes up the PC by sending a Magic Packet and passes the request. Thus, communications are handled in a completely transparent way to the external network (and users). The approach known as *context-aware power management*, instead, exploits some context information (for instance, the presence/absence of the user) for a finer grained power management. This approach can be used not only for PCs, but for

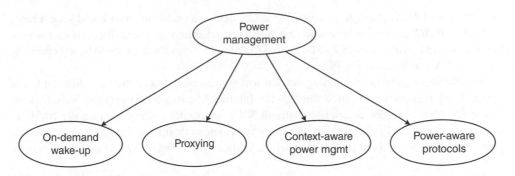

Figure 18.1   General approaches to power management of Internet edge devices

other edge devices as well. For example, displays used for the diffusion of information can be turned off when no one is in the nearby area. Finally, the use of *power-aware* protocols and applications is yet another approach to save power for edge devices.

In the next sections, the above-mentioned approaches and their potential to save energy are analyzed and explored in detail. It is important to emphasize that these approaches are not necessarily alternative but may also be used jointly.

18.3   Remotely Controlled Activation and Deactivation

Remotely controlled activation and deactivation is essentially based on the *Wake-on-LAN* (WoL) mechanism, widely used in Ethernet networks, which allows a PC in *sleep* state to be awakened – remotely – by sending a special message called *Magic Packet* [11]. The *Magic Packet* is typically sent from the same LAN of the target PC; however, it could also be sent by any device on the Internet. The WoL mechanism requires that part of the Ethernet network interface card (NIC) remains always active. Therefore, the PC cannot be disconnected from the power source or should use an alternative source of supply (i.e., a battery). The NIC component that remains active introduces, of course, a standby power consumption that is much smaller than the power consumed by the PC in the active mode.

The Magic Packet is a particular MAC frame that contains 16 repetitions of the MAC address of the target PC. It is usually sent as a UDP message destined to a specific port (9). After receiving a Magic Packet, the WoL component of the NIC wakes up the PC. There are two ways to transmit a Magic Packet over the network. It can be sent to the broadcast address of the subnet of the target PC (*subnet-directed broadcast*), or directly to the target PC (*unicast wake-up packet*).

The former case is the most common one since *subnet-directed broadcast* was the original transmission method for sending wake-up packets. With this technique, the Magic Packet is received by all the NICs in the network; however, it is discarded by all the NICs but the one whose MAC address matches the specified address. The main drawback is that, since Magic Packets are sent to a broadcast address, typically, they are not forwarded by routers. However, this limitation can be easily overcome in several ways, allowing a PC to be woken up by any computer on the Internet. This can be achieved, for instance, by configuring routers in such a way to allow them to forward Magic Packets. This solution, however, makes the network vulnerable to *DDoS attacks* (e.g., *Smurf Attacks*), that is, a malicious user could send a large amount of *ICMP* packets in broadcast, causing a remarkable response traffic. Another way is using a *virtual private network* (VPN) so that the remote computer appears to be a member of the same LAN as the sleeping PC.

The alternative solution is sending *unicast wake-up packets*. Since they are directed to the target IP address, they are routed through the Internet like regular datagrams. Nevertheless, this approach may not be compatible with all NICs, especially with oldest ones. In addition, such packets will not be delivered to PCs that have changed their IP address (e.g., via dynamic host configuration protocol (*DHCP*)) or whose address is no longer present in the *ARP cache* of the router.

Several extensions to the basic WoL mechanism described above have been proposed. For example, some Intel NICs allow several options, namely, *Wake on Directed Packet, Wake on Magic Packet, Wake on Magic Packet from power-off state,* and *Wake on Link*. In particular, *Wake on Directed Packet* [12] is an extension that makes the wake-on-demand mechanism much more flexible. Basically, a sleeping PC can be woken up by any packet directed to it, for example, by the request to open a TCP connection. Obviously, spurious wake-ups may occur with such mechanism, thus resulting in energy waste. Moreover, a PC consumes much more energy during the wake-up transition than during normal operating conditions. Therefore, spurious wake-ups should be avoided by filtering wake-up requests (see also Section 18.4).

Several power management systems for large-scale distributed systems have been proposed that make use of the wake on-demand mechanisms. *Polisave* [13] is a client–server system that allows to schedule actions for PCs associated with the service. In order to avoid energy waste, it is possible to specify the time when a client PC must be turned on/off, or must go into *standby* or *hibernation* state (i.e., the energy states defined in the *ACPI standard* [14]). In Polisave the client PC periodically queries the server in order to find out if there are actions scheduled for it. If a shutdown (or hibernation) is planned, the PC turns off. Conversely, if a PC is scheduled to be switched on, the server sends a Magic Packet to the NIC of the target PC.

Gicomp [15] is another similar system that allows to install/modify power management policies on controlled PCs. In particular, it allows to define the time to dim and turn off the display, the disk spin-down time-out, the suspend time-out, the hibernate time-out, and other options, according to the features offered by the operating system. Like Polisave, it follows a client–server paradigm. Clients and server communicate through the *XMPP protocol*, which guarantees confidentiality and authentication.

Both *Polisave* and *Gicomp* are able to work in the presence of network address translation (*NAT*) servers and *Firewalls*. They allow users to remotely control (through a web interface) all PCs associated with their personal account. PCs can be turned off, suspended, or turned on (through WoL). Finally, *Gicomp* solves the problem of Magic Packets' routing, using a specific *Waker* in each served IP subnet. *Wakers* are proxies, acting on behalf of the main server, that take care of waking up PCs on their IP subnet.

The on-demand wake-up technique has some limitations. First, it can be used only when the PC has a NIC with WoL support, that is, an Ethernet card. It also requires a proper configuration of both *BIOS* and operating system. It also suffers from security limitations. An attacker could turn on a sleeping PC through WoL, provided that she/he is on the same IP subnet of the sleeping PC. This is because the wake-up procedure does not require authentication and the content of the magic packet is transmitted as plaintext. To mitigate this problem, some NICs allow to insert a password in the Magic Packet, in addition to the MAC address.

However, this method can be easily overcome by sniffing the network traffic, as Magic Packets are not encoded. For these reasons, some PCs have an improved chipset to provide security for WoL. For instance, *Intel AMT* (a component of *Intel vPro technology*) supports *transport layer security* (TLS) encryption in order to secure an *out-of-band* communication tunnel for remote management commands such as WoL [16, 17].

18.4   Proxying

In this section, we analyze solutions on the basis of the use of a *proxy*. A proxy is an entity capable of responding to requests coming from the network on behalf of a sleeping device (e.g., a PC). The idea of using a proxy for energy conservation is not new. Indeed, proxy-based architectures have been proposed to guarantee an energy-efficient Internet access from mobile devices [18, 19]. However, in that case, the proxy architecture was designed to support a mobile device running standard client–server applications [19]. More recently, the idea of a proxy-based architecture has been extended to implement energy-aware solutions in the Internet [20]. In this case, the proxy acts on behalf of a host to respond to minimal network interactions and wakes up the host if needed.

A power management proxy works as shown in Figure 18.2. Initially, the edge device (e.g., a PC) has to associate with the proxy. Then, when the device is in sleep state and a request arrives from the network, there are two possible options. If the proxy is able to manage the received request by itself, it immediately serves it on behalf of the device, which can thus remain in sleep state. Otherwise, the proxy sends a wake-up message to the device and, then, forwards the received request to it.

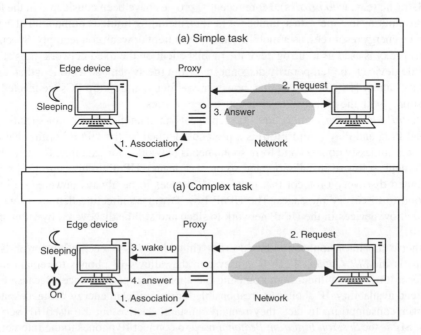

Figure 18.2   Power management proxy operation scheme

Table 18.1   List of proxy-based solutions

Category	Research work
Application-specific proxy	UPnP low power [21]
	SIP catcher [22]
	Proxy for Gnutella [23]
	EE-BitTorrent [24, 25]
Network connectivity proxy	Concept, design and
	implementation
	[26, 20, 27–32]
	Somniloquy [33]
	SleepServer [34]
	ECMA-393 [35]

Proxy-based solutions can be further divided into two categories, depending on the kind of proxy they rely upon. We can distinguish between *application-specific proxy* and *network connectivity proxy*. The main proxy-based solutions proposed in the literature are listed in Table 18.1 and are discussed in the following subsections.

18.4.1   Application-Specific Proxy

Proxying is a very common technique in distributed computing. Traditionally (server) proxies have been used in distributed applications to improve the system performance and reduce the network traffic (e.g., *web proxies*). More recently, proxies have been considered in the field of mobile computing to cope with a number of factors, including limited computational capabilities, scarce energy resources, user mobility, and intermittent or weak connectivity. Specifically, a (client) proxy is used as a surrogate of the mobile client on the fixed network, thus allowing the mobile device to be temporarily disconnected from the system, so as to lengthen the lifetime of its battery [18]. In this section, however, we focus on desktop PCs, connected to the power supply with the objective to eliminate energy wastes.

The *UPnP Low Power* architecture [21] represents an example of protocol-specific proxy. Universal Plug and Play (UPnP) [36] is a protocol, defined by the UPnP Forum, that allows devices to seamlessly connect and form spontaneous networks, for example, for data sharing, entertainment, software installation, and so on. The legacy UPnP architecture [36] relies on a distributed discovery protocol that requires all devices to be always powered on in order to respond to discovery messages. The UPnP Low Power architecture defines a low-power proxy to allow devices in the UPnP network to sleep and still be discovered by UPnP control points.

Another example of application-based proxy (in addition to a mechanism for on-demand wake-up) is the *SIP Catcher* [22]. It is a system that allows IP phones to remain in sleep mode for a long time without compromising the application performance. Because of their widespread availability, IP phones are responsible for a significant energy waste, despite their low power consumption. In fact, they remain constantly active but are used for very short periods. SIP is the *Session Initiation Protocol* used to connect IP phones to the Internet. Basically, a user registered with the SIP server and sends an *invite* message to the SIP server when she/he wants to call another IP phone. This locates the recipient IP phone over the Internet and forwards the invite message from the caller. The IP phone responds with a *trying* and a *ringing* message and then rings. At this point, the caller and the called party can start communicating. However, if the IP phone is in sleep mode, it will be obviously unreachable. In this case, a SIP catcher can overcome the drawback. The SIP catcher is a system that runs on the last hop router and acts as a proxy for SIP calls. Specifically, when it detects an *invite* message directed to the sleeping IP phone, it wakes the IP phone up and, once reactivated, transmits the invite message. Meanwhile, the catcher responds to the caller sending the *trying* message. Once reactivated, the IP phone sends the *ringing* message, thus completing the SIP protocol handshake, while being completely transparent to the caller.

Proxying techniques have been also proposed to increase the energy efficiency of *peer-to-peer* (P2P) applications, such as *file sharing/distribution*. Recent studies indicate that a large amount of the overall Internet traffic originates from P2P applications [37]. Nevertheless, P2P file sharing protocols – such as *BitTorrent* and *Gnutella* – have been designed assuming that PCs are always on and, thus, they are not energy efficient. To this end, various solutions have been proposed, including proxy-based architecture (other proposals taking different approaches will be presented in the subsequent sections).

In Ref. [23] the authors present a proxy-based solution for *Gnutella* that also exploits the WoL mechanism. Gnutella uses a flooding mechanism to find files over the overlay network. It defines five different messages, namely *Ping, Pong, Query, Query Hit,* and *Push*. The *Query* and *Query Hit* message are used to find files and respond to query messages, respectively. In the solution presented in Ref. [23] the power management proxy is a microcontroller with limited storage capacity and low energy consumption (much less than the host) and it is positioned on the Ethernet NIC of the host or in a LAN switch. The proxy detects requests for files directed to the sleeping host and wakes it up through a Magic Packet. Thus, the host can serve the requested files. In order to take over for the sleeping host, the proxy shares information with it, about the power state of the host (sleeping or fully powered-on), the IP list of its *neighbors,* and the list of the shared files. The P2P proxy supports only a subset of *Gnutella* functionalities. Specifically, it can start and accept neighbor connections, receive and forward *Query* messages, send *Query Hit* messages, and wake up the sleeping host. Instead, it cannot serve files or store them. When the PC goes to sleep, its TCP connections with neighbors are terminated and established again by the proxy. When the proxy wakes the PC up, the opposite occurs. In both cases, everything happens transparently to the user.

Still in the framework of P2P file sharing, *EE-BitTorrent* [24, 25] is another proxy-based solution for making the *BitTorrent* protocol energy efficient. *BitTorrent* is the most commonly used protocol for P2P file sharing; however it was not designed with energy efficiency in mind. In particular, it requires that a *peer* is always active while downloading a file and remains active, for some time, after completing the download, so as to provide the same file to other peers. *EE-BitTorrent* relies on a *BitTorrent* proxy that serves a large number of *BitTorrent* peers. When a user requests a file, the query is sent to the proxy in a transparent way to the user. The proxy also implements a caching mechanism. Hence, in most cases the requested file is already available on the proxy cache and can be immediately downloaded, thus reducing the download time and energy consumption at the user PC. When the file is not immediately available, the download service is undertaken by the proxy and the user can, thus, turn off her/his PC (Figure 18.3(a)). The proxy gets a copy of the file from the overlay network, acting as a regular *BitTorrent* peer (Figure 18.3(b)). When the copy is available, the proxy wakes the

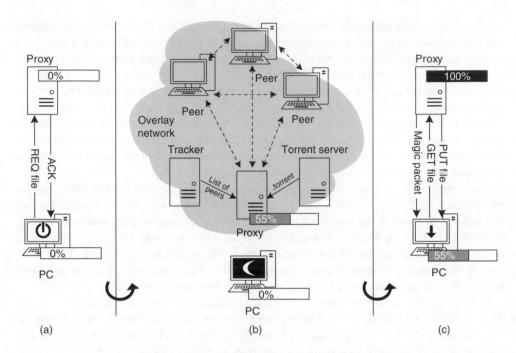

Figure 18.3   EE-BitTorrent operation scheme

PC up, through a Magic Packet, and transfers the file to it (Figure 18.3(c)). If the on-demand wake-up mechanism is not available, the PC can explicitly require the file to the proxy after its reactivation. An important aspect of the system concerns the proxy location. Since the behavior of EE-BitTorrent does not depend on the proxy location, there are many options for placing the proxy, mainly driven by the specific deployment scenario. In an enterprise environment (e.g., a university or business department) the proxy could be located in the same LAN of the served PCs. In a residential scenario, instead, this is not a reasonable option as the proxy would serve only a limited number of PCs (those in the same home). In such a scenario, it could be located in the ISP network and offered as a (free) service to users, or it could be a cloud proxy. Also, a group of users could manage a social (i.e., shared) proxy – connected to the Internet through a high-speed network – for reducing energy consumptions at their home PCs.

Another proxy-based solution, similar to EE-BitTorrent, is presented in Ref. [38], taking into account the efficient sharing (in terms of energy consumption) of files for which only a very limited number of copies are available on the Internet. Downloading such a file often results in a client–server transfer from the peer that provides the file to the peer that requests it, at a very low bit rate. Hence, the benefits of the P2P paradigm are lost, resulting in an increased energy consumption. In the proposed solution, peers are coordinated in such a way that only a limited number of them remains active and act as proxies for the other peers that go in sleep mode.

For the sake of completeness, we also mention here some proxy-based solutions for mobile devices that use a simplified version of the *BitTorrent* protocol [39, 40]. In this case, however, the main objective is to increase as much as possible the battery lifetime of the mobile device.

18.4.2   Network Connectivity Proxy

The proxy-based solutions presented above are *application-specific* as they refer to a particular application or protocol. Therefore, it is necessary to use a different proxy for each specific application. In addition, they typically require the user intervention, that is, they are not transparent. Ideally, a PC should transparently enter sleep mode, whenever it is idle, in order to save energy. At the same time, it should still appear connected and fully operational to the other network devices. This would maximize energy savings while minimizing the impact on the performance of network applications. This goal can be achieved by using a *network connectivity proxy* (NCP), that is, an entity that is capable of maintaining the network presence on behalf of a sleeping PC, managing all packets destined to that PC. The concept of NCP was originally proposed in Ref. [27] for Ethernet networks and, then, extended in subsequent papers [20, 28, 29] for IP networks in general. Key challenges in the design and implementation of an NCP have been addressed in Ref. [30], where the authors propose some possible solutions and show that using an NCP can result in significant energy savings, up to 70%. A sleep proxy similar to the NCP proposed in Ref. [27] is also proposed in Ref. [31].

In order to design an NCP, it is necessary to have a detailed knowledge of the activities carried out by a host (e.g., a PC) to maintain network connectivity. Basically, a host performs a series of actions. Specifically, it replies to periodic ARP requests, generates periodic DHCP requests to maintain the IP address, replies to ICMP messages (e.g., *ping* requests), accepts TPC connections by replying to TCP SYN segments, and, more generally, manages all incoming packets appropriately. A detailed analysis about the packets received by a PC during idle periods, and the related protocols, was carried out in Ref. [29] and, more recently, in Ref. [32]. The latter considers both home and office environments. Once the type and fraction of received packets are known, they can be classified according to the class of actions they require. Hence, NCP requirements can be defined accordingly [26, 32]. In summary, when a packet directed to the sleeping PC is received, the NCP should perform one of the following actions:

(a) Directly respond to the packet.
(b) Discard the packet.
(c) Redirect the packet to another (active) PC for further processing.
(d) Wake up the host and pass the packet for appropriate processing.
(e) Put the packet in a queue to transfer it to the host when this is reactivated.

In addition, the NCP could be instructed to generate periodic requests on behalf of the sleeping PC (e.g., DHCP lease requests for maintaining the IP address) [26].

According to the NCP model outlined above, several practical variants can be envisaged [32]. In fact, the design space is quite large since different solutions may vary in many aspects, including *complexity* (the set of functionalities implemented by the proxy), degree of *transparency* (the possible differences in the user/application behavior with and without the proxy), *deployment* (the place where the proxy is physically located, e.g., individual PC, router/firewall, separate PC), and *implementation* (e.g., device attached to the NIC/motherboard of the PC, external USB-connected device).

In Ref. [32] the authors consider four different proxies with different complexity and compare their performance in terms of energy efficiency. They also propose a general and flexible NCP architecture that can accommodate different design choices and present a simple implementation, wherein the NCP is assumed to be a standalone machine and, thus, it is in charge of maintaining the network connectivity of several PCs in the same LAN. The following NCP variants are considered in the analysis.

- *Proxy-1*: drops all packets classified as *ignorable* and wakes up the PC for handling all other packets.
- *Proxy-2*: drops all packets classified as ignorable, responds directly to protocol packets that require a minimum handling, and wakes up the PC for all other packets.
- *Proxy-3*: performs the same actions as *Proxy-2*, but it is more selective as it wakes up the sleeping PC only when the received packet belongs to a set of user-specified applications.
- *Proxy-4*: performs the same actions as *Proxy-3* with respect to incoming packets. In addition, it can be instructed to wake up the PC to perform scheduled task such as network backups, antivirus updates, software updates, and so on.

The performance comparison is based on traces derived from real measurements, carried out both in home and office environments. The obtained experimental results show that *Proxy-1* is inadequate in office environments and only marginally adequate in home environments. *Proxy-3* provides significant energy savings in both home and office scenarios (it allows the PC to *sleep* for most of the idle time). Instead, the performance of *Proxy-2* largely depends on the specific environment. Specifically, the additional complexity, compared to Proxy-1, makes it a good choice in home environments, although it is not a good candidate for office environments, where the amount of traffic to manage is much higher. Finally, the performance of *Proxy-4* is close to that of *Proxy-3* since scheduled tasks are typically infrequent.

Somniloquy [33] is a private NCP, that is, it is supposed to serve just one PC. It is conceived as an external device, connected to the PC through a USB port, which includes a *low-power* processor capable of running an embedded operating system, flash memory to store data (e.g., files) while the PC is sleeping, and one or more network interfaces to communicate with the external network. When the PC is sleeping, network connectivity is maintained through the NIC of the proxy. The latter uses a packet filter – defined in the form of regular expressions – to select packets and wake up the PC in case of a match. This allows to respond to network applications, such as remote *Secure Shell (SSH)*, file access requests, and VoIP calls, even when the PC is sleeping. In addition, *Somniloquy* can act as an application proxy for some common network applications such as *instant messaging* and *P2P file sharing*. This is accomplished by implementing a lightweight version of the specific application (*stub*) on the proxy. The application stub allows the proxy to manage autonomously the majority of actions required by the application and wakes up the PC only on complex events. A prototype implementation of *Somniloquy* based on the Gumstix[1] platform is described in Ref. [33]. The prototype includes a 200 MHz XScale processor with 2 GB of flash memory and 64 MB of RAM, a wired Ethernet (or wireless WiFi) NIC for connectivity, and two USB ports (one for sleeping/waking up the PC, the other one for relaying data received from NIC to the PC), and runs a version of Embedded Linux that supports a full TCP/IP protocol stack. To give an idea of the amount of energy saved by *Somniloquy* we can consider that a common PC consumes approximately 100 W in normal operating conditions, while the total power consumed by the PC in sleep mode and the external device implementing *Somniloquy* is approximately 5 W.

Somniloquy and the different proxy variants considered in Ref. [32] are not able to preserve TCP connections when the PC is in sleep state. Instead, this issue is specifically addressed in

---

[1] http://www.gumstix.com/

Ref. [26]. In addition to the outlined tasks (i.e., discarding ignorable packets, directly replying to packets that require minimal actions, and waking up the host when needed), the NCP proposed in Ref. [26] is also able to maintain TCP connections and UDP data flows. This is achieved by splitting the TCP connection at the proxy (see Section 18.6.1 for details about *splitting*). We assume that the proxy runs in the same network of the PC (e.g., on a router) and can, thus, cover several hosts. TCP packets destined to a given PC are buffered locally at the NCP when the PC is sleeping, and are later relayed, when the PC is awake again. Queuing of packets for later processing may actually make sense for some network applications such as *instant messaging* and *SSH*.

SleepServer [34] is another *proxy-based* solution that allows a host (e.g., PC, printer) to go in sleep mode while remaining reachable at the application layer. It does not require any change to the network infrastructure or any additional hardware, but only software agents installed on hosts. Indeed, it is completely implemented in software, using virtualization techniques. Specifically, the proposed architecture is physically composed of one or more SleepServer (SSR) machines on the same subnet as the hosts (but it can also work for hosts on different subnets using the VLANs mechanism). Each SSR serves a set of hosts and contains their images, in the form of virtual machines (VM). An SSR maintains the presence of the hosts in the network when they are in a state of sleep. On each SSR, a *SSR-Controller* is installed. It is a software component that manages (i) the creation of host images, (ii) the communication between hosts and their images, (iii) the resources allocation, and (iv) sharing among the images, providing isolation between images. Each host has also a software component, the *SSR-Client*, that connects the host to the SleepServer, passing its MAC and IP addresses and its firewall configurations. Before going to sleeping state, it sends its applications' state and all its open TCP and UDP ports to the SSR-Controller that creates the host image on the SSR machine. The image uses the same configuration parameters as the corresponding host. Hence, while the host is asleep, the image interacts with the network on behalf of it. If the host interaction is required, the SSR-Controller wakes up the host and disables its image on the SleepServer. In Ref. [34] it is shown that this solution allows significant energy savings (about 60–80%) and is able to support heterogeneous operating systems. Due to its high scalability, it is especially suitable for enterprise LAN environments with a large number of hosts (PCs, printers, etc.) connected to the network.

Finally, ECMA-393 [35] is a standard, adopted in February 2010, that specifies the maintenance of network connectivity and presence by proxies in order to extend the sleep duration of hosts, so as to save energy. The standard defines the behavior and the architecture of the proxy. In particular, it specifies the capabilities that a proxy may expose to a host, the information that must be exchanged between a host and a proxy, the proxy behavior with 802.3 (Ethernet) and 802.11 (WiFi) NICs and, more generally, the behavior of a proxy, including responding to packets, generating packets, ignoring packets, and waking up the host.

18.5   Context-Aware Power Management

In order to minimize energy consumption, PCs and other edge devices should be ideally turned off, or put in sleep mode, whenever they are not used. Also, they should be switched on again as soon as the user needs to use them. However, manually managing the power state of edge devices could be too onerous and frustrating for users. For these reasons, *Context-Aware Power Management (CAPM)* strategies have been developed, that is, strategies that use context information to automatically manage the power state of a device at runtime. Essentially, they aim to determine – by means of proper sensors – if the user *is using, is not using,* or *is about to use* a device. Through this information, the system is able to change the power state of the device, thus resulting in energy savings, while providing, at the same time, an acceptable service quality to the user.

CAPM strategies can also be used to optimize the energy consumption of specific components of a PC – such as hard drive, NIC, CPU, or display – according to the actual usage by the user. For example, the authors of Ref. [41] propose a solution that relies on a camera to determine if a user is looking at the display and turns off the display if the user is not present. Specifically, the CAPM system periodically acquires images by the laptop camera, which are processed by a *face detector* algorithm. If the user's face is not found, the system turns off the display.

However, more significant energy savings can be achieved by switching the entire machine to a low-power state during idle periods. Obviously, the power management system cannot turn off the PC as soon as the interaction with the user ceases, but it must infer from the context whether the user has actually stopped using the computer. In fact, transitions from one state to another have a significant cost in terms of

(i) *energy*: switching on a sleeping PC causes a considerable power consumption (much higher than the consumption in idle state);
(ii) *time*: in order to resume a PC to a fully operational state, tens of seconds could pass and, during this time interval, the user is unable to use her/his PC;
(iii) *lifetime of the device*: switching off and on frequently a device may reduce its lifetime.

The remarks above suggest that a too aggressive power management strategy, characterized by frequent shutdowns, would not lead to significant energy savings (due to the consumption during the wake-up phase) and could be highly frustrating for the user (forced to wait long resume periods). Therefore, it is possible to define a *break-even* time, that is, the minimum time interval that a device must spend in a sleep state in order to justify the passage to the low-power state and the subsequent reverse transition. For PCs, this time is in the order of minutes. Hence, a CAPM strategy requires accurate information to predict whether an idle period is long enough to justify the cost of the state change.

The required information can be obtained by means of a *location-aware* system, that is, a system able to determine the user's position with respect to the PC. Depending on the distance from the machine, the system can, thus, evaluate if the user is leaving her/his workplace or has temporarily discontinued the use of the PC. Hence, it can decide whether or not to put the PC into standby mode. Similarly, it can recognize when the user is back to the PC and switch it on, if necessary. In the literature there are various power management approaches that rely on the user's location estimation. They can be classified into two main categories that are discussed below.

The first category includes CAPM mechanisms that rely on very accurate information about the user's location, obtained through sophisticated location systems. For instance, the solution presented in Ref. [42] exploits an ultrasonic system that provides the user's location with high accuracy. Alternatively, a ultra-wideband (UWB) radio system can offer a good compromise between accuracy (below 1 m) and deployment costs. This kind of approach can be defined as *reactive*, since the system defines some *spatial zones* and triggers special events (i.e., switch on or suspend a PC) when a user enters or leaves a specific zone. Obviously, although these solutions provide excellent performance, their complexity and costs are typically very high.

The second category includes solutions that do not rely on very accurate location information. They typically exploit low-power sensors that provide only approximate information about the user's position. For instance, they check the radio connectivity of personal mobile devices, such as smartphones, to infer the presence/absence of the user in the working area [43, 44]. The solution proposed in Ref. [44] uses a policy called *Sleep/Wake-up on Bluetooth*. Specifically, the PC periodically runs the *Bluetooth discovery* procedure to detect the presence of the user's Bluetooth phone. If the latter is not discovered, the PC enters the standby mode. Then, when the user comes back within the Bluetooth coverage area, a nearby server detects the phone and wakes up her/his PC. Approaches falling in this second category can be defined *proactive*, as the system guesses the user's intentions and changes the power state of the PC accordingly. They are easy to implement; however, they are much less accurate than the previous ones. Hence, they may cause undesired shutdowns or activations of the PC.

When using low-power low-accuracy sensors, data provided by different sensors can be combined together, using AI techniques, in order to get more accurate location information. For instance, in Ref. [45] the authors exploit not only the information provided by Bluetooth phones (as in Ref. [44]), but also other context information provided by acoustic sensors and software sensors that monitor the user's activity on the PC. All these data are then processed using *Bayesian inference techniques*, so as to infer the user's position and activity, and act on the power state of the PC accordingly. Another interesting solution, called *non intrusive location-aware power management scheme* (*NAPS*) is presented in Ref. [46]. NAPs is specifically designed for PCs and does not require very accurate location information. It divides the space around a PC in some *virtual zones* and then uses the *received signal strength indicator* (RSSI) of a sensor node, carried on by the user, to estimate the zone where the user is located. According to the estimated distance from the computer, the system performs different actions. If the user moves away from her/his PC, the system acts first at the application level, closing unnecessary applications that use the CPU (e.g., web browsers) and, then at the device level, switching off some components, such as video and hard disk. If the user moves further away, the PC is put into the sleep state.

Obviously, the efficiency of a CAPM solution depends on several factors. For instance, the use of very accurate location methods, or the addition of context information, can increase the performance of the power management system. On the other hand, we also need to consider the additional cost of sensors and their energy consumption. The study in Ref. [47] analyzes the costs and potentialities of CAPM systems. The obtained results show that context information and location methods must be carefully chosen in order to maximize the ratio between energy saved through power management and energy consumed by sensors. At the same time, the system should be capable not only of saving energy, but also be transparent to the user. Forcing the user to manually turn on her/his PC, or to wait long resuming times, could be irritating and lead the user to disable the power management system. It is thus preferable to reduce the intrusiveness of the system, at the cost of a slightly higher energy usage. Anyway, as shown in Refs. [42, 45, 47], CAPM systems are able to produce significant energy savings, although occasional unnecessary shutdowns or switches-on are unavoidable. Obviously, CAPM systems can never achieve an optimal performance, because their effectiveness strictly depends on the particular use that the user makes of her/his PC and requires an accurate prediction of the user's future intentions.

18.6 Power-aware Protocols and Applications

In this section we describe some techniques for designing *power-aware* protocols and applications. All common protocols (e.g., TCP) have been implicitly designed assuming that the hosts on which they run are continuously active and, therefore, they are not power-aware. To be used in hosts that can switch between sleep and active modes, these protocols (applications) must be properly adapted. To this end, there are two viable approaches: (i) modifying existing protocols (applications) in order to make them power-aware, or (ii) defining new protocols and applications from scratch.

To implement new energy-efficient protocols and applications, it is recommended to use specific software tools, such as the ones presented in Refs. [48, 49]. These tools perform an energy profiling of the protocol/application in all its parts and assist the developer during programming, so that she/he can make design choices aimed at reducing energy consumption. However, the approach (i) is the most widely used for very common protocols and applications. Therefore, in the following we refer to techniques to modify existing protocols and applications, in an energy-efficient perspective. According to Ref. [49], it is possible to act both at the transport and the application layer. The two approaches are discussed below.

18.6.1 Transport Protocols

In this section we survey solutions working at the transport layer of the networking protocol stack. The main advantage of this approach is that energy-aware capabilities added to the transport protocol (e.g., TCP) can be exploited by all the applications running on top of it.

With reference to TCP, many research activities have been carried out to make the protocol energy efficient in the context of mobile/wireless networking [50–54], where an improvement in the efficiency of the networking subsystem can significantly increase the lifetime of the mobile computer. Although the above-mentioned solutions could also apply to stationary PCs, in the following we focus mainly on solutions specifically targeted at energy efficiency in stationary/wired environments.

In Ref. [55] the authors analyze the energy cost of TCP, due to computational activities, by investigating the protocol consumption on different hardware platforms and with different (Unix and Linux) operating systems. In addition, they propose solutions to improve the energy efficiency of the existing TCP implementations.

The computational energy cost of TCP includes the energy consumption due to the following activities:

1. moving data from the user space into kernel space (*user-to-kernel copy*), as data sent through a TCP socket is first queued in the socket buffer and then copied into kernel space for further processing;
2. copying packets to the network card (*kernel-to-NIC copy*);
3. processing in the TCP/IP protocol stack, including the cost for computing the checksum, preparing ACKs, responding to time-out events and to triple duplicate ACKs, and other costs (window maintenance, estimation of Round Trip Time, interrupt handling, etc.).

The experimental results clearly show that a large part of the energy consumption associated with TCP is because of the copy operations, while only about 15% of the consumed energy is linked to TCP processing. Considering a more detailed breakdown of the energy costs, the authors estimate the consumption for each phase of the protocol. In particular, the step of computing the checksum alone covers about 30% of the energy consumption of the entire TCP processing. In order to reduce the energy consumption, the authors present some solutions. For instance, using *zero copy* technique – that is, directly copying the user data from the user buffers to the NIC – it is possible to skip the user-to-kernel copy phase. Instead, in order to reduce the kernel-to-NIC copy cost, two methods are proposed, aiming at decreasing the number of copies: (i) maintain the TCP send buffer on the NIC itself and (ii) maximize the data transfer size from the kernel to the NIC. Using all of these mechanisms, the authors showed that it is possible to achieve a reduction in the overall cost at a sender of about 30%.

A significant problem concerning energy efficiency in TCP is the maintenance of a connection when a PC goes to sleep. Many applications (e.g., *SSH*, *Instant Messaging*) need a permanent TCP connection between the client and the server. To maintain the persistence of the connection, hosts must generate (and respond to) periodic *keep-alive messages* even when the TCP connection is idle, that is, when neither the client nor the server needs to send data. These messages can be generated directly by the TCP protocol (at least once every 2 hours) or by the application. Nevertheless, when a PC is in the sleep state, its processor is stopped and, thus, it cannot process incoming packets and reply with ACK packets. After a predetermined time-out period, data packets (e.g., keep-alive messages) are assumed to get lost and retransmitted. If no answer is received after a certain number of attempts, the sending host drops the connection, cleaning up all the resources associated with the connection. The application is then notified that the connection is closed, resulting in an error, since the application expected a persistent connection. In the literature, two solutions have been proposed to modify the TCP protocol in order to overcome this problem, namely, *Green TCP/IP* [56] and *splitting* [29].

Green TCP/IP, proposed in Ref. [56], adds the concept of *connection sleep state* to the legacy TCP protocol. Basically, the Green TCP/IP client notifies the Green TCP/IP server that it is going to sleep. Thus, the server logically keeps the connection alive but does not send any data or ACK packet to the sleeping client. Besides, the *socket* associated with the connection is blocked in the server so as to avoid excessive queuing of data to send. When the client wakes up, it has to notify the server – simply sending a data packet on the sleeping connection – and, thus, the data flows between the server and client can be immediately resumed. Obviously, the Green TCP/IP must be compatible with legacy TCP, in order to allow the coexistence with regular TCP/IP hosts. To this end, a new *TCP_SLEEP* option has been introduced into the header of the segment. This field is used to inform the server that the client is entering the sleep mode. When the server receives a packet containing the *TCP_SLEEP* option, it will avoid dropping the connection for that client. Considering that – according to the TCP protocol – a host ignores each option that it does not understand, this solution is, therefore, backward compatible.

An alternative approach to Green TCP/IP is proposed in Ref. [29] and is based on *splitting*. The main goal of this solution is to allow an application to receive replies for keep-alive messages even if the host at the other end of the TCP connection is sleeping. In addition, it provides a quick way to fully resume the TCP connection toward a sleeping host in case the application needs data to be transmitted. In order to realize this solution, the TCP connection at each host can be split in two parts, adding a *shim layer* between the *socket interface* and the application. This shim layer presents a socket interface to the application (so the application does not need any changes) and uses the existing socket layer of the TCP software implementation. The shim layer 'deceives' applications making them see an established connection at

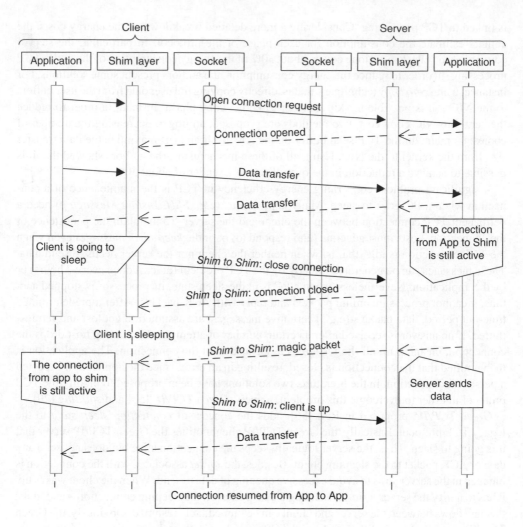

Figure 18.4 TCP connection splitting mechanism

all times. Actually, splitting the TCP connection, applications continue to see the connection with the shim layer, even when the TCP connection between the client and the server hosts has been closed. The shim layer's functionality is used only when power management is enabled (i.e., the client goes into sleep mode) and it works transparently to the application. Figure 18.4 shows how the splitting mechanism works. When the client enters the sleep state, its shim layer notifies the corresponding layer in the server, so as to drop the current TCP connection. If the server wants to send data to a sleeping client, its shim layer preliminarily wakes up the client, through a wake-up message. When the client is up again, the shim layer reestablishes the TCP connection with the opposite shim layer so that data transfer can start.

18.6.2   Application-Layer Protocols

In addition to improving transport-layer protocols at kernel level, it is also possible to design energy-efficient applications at user level. In this section we review the main energy-aware application-layer protocols proposed in the literature.

Green Telnet [57] is a very relevant example of power-aware application-layer protocol, and the approach used in *Green Telnet* can be easily extended to other client/server applications. The goal is to allow clients to go into sleep state without losing their session with the server. For this purpose, the authors propose some changes both at server and client side. Basically, *Green Telnet* decouples the state of the TCP connection from the state of the Telnet server, by an abstraction process. The application server (*gtelnetd*) does not operate directly on a socket but uses an intermediate buffer. In practice, when a client goes to the sleep state, it notifies the server – with a special message – of its intention to interrupt the communication. While the client is sleeping, the application at the server side continues to write data. Nevertheless, such data is not sent immediately to the client but is stored in the buffer. When the client wakes up, it creates a new TCP connection with the server (on a proper port communicated by the server before the power state change) and receives all the data stored during the inactive period. The software has been implemented through three processes at the server side that control the phase of client reconnection, the sending and the receiving of data to and from the buffer shared with the *gtelnetd demon,* respectively. Since *gtelnetd* acts on the intermediate buffer, all the sleep/wake-up operations occur transparently. The original Telnet protocol has been modified to implement Green Telnet by inserting a new field in the packet header and defining the new control messages that regulate the communication of the power state change.

Various methods for reducing the energy consumption have also been proposed for P2P protocols. A detailed survey of energy-efficient P2P systems and applications is available in Ref. [58]. We describe below the main solutions proposed in the literature, from a protocol perspective.

Significant energy savings can be achieved in P2P systems by properly allocating tasks to peers. A "client" peer that needs a service has to find a "server" peer that can satisfy its request. Hence, this "server" peer can be suitably chosen so as to minimize the energy consumption. In Refs. [59] and [60] the authors consider a Web application on P2P overlay networks. This is a typical example of a transaction-based application, in which the main cost for the fulfillment of the request consists in the consumption of CPU resources (i.e., in processing) while the expense for the content distribution is absolutely marginal. Thus, the authors present a computation model and a power consumption model in order to describe how processes run on a server peer and how much energy their execution consumes. In particular, two versions – a simple and a multilevel version – of the power consumption model are proposed. The simple model is more suitable for PCs with one CPU, whereas a server computer with multiple CPUs follows the multilevel model. Exploiting these models, the authors propose an algorithm by which a client peer can select a server peer in a set, so as to satisfy some constraints (e.g., temporal constraints) and reduce power consumption. Simulation results show energy savings up to 12.2%, compared to traditional Round Robin algorithms.

The most common use of peer-to-peer applications consists in file sharing. In this context, *BitTorrent* is the most commonly used protocol over the Internet. Hence, a number of solutions has been proposed to make it energy efficient. *Green BitTorrent* [61] is a modified version of the

BitTorrent protocol that allows *peers* that have completed their download process (*seeds*) and that are not currently involved in any upload operation, to go into sleep mode. From the point of view of a generic peer *P*, the other peers in the same *swarm* can be in one of the following states: *connected*, if the TCP connection is active; *sleeping*, if the peer is disconnected but the TCP connection could be reestablished and *unknown*. When *P* detects a peer disconnection, it sets, in a list, the state of that disconnected host to "sleeping." However, the TCP connection is not dropped. When the number of connected peers is less than a predefined threshold, peer *P* can explicitly wake up a sleeping peer, by sending a special wake-up message (i.e., a Magic Packet or another WoL mechanism). Once *P* completes its download, it starts an *inactivity timer* to clock the idle periods (i.e., without upload operations). After the timer expiration, the peer can go to sleep. Basically, it communicates its intentions sending a message to all the connected peers and, then, it enters the sleep state. While *P* is in this state, it can receive a wake-up message from another peer. In this case, it establishes a TCP connection with the peer that sent the message and starts exchanging data with it. Green BitTorrent is compatible with the legacy version of the protocol, even if peers using legacy BitTorrent protocol experience a slight performance degradation, in terms of higher average download time. It allows to obtain a considerable reduction of the energy consumption [61].

18.7   Conclusions

In this chapter we have addressed the problem of energy-efficient protocol design for reducing energy wastes because of edge devices (i.e., PC, printers, IP phones, etc.) that are typically left on even when they are not needed. Specifically, we have surveyed the main solutions proposed in the literature to address this problem and presented a taxonomy. According to the proposed taxonomy, we can classify these solutions into four main categories, namely, *on-demand wake-up, proxying, context-aware power management,* and *power-aware protocols.*

Solutions exploiting on-demand wake-up allow to turn on a host remotely, by sending a special packet called Magic Packet. These solutions are very common and used in many power management systems, also for large-scale distributed systems. However, they exhibit a number of limitations, mainly in terms of security and privacy. Some proprietary solutions have been proposed to overcome these issues. However, they have not yet reached the same degree of diffusion and compatibility as the original solution.

Solutions based on proxying allow a host to delegate the answer to certain types of requests to a proxy and go to sleep. Both application-specific and network connectivity proxies are available. Network connectivity proxies are application independent and allow significant energy savings, up to 70%, depending on the usage model of the host, while guaranteeing continuous network connectivity. A network connectivity proxy can be either a *private* proxy serving just one host or a *shared* proxy capable of serving many hosts. The former solution is suitable for PCs, while the latter one is more appealing for large distributed systems.

Context-aware power management strategies allow to manage the power state of a host by exploiting proper context information. They rely on specific sensors to determine the user's position and turn off the PC when the user is far from it. Several approaches can be used to obtain the required context information. They can make use of accurate and expensive sensors or low-cost general devices (such as Bluetooth phones). Also, it is necessary to consider the additional energy consumed (e.g., by sensors), and the intrusiveness of the power management system. Ideally, power management should be totally transparent to the user. Instead, forcing the user to manually turn on her/his host or to wait long resuming times due to wrong decisions could be irritating and lead the user to disable power management. Currently, to the best of our knowledge, no such strategy is used in commercial systems. This is mainly because it is very difficult to predict the users' intentions. However, context-aware power management is a stimulating research field.

Finally, the last approach consists in designing power-aware protocols and applications. In particular, we have focused on methods to modify existing protocols and applications in order to make them energy efficient. While designing new energy-efficient solutions from scratch would be more effective, some protocols (e.g., TCP, BitTorrent) are so common that it is almost impossible to reimplement them from scratch. Hence, modifying existing protocols and applications in a green perspective is the only way to achieve energy efficiency while preserving backward compatibility.

Before concluding this chapter, it may be worthwhile emphasizing that the previous approaches are not necessarily alternative. Instead, some of them can coexist. For instance, network connectivity proxies typically rely on magic packets to wake up the served host, when necessary. Also, power-aware protocols and applications can coexist with context-aware power management strategies or proxy-based solutions.

References

[1] K. Kawamoto, J. Koomey, B. Nordman, R. Brown, M. Piette, M. Ting, and A. Meier, "Electricity used by office equipment and network equipment in the U.S.: detailed report and appendices," Technical Report LBNL-45917, Energy Analysis Department, Lawrence Berkeley National Laboratory, 2001.

[2] B. Raghavan, J. Ma, "The Energy and emergy of the Internet", Proceedings of ACM Workshop on Hot Topics in Networks (Hotnets 2011), Cambridge, Massachusetts, USA, November 14–15, 2011.

[3] P. Bertoldi, B. Hirl, N. Lab, "Energy efficiency status report 2012 – electricity consumption and efficiency trends in the EU-27", JRC Scientific and Policy Reports, European Commission, Joint Research Centre, Institute for Energy and Transport, 2012.

[4] S. Ruth, "Green IT – more than a three percent solution", IEEE Internet Comput. Mag., vol. 13, no. 4, pp. 74–78 2009.

[5] R. Bolla, R. Bruschi, F. Davoli, F. Cucchietti, "Energy efficiency in the future Internet: a survey of existing approaches and trends in energy-aware fixed network infrastructures", IEEE Commun. Surveys Tutorials, vol. 13, no. 2, 2011.

[6] R. Bolla, R. Bruschi, K. Christensen, F. Cucchietti, F. Davoli, S. Singh, "The potential impact of green technologies in next generation wireline networks – is there room for energy savings optimization?", IEEE Commun. Mag., vol. 49, no. 8, pp. 80–86, 2011.

[7] S. Karayi, "The PC energy report, 1E", National Energy Foundation (NEF), London [Online], 2007. Available at: http://www.1e.com/energycampaign/downloads/1E_reportFINAL.pdf

[8] G. Newsham, D. Tiller, "A case study of the energy consumption of desktop computers", Proceedings of IEEE Industry Applications Society Annual Conference, Houston, Texas, USA, October 4–9, 1992.

[9] K. Christensen, "The next frontier for communications networks: power management", Proceedings of SPIE - Performance and Control of Next-Generation Communications Networks, Vol. 5244, pp. 1–4, 2003.

[10] K. Christensen, P. Gunaratne, B. Nordman, A. George, "The next frontier for communications networks: power management", Comput. Commun., vol. 27, no. 18, pp. 1758–1770, 2004.

[11] Magic Packet Technology, White Paper, Publication# 20213, Rev: A, Amendment/0, 1995.

[12] "Remote Wake-Up: Intel® Network Adapters User Guide", Intel Corporation [Online], 2008, Available at: http://driveragent.com/archive/17228/image/7-0-93.

[13] L. Chiaraviglio, M. Mellia, "Polisave: efficient power management of campus PCs", Proceedings of International Conference on Software, Telecommunications and Computer Networks (SoftCOM 2010), Split, Dubrovnik, Croatia, September 23–25, 2010.

[14] "Advanced Configuration and Power Interface Specification, revision 5.0", Hewlett-Packard, Intel Corporation, Microsoft, Phoenix Technologies, Toshiba, December 6 [Online], 2011. Available at: http://acpi.info /DOWNLOADS/ACPIspec50.pdf.

[15] K. Kurowski, A. Oleksiak, M. Witkowski, "Distributed power management and control system for sustainable computing environments", Proceedings of International Conference on Green Computing (IGCC 2010), Chicago, Illinois, USA, August 15–18, 2010.

[16] "Hardening Measures Built into Intel® Active Management Technology", Intel ®, 2010 [Online]. Available at: http://software.intel.com/en-us/articles/hardening-measures-built-into-intel-active-management-technology/.

[17] "Intel® Core™ vPro™ Technology: Intelligence Adapts to Your Needs", White Paper, Intel® [Online]. Available at: http://www.intel.com/content/dam/www/public/us/en/documents/white-papers/remote-support-vpro-intelligence-that-adapts-to-your-needs-paper.pdf

[18] E. Pitoura, G. Samaras, "Data Management for Mobile Computing", Norwell, Massachusetts, USA: Kluwer Academic Publishers, 1997.

[19] G. Anastasi, M. Conti, E. Gregori, A. Passarella, "Performance comparison of power saving strategies for mobile web access", Perform. Eval., vol. 53, no. 3–4, pp. 273–294, 2003.

[20] B. Nordman, K. Christensen, "Improving the energy efficiency of ethernet-connected devices: a proposal for proxying", White Paper, Version 1.0, Ethernet Alliance, 2007.

[21] UPnP Low Power Architecture V1.0, UPnP Forum, August 27, 2007 [Online]. Available at: http://www.upnp .org/specs/lp.asp.

[22] M. Jimeno, "The SIP Catcher: a Service to Enable IP Phones to Sleep" [Online]. Available at: http://www .youtube.com/watch?v=KdAm4olcVoo.

[23] M. Jimeno, K. Christensen, "A prototype power management proxy for Gnutella peer-to-peer Ffile sharing" Proceedings of IEEE Conference on Local Computer Networks (LCN 2007), Dublin, Ireland, October 15–18, 2007.

[24] G. Anastasi, M. Conti, I. Giannetti, A. Passarella, "Design and evaluation of a BitTorrent proxy for energy saving", Proceedings of IEEE Symposium on Computers and Communications (ISCC 2009), Sousse, Tunisia, July 5–8, 2009.

[25] G. Anastasi, I. Giannetti, A. Passarella, "A BitTorrent proxy for green Internet file sharing: design and experimental evaluation", Comput. Commun., vol. 33, no. 7, pp. 794–802, 2010.

[26] M. Jimeno, K. Christensen, B. Nordman, "A network connection proxy to enable hosts to sleep and save energy", Proceedings of IEEE International Performance Computing and Communications Conference (IPCCC 2008), Austin, Texas, USA, December 7–9, 2008.

[27] K. Christensen, F. Gulledge, "Enabling power management for network-attached computers", Int. J. Netw. Manag., vol. 8, no. 2, pp. 120–130, 1998.

[28] K. Christensen, B. Nordman, R. Brown, "Power management in networked devices" IEEE Comput., vol. 37, no. 8, pp. 91–93, 2004.

[29] C. Gunaratne, K. Christensen, B. Nordman, "Managing energy consumption costs in desktop PCs and LAN switches with proxying, split TCP connections, and scaling of link speed", Int. J. Netw. Manag., vol. 15, no. 5, pp. 297–310, 2005.

[30] R. Khan, R. Bolla, M. Repetto, R. Bruschi, M. Giribaldi, "Smart proxying for reducing network energy consumption", Proceedings of International Symposium on Performance Evaluation of Computer and Telecommunication Systems (SPECTS 2012), Genoa, Italy, July 8–11, 2012.

[31] S. Cheshire, "Method and apparatus for implementing a sleep proxy for services on a network", United States Patent N. 7,330,986, February 12, 2008.

[32] S. Nedevschi, J. Chandrashekar, B. Nordman, S. Ratnasamy, N. Taft, "Skilled in the art of being idle: reducing energy waste in networked systems", Proceedings of USENIX Symposium on Networked System Design and Implementation (NSDI, 2009), Boston, Massachusetts, USA, April 22–24, 2009.

[33] Y. Agarwal, S. Hodges, J. Scott, R. Chandra, P. Bahl, R. Gupta, "Somniloquy: augmenting network interfaces to reduce PC energy usage", Proceedings of USENIX Symposium on Networked System Design and Implementation (NSDI, 2009), Boston, Massachusetts, USA, April 22–24, 2009.

[34] Y. Agarwal, S. Savage, and R. Gupta, "SleepServer: energy savings for enterprise PCs by allowing them to sleep", Proceedings of USENIX Annual Technical Conference, Boston, Massachusetts, USA, 2010.

[35] Standard ECMA-393 "ProxZzzy for Sleeping Hosts, 1st edition," 2010.

[36] UPnP Device Architecture, UPnP Forum [Online]. Available at: http://www.upnp.org/standardizeddcps/ default.asp.

[37] H. Schulze, K. Mochalski, IPOQUE – Internet Study 2008/2009, Leipzig, Germany, 2007.

[38] H. Hlavacs, R. Weidlich, T. Treutner, "Energy efficient peer-to-peer file sharing", J. Supercomput., vol. 62, no. 3, pp. 1167–1188, 2012.

[39] I. Kelenyi, A. Ludanyi, J. Nurminen, "BitTorrent on mobile phones-energy efficiency of a distributed proxy solution", Proceedings of International Green Computing Conference (IGCC 2010), Chicago, Illinois, USA, August 15–18, 2010.

[40] I. Kelenyi, A. Ludanyi, J. Nurminen, I. Pusstinen, "Energy-efficient mobile BitTorrent with broadband router hosted proxies", Proceedings of IFIP Wireless and Mobile Networking Conference (WMNC 2010), Budapest, Hungary, October 13–15, 2010.

[41] A. Dalton, C. Ellis, "Sensing user intention and context for energy management", Proceedings of Workshop on Hot Topics in Operating Systems (HotOS IX), Lihue, Hawaii, USA, May 18–21, 2003.

[42] R. K. Harle, A. Hopper, "The potential for location-aware power management", Proceedings of International Conference on Ubiquitous Computing (UbiComp 2008), Seoul, South Korea, September 21–24, 2008.

[43] M. Youssef, A. Agrawala, "The Horus WLAN location determination system", Proceedings of International conference on Mobile Systems, applications, and services (MobiSys 2005), Seattle, Washington, USA, 6–8 2005.

[44] C. Harris, V. Cahill, "Power management for stationary machines in a pervasive computing environment", Proceedings of Hawaii International Conference on System Sciences (HICSS 2005), Hawaii, USA, January 3–6, 2005.

[45] C. Harris, V. Cahill, "Exploiting user behaviour for context-aware power management", Proceedings of IEEE International Conference on Wireless And Mobile Computing, Networking And Communications (WiMob 2005), Montreal, Canada, August 22–24, 2005.

[46] Z.-Y. Jin and R. K. Gupta, "RSSI based location-aware PC power management", In Workshop on Power Aware Computing and Systems (HotPower 2009), Big Sky, Montana, USA, October 10, 2009.

[47] C. Harris, V. Cahill, "An empirical study of the potential for context-aware power management", Proceedings of International Conference on Ubiquitous computing (UbiComp 2007), Innsbruck, Austria, September 16–19, 2007.

[48] A. Kansal, F. Zhao, "Fine-grained energy profiling for power-aware application design", ACM SIGMETRICS Perform. Eval. Rev., vol. 36, no. 2, pp. 26–31, 2008.

[49] W. Baek, T. Chilimbi, "Green: a framework for supporting energy-conscious programming using controlled approximation", Proceedings of ACM SIGPLAN conference on Programming language design and implementation (PLDI 2010), Toronto, Canada, June 5–10, 2010.

[50] A. Ayadi, P. Maille, D. Ros, "TCP over low-power and lossy networks: tuning the segment size to minimize energy consumption", Proceedings of IFIP International Conference on New Technologies, Mobility and Security (NTMS 2011), Paris, France, February 7–10, 2011.

[51] C. Song, S. W. Turner, H. Sharif, "An energy-efficient TCP quick timeout scheme for wireless LANs", Proceedings of IEEE International Performance, Computing, and Communications Conference (IPCCC 2003), Phoenix, Arizona, USA, April 9–11, 2003.

[52] F. Keceli, I. Inan, E. Ayanoglu, "Fair and efficient TCP access in IEEE 802.11 WLANs", Proceedings of IEEE Wireless Communications and Networking Conference (WCNC 2008), Las Vegas, Nevada, USA, March 31–April 3, 2008.

[53] N. Cho, K. Chung, "TCP-New veno: the energy efficient congestion control in mobile ad-hoc networks", Proceedings of International Conference on Embedded and Ubiquitous Computing (EUC 2006), Seoul, South Korea, August 1–4, 2006.

[54] A. Seddik-Ghaleb, Y. Ghamri-Doudane, S. Senouci, "A performance study of TCP variants in terms of energy consumption and average goodput within a static ad hoc environment", Proceedings of International Conference on Wireless Communications and Mobile Computing (IWCMC 2006), Vancouver, Canada, July 3–6, 2006.

[55] B. Wang, S. Singh, "Computational energy cost of TCP", Proceedings of Annual Joint Conference of the IEEE Computer and Communications Societies (INFOCOM 2004), Hong Kong, China, March 7–11, 2004.

[56] L. Irish, K. Christensen, "A 'Green TCP/IP' to reduce electricity consumed by computers", Proceedings of IEEE Southeastcon 1998, Orlando, Florida, USA, April 24–26, 1998.

[57] J. Blackburn, K. Christensen, "Green telnet: modifying a client-server application to save energy", Dr. Dobb's J., vol. 414, pp. 33–38, 2008.

[58] A. Malatras, F. Peng, B. Hirsbrunner, "Energy-efficient peer-to-peer networking and overlays", Chapter 20 in Handbook of Green Information and Communication System, M. S. Obaidat, A. Anpalagan, I. Woungang, Eds, Elsevier: Academic Press, 2012.

[59] T. Enokido, A. Aikebaier, M. Takizawa, "A model for reducing power consumption in peer-to-peer systems", IEEE Syst. J., vol. 4, no. 2, pp. 221–229, 2010.

[60] T. Enokido, A. Aikebaier, M. Takizawa, "Process allocation algorithms for saving power consumption in peer-to-peer systems", IEEE Trans. Ind. Electron., vol. 58, no. 6, pp. 2097–2105, 2011.

[61] J. Blackburn, K. Christensen, "A simulation study of a new green BitTorrent", Proceedings of International Workshop on Green Communications (GreenComm 2009), Dresden, Germany, June 18, 2009.

19

Information-Centric Networking: The Case for an Energy-Efficient Future Internet Architecture

Mayutan Arumaithurai[1], Kadangode K. Ramakrishnan[2] and Toru Hasegawa[3]

[1]*Institute of Computer Science, Computer Networks Group, University of Goettingen, Goettingen, Germany*
[2]*Riverside Computer Science and Engineering, University of California, Riverside, USA.*
[3]*Information Networking, Osaka University, Osaka, Japan*

19.1   Introduction

The Internet was traditionally designed in a host-centric manner with its primary focus being the establishment of end-to-end connectivity between them. It was designed on the assumption that the network elements are always available (i.e., switched on and connected) and that end-to-end connectivity such as that provided by TCP/IP is sufficient to facilitate data transfer between two nodes. If an established end-to-end connectivity between two nodes is broken, the infrastructure primarily focuses on re-establishing the broken connection. In case the nodes are mobile, Mobile-IP [1] and related protocols further ensure that the communication (via TCP connection) is maintained even as the nodes move.

In reality, such a heavy focus on connection establishment is not necessary for all usage scenarios, especially in the case of data delivery and could result in inefficient use of network resources such as bandwidth and power. Moreover, the presence of intermediate nodes that facilitate end-to-end connectivity such as proxies, Home-Agents, and Network Address Translation (NAT) devices further complicate energy-saving approaches such as shutting down routers, network interfaces, or even switching off certain routing paths based on the network

Green Communications: Principles, Concepts and Practice, First Edition.
Edited by Konstantinos Samdanis, Peter Rost, Andreas Maeder, Michela Meo and Christos Verikoukis.

load. Instead, the focus should be on the data regardless of where it is obtained from. This would allow the requester to obtain the data from a cache/node that is fewer hops away from the original source of the data if it becomes available during an ongoing transfer. Moreover, in case the source of the content becomes unavailable (because it moved or switched off to save power) during an ongoing transfer, the requester could continue to download the remaining content from other sources. Predominantly battery powered hosts such as laptops, smartphones, and tablets that connect intermittently to the Internet and are not powerful enough to support too many parallel requests cannot be considered as reliable sources for content, especially a large piece of data. The influx of such devices further strengthens the need to focus on content instead of establishing and maintaining connection.

In short, users are primarily interested in obtaining content and do not care much about where they obtain the content from. But, the Internet as it is currently designed focuses on establishing end-to-end connectivity. In this chapter, we argue that energy optimization techniques applied on the current Internet infrastructure will not result in orders of magnitude increase in energy efficiency and that we need to consider the possibility of deploying future Internet architectures, namely Information-Centric Networking (ICN) [2–5]. We will first look at popular enhancements to the current design of the Internet that primarily focuses on the actual content retrieval, discuss the energy saving potential they pose and the reasons behind why they are limited in their functionality as a pure information-centric network and the lessons learnt. We will then shift our focus to the ICN paradigm that focuses primarily on content and helps to deliver content in a more intelligent manner. Finally, we discuss the potential it has in terms of energy efficiency, the challenges it imposes and present some use-scenarios.

19.2   Popular Content-Centric Enhancements

Though the Internet has been designed with a focus on end-to-end connectivity, many popular solutions exist that try to overcome this and turn the focus on content.

19.2.1   Peer-to-Peer

Peer-to-peer is a prime example of a content-centric approach where users interested in a particular content attempt to obtain it from other peers. Popular peer-to-peer services such as BitTorrent make use of a Tracker Server to store the mapping between available content and which of the peers have it. A peer interested in a particular content contacts a Tracker Server and obtains a list of peers that are serving that particular content. The advantage of peer-to-peer solutions is that the peers that are downloading a particular content can also choose to serve that content, thereby increasing the number of sources for a particular content. Peer-to-peer solutions thus provide users a wide range of options from where one could obtain the content. Peer-to-peer services also facilitate the possibility for a requester of content to obtain it from multiple sources simultaneously.

19.2.1.1   What is the Energy Saving Potential?

Peer-to-peer facilitates multiple nodes to participate in the redistribution of content and therefore make the content simultaneously available in multiple nodes. This increases the likelihood that a requester of content finds one or more optimal sources to download the data. The optimal source could be close enough to the requester compared to the original source, thereby providing the possibility to reduce the number of hops the data has to traverse. Moreover, the original source for the data need not be available 24 hours on-line because other peers that are available contribute their uplink capacity to support redistribution. Moreover, unlike end-to-end connectivity based data retrieval, peer-to-peer systems are resilient to churn, and, therefore, switching off hosts or network routers based on energy saving plans does not affect the data delivery significantly. Peer-to-peer-based data delivery is also resilient to changes in paths.

19.2.1.2   Why They are not Completely Effective as a Content-Centric Alternative?

The disadvantage of a peer-to-peer-based solution is that it is not topology aware and, therefore, proximity in terms of hops in the peer-to-peer topology does not in reality mean that they are close to each other in the routing topology. For instance, three peers that appear close to each other on the peer-to-peer topology could in fact be world apart with one of them being in the United States, another in Europe, and another in Japan. Therefore, in terms of the actual distance the content has to traverse, it might have to traverse a larger number of hops, thereby increasing energy consumption. To overcome the problem of topology unawareness, solutions such as Application Layer Transport Optimization (ALTO) [6] have been proposed. The ALTO servers are envisioned to have information about the network topology and other factors and, therefore, support the clients in the peer selection process. The ALTO-based solution looks promising, but cannot operate at small timescales because the updates it receives are usually averaged over larger timescales. Furthermore, the effectiveness of the ALTO solution depends on the level and accuracy of the information it obtains from the various network operators.

19.2.2   Content Delivery Network (CDN)

Content Delivery Networks or Content Distribution Networks (CDNs) are a distributed network of large storehouses for content and support the redistribution of content. The goal of a CDN is to serve content to end users with high availability and high performance. CDNs help users to obtain their content faster and reduce the load on the original source of content as well as on the network. CDNs are in fact a group of servers present in data-centers that cache and serve content such as downloadable files (movies, software, documents), web-objects (images,scripts, text), location-specific advertisements, and other static content. They are also used by content providers to serve live-streaming and video on demand. CDNs were usually deployed in backbone networks, but recently, network operators have been deploying smaller scale CDNs closer to the edge to optimize traffic in their network as well as to provide content providers an alternative CDN service.

19.2.2.1   What is the Energy-Saving Potential?

As CDNs are essentially data-centers serving content, energy optimization techniques used in data-centers are applicable. Cheaper/renewable energy sources, efficient cooling mechanisms, and efficient load distribution could be used for energy efficiency, which is difficult and infeasible for smaller/individual content providers. Moreover, due to the concentration of content, servers can serve multiple contents instead of having to be available 24/7 just serving a single piece of content regardless of the load. The advantage is that end-nodes that serve only one type of content can be switched off because their content is being served by a CDN server. Furthermore, based on load, multiple clusters could be switched off. Content could also be served from a closer CDN, thereby reducing latency and the number of hops.

19.2.2.2 Why They are not Completely Effective as a Content-Centric Alternative?

CDNs are application layer solutions, and, therefore, the client will have to establish connection to the content provider (e.g. HTTP), in order to receive a list of content and the corresponding CDN cache server where the content can be obtained from. Therefore, content providers might need to have their servers available for initial connection establishment and depend on CDNs to increase efficiency. Moreover, CDN-based solutions can be suboptimal because the CDN source has to be decided prior to the actual data transfer. A web server might place data in a CDN and request users to go there, but a CDN server close to the user might not be used because the web server did not store content there. Dynamically deciding which content to cache in which CDN server is not straightforward. In the case of mobile nodes, this could result in larger inefficiency because a Uniform Resource Identifier (URI) that has been resolved earlier to a particular CDN might not be the optimal one once a node moves and because no URI resolution is involved after movement, the nonoptimal CDN is being used, which could be a larger number of hops away.

19.2.3 Domain Name Systems (DNS)

DNS is another example of a content-centricity approach. The DNS stores mapping between a URL and the IP address where the content can be obtained. For instance, a user searching for "Google.com" can be redirected to any of the Google servers based on how the DNS is configured. The configuration could be such that the load is balanced or the request is redirected to the server closest to where the request was made. When an end user moves, he can renew his DNS request to receive a server close to him.

19.2.3.1 What is the Energy-Saving Potential?

The separation between end users to request for information and a mapping service that maps information to location/node has the possibility to save energy in terms of the number of hops the data has to traverse.

19.2.3.2 Why They are not Completely Effective as a Content-Centric Alternative?

Nevertheless, the problem with DNS is that, it is performed at the beginning and is rarely updated during the session. Moreover, DNS updates are not possible in shorter time frames and make sense only for big content providers.

19.3   ICN: Motivation

ICN, also known as Content-Centric Networking (CCN) [2, 3, 5], is a new paradigm for the future Internet where the network provides users with named content, instead of communication channels between hosts. It places the content at the forefront of the design of the architecture. Unlike current IP networks that support end-to-end connectivity, ICN networks route traffic based on content names. Therefore, instead of focusing on end-to-end connection and maintaining such connections even when data is available on a router/node along the path, ICN allows us to focus on obtaining content. In the case of an IP-based end-to-end connection scenario where multiple recipients are interested in the same content, the content is delivered only from the source on a per flow basis even if the flows traverse the same intermediate nodes. Intermediate routers will therefore end up processing the same request and data multiple times in the case of IP. In the case of mobile nodes, as a mobile node moves, it can continue to obtain the data from the same source or some other efficient source with the help of ICN. It does not have to use a complex protocol like Mobile IP that involves a lot of signaling just to maintain end-to-end connectivity. To summarize, ICN is content-centric; facilitates easy caching and obtaining data from closer sources; and supports obtaining parts of the data from multiple sources at the same time.

ICN with its focus on content changes routing completely. The routing is done such that the request is sent to the closest source. Such an optimization at the routing level facilitates the introduction of drastic changes in energy efficiency. At the routing layer, the decisions made are topology-aware unlike peer-to-peer or CDN-based solutions. Here, in-path caches that serve the content can further improve latency, reduce unnecessary usage of network resources, and thereby result in energy efficiency. Dedicated resources such as CDNs and routers with a larger cache could compliment ICN and help further improve energy efficiency.

As explained in Section 19.5, ICN has the potential to introduce significant changes in energy consumption and energy saving mechanisms. The energy saving mechanisms described in other chapters of this book for the current Internet might also be applicable in an ICN-based environment. For instance, shutting down unused network cards or devices could be used in ICN too. In fact, ICN might make it easier to deploy many of these energy saving mechanisms. As ICN is a new and evolving technology, there is a lot of scope to include energy efficiency in the design phase as being done in the GreenICN joint project between the European Union and Japan [7].

19.4   ICN: Background and Related Work

Research on ICN is at an early stage, with many key issues still open, including naming, routing, resource control, security, privacy, and a migration path from the current Internet. Next, we list some of the interesting ongoing works related to ICN.

19.4.1   Named Data Networking (NDN)

Named Data Networking (NDN)[1] [3], originally known as Content-Centric Networking (CCN) [2], is a popular ICN protocol where content/information is looked up and delivered

---
[1] http://named-data.net/

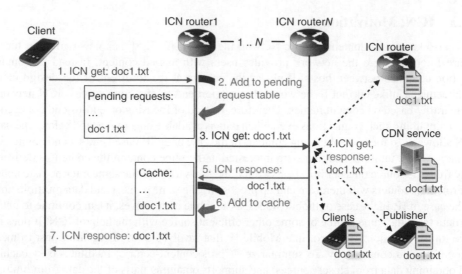

Figure 19.1 Message flow highlighting the name-based data retrieval in ICN. Requested data can be obtained from one of the multiple sources of the data. In this case, the data can be obtained from the cache of an ICN router, from a CDN service, from other clients, or directly from the original publisher (see Step-4). Step-2 shows that the `Interest` is added to the Pending Interest/Request Table (PIT), and in Step-6 we can observe that the data is stored in the cache of the router before being forwarded

according to its name without knowing the identity and location of the sender. NDN uses two packet types, `Interest` and `Data`. A consumer queries for named content by sending an `Interest` packet; a provider in turn responds with a `Data` packet. NDN requires a new forwarding engine instead of IP, which contains the Forwarding Information Base (FIB), Content Store (buffer memory which caches content), and Pending Interest Table (PIT). FIB is used to forward `Interest` packets toward potential source(s) of matching data. PIT keeps track of bread crumbs of `Interest` (i.e., to support reverse-path forwarding), which the `Data` packets follow to reach the original requester(s). If multiple `Interest` packets arrive for the same data from multiple end-nodes, they will be aggregated in the PIT and served when the data arrives. The Content Store maintains a cache of the data in order to satisfy potential future requests for that data.

Figure 19.1 shows a simple message flow in an NDN architecture. Let us assume that the requester would like to get a movie file (doc1.txt) that is published by the publisher. In an IP network, the requester would make a DNS request to identify the IP address of the publisher and issue a request that would go all the way to the publisher. On the other hand, in an ICN network, the requester would issue a request (Step-1, i.e., ICN Get) for that movie file to the ICN network. The first hop router has multiple options to deal with this request. If the movie is present in its cache, it could deliver it directly from it. Else, an ICN router could choose to forward it via any one of the paths such as those depicted in Figure 19.1 where one path leads to a CDN like content store, another to other clients that have the data, another to an ICN router cache, and another to the publisher.

19.4.2   Content-Oriented Publish/Subscribe System (COPSS)

Content-Oriented Publish/Subscribe System (COPSS) [5] was proposed as an enhancement of NDN to provide push-based pub/sub multicast capability in ICN. COPSS allows subscribers interested in a particular type of content or topic to issue a request for subscription. The COPSS network handles the delivery of the content to the interested subscribers when publishers publish content for these topics. The advantage of COPSS compared to NDN is that the routers need not maintain per packet state as being done by the PITs. Instead, they have to maintain a state per subscription in the Subscription Table (ST). A Rendezvous Point (RP) forms the root of the tree to which subscriptions are sent. Publishers forward the data toward the RP, which in turn forwards it to the subscribers via the ST. COPSS extends the naming framework with the introduction of the concept of hierarchical content descriptors to enable efficient large-scale information dissemination. COPSS also provides support for a hybrid environment [8, 9] that comprises nodes belonging to both IP and ICN nodes that can choose to function with full or partial ICN capability.

19.4.3   Projects Supported by the European Union

The Publish Subscribe Internet Routing Paradigm (PSIRP) [10] project developed an information-centric network architecture based on a publish/subscribe paradigm. PSIRP proposed to replace the current Internet protocols entirely, applying a layer-less clean-slate architecture for routing, security, mobility, and other basic network services. This was followed by the PURSUIT project [11] to address open issues such as resource control and advanced concepts for information scoping. PSIRP/PURSUIT introduced several contributions on several aspects of ICN, for example, publish/subscribe architecture (e.g., Refs. [11, 12]), fast-forwarding strategies (e.g., Ref. [13]), a new transport layer protocol (e.g., Ref. [14]), and mobility support (e.g., Ref. [15]).

The 4WARD project [16] developed an ICN architecture called Networking of Information (NetInf) [17]. It has an object model that can handle information at different abstraction levels, enabling the referencing of information independent of its encoding. The NetInf naming scheme provides name–data integrity and name persistency. The SAIL project [18] continued developing NetInf where 4WARD left off. For instance, the naming scheme was revised, the object model was simplified, and the routing and name resolution framework has become more concrete with, for example, an inter-domain interface.

The Architecture and Applications of Green Information Centric Networking (GreenICN) project [7] is a relatively new project that was started in April 2013 and is supported by both the European Union and Japan. GreenICN plans to build an energy-efficient content-centric network from the onset instead of treating it as an after-thought. GreenICN aims to perform an application-driven design, with disaster and large-scale video delivery as the chosen key application scenarios. The aftermath of a disaster introduces challenges in terms of data delivery as well as efficient use of scarce resources such as power. Many of the functioning devices such as base stations and mobile terminals are primarily driven with the use of batteries. GreenICN aims to efficiently distribute disaster notification and critical rescue information, with its ability to exploit fragmented networks with only intermittent connectivity, while ensuring that energy is efficiently consumed. Video delivery on the other hand introduces issues of scale in terms of network traffic, efficient use of caching, and load on the various nodes. GreenICN also aims to investigate migration from IP to a pure ICN architecture.

19.4.4   Internet Research Task Force (IRTF)

The IRTF [19] is associated with the Internet Engineering Task Force (IETF)[20]. It promotes research of importance to the evolution of the Internet by creating focused, long-term Research Groups working on topics related to Internet protocols, applications, architecture, and technology. Recently, Information-Centric Networking Research Group (ICNRG)[21] was started with the main objective of coupling ongoing ICN research in the aforementioned areas with solutions that are relevant for evolving the Internet at large. The ICNRG serves as a forum for exchange and analysis of ICN research ideas. Its current goals are to produce documents that further the understanding of the current state of the art and identify the research challenges. In-network caching techniques have also been investigated in the IETF for potential standardization in the Decoupled Application Data Enroute (DECADE) Working Group [22] and might be pursued in the ICNRG Working Group too. The Light-Weight Implementation Guidance (lwig) [23] focuses on small devices and work related to energy efficiency.

19.4.5   ICN-Related Research papers

DONA [4] was one of the first, clean-slate, ICN proposals. DONA uses flat, self-identifying and unique names for information objects and binds the act of resolving requests for information to locating and retrieving information. The authors of Ref. [24] investigate the use of CCN in the case of real-time applications such as audio-conference, while the authors of Ref. [25] investigate the use of COPSS for a gaming application with stringent requirements. CONIC [26] is a network architecture designed for efficient data dissemination using storage and bandwidth resources in end-systems (i.e., available storage located in end hosts is used for caching). A similar approach, where content is cached in routers, is the Cache and Forward architecture [27]. MultiCache [28] is an information-centric overlay network architecture aiming to improve network utilization via resource sharing. In MultiCache, network operators deploy and control proxy overlay routers that enable the joint provision of multicast and caching, targeting both synchronous and asynchronous requests.

19.5   ICN: Energy Efficiency

19.5.1   Content-Centric Routing

Content-centric routing ensures that the routers are able to make an informed decision on what the user is interested in. Based on the request, the routers are able to forward it to the closest source possible. Such a service is naturally resilient to energy saving techniques such as switching off network elements and can react to changes in the routing path at quicker time scales. It provides services similar to peer-to-peer at the routing layer by facilitating the recipient to receive from one or more of the multiple sources (including in-path) caches; thereby enhancing efficiency, reducing latency, and control by relying on topology-based decisions.

Furthermore, it allows easy integration of CDN like services at the routing layer. It will reduce the overhead and complexity involved in unnecessarily re-routing the data, especially in the case of Mobile IP, to maintain existing connections. This could result in the reduction of latency and also help find optimal routes that need not go via predetermined proxy nodes. Similar to peer-to-peer, nodes that are currently downloading content could also double up as content providers because they are in any case switched on.

19.5.2   Reduction in the Number of Hops

Reduction in the number of hops implies that less nodes are used to process the request as well as data, thereby reducing the overall energy consumption in the network on a per request/data level. The ICN architecture, as mentioned earlier, has in-path caching, that is, routers along the way are capable of caching content. Therefore, popular content that has multiple requests at the same time might find a close enough cache, thereby reducing the number of hops the request and the data have to traverse. The in-path caching support could be more efficient than a CDN like solution because the data storage is available on the path from the requester to the data source and need not take a different path. The data source in this case could be either another peer or a CDN service. In case there are multiple sources, a closer source can be utilized to serve the data, thereby reducing the number of hops.

In-path delivery also ensures that in the case of mobility when a node moves, the request would find a closer source from the requester at the new location to the previous data source. Moreover, when a mobile node moves, one need not worry about maintaining the same IP, Mobile IP, and so on. What one needs to do is find a closer source for that data and continue delivering it. Even if there is only one source, signaling can be avoided and a new request from the new location is sent. Reduction in signaling messages after node movement also has energy saving potentials.

Figure 19.2 illustrates an example scenario where a client is seen to be requesting for the data doc1.txt in an IP as well as an ICN scenario. In the case of IP (Figure 19.2a), we can observe that the client receives the URL of the server in which doc1.txt could be found from the Google search engine (Step-2) and uses this URL to obtain the IP address of that server

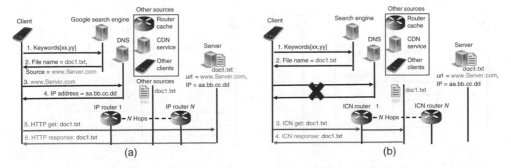

Figure 19.2   Message flow for obtaining data in a standard IP and ICN environment. (a) IP message flow: A standard HTTP Get command for data doc1.txt issued from a client to a server. The server receives the request even if other sources of that data exist closer to the client. (b) ICN message flow: A standard ICN Get command for data doc1.txt being issued to the ICN network and the first hop ICN router directing it to a closer source for that data

from the DNS server (Step-3, Step-4). It then issues an HTTP Get command for `doc1.txt` to the IP address returned by the server. This request is forwarded by the IP routers to the server mentioned in the IP header. It can be seen that even in the presence of other sources for the same data at a closer location (lesser number of hops) to the requester, the HTTP Get message is forwarded to the server. In the case of ICN (Figure 19.2b), we can observe that the search engine just returns the name of the data, that is, `doc1.txt` and does not provide a location or a URL. The client then issues an ICN Get command with the name of the data to the first hop ICN router that in turn forwards it to one of the sources that is closer (in terms of hops) to the requester, instead of forwarding it to the original source of the data. This is feasible because the ICN router does name-based forwarding and is aware of the name of the data required and where in the network is that data available. The energy efficiency achieved in this case can be represented by:

$$\eta \propto (n_1 rD + nrS + nrs) \tag{19.1}$$

where η is energy saved, n_1 is the number of hops to the DNS server, D is the cost at the DNS server to process the query, r is the cost incurred at every router to process the request or the data, n is the difference in the number of hops that the HTTP Get traverses to reach the server versus the number of hops that the ICN request travels to reach a closer source, s is the size of the request, and S is the size of the data. If we assume that the requested data file is a large video file or that the number of hops (n) is large, the gain could be considerably high.

Figure 19.3 illustrates an example scenario where multiple clients are seen to be simultaneously requesting for the same data `doc1.txt` in an IP as well as an ICN scenario. In the case of IP (Figure 19.3a), we can see that all the simultaneous requests are being forwarded by the IP routers to the server mentioned in the IP header. It can also be seen that even in the presence of other sources for the same data at a closer location (lesser number of hops) to the requester, the HTTP Get message is forwarded to the server. In the case of ICN (Figure 19.3b), we can observe that the first hop ICN router is adding all the requests to a `Pending Request Table` and forwarding just one of the requests to a source that is closer (in terms of hops) to the requesters. This is again feasible because the ICN router does name-based forwarding and is aware that the multiple requests from different clients are for the same data by looking

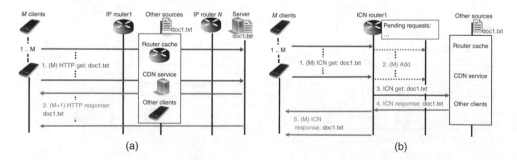

(a)   (b)

Figure 19.3 Message flow comparing the case of multiple simultaneous HTTP Get versus ICN Get. (a) IP message flow: Multiple simultaneous HTTP requests for the same data `doc1.txt` is forwarded to the server. Therefore the server responds with M+1 copies of the same data. (b) ICN message flow: Multiple simultaneous ICN requests for the same data `doc1.txt` is added to the `Pending Request Table` and only one ICN request is forwarded to the closest source at the name of the required data. Energy efficiency represented by Eq. (19.1) can therefore be updated to:

$$\eta \propto (n_1 rD + (M-1)nrS + (M-1)nrs - (M-1)p) \tag{19.2}$$

where η is energy saved, M refers to the number of parallel requests, p is the additional cost incurred to add the M requests to the Pending Request Table, n_1 is the number of hops to the DNS server, D is the cost at the DNS server to process the query, r is the cost incurred at every router to process the request or the data, n is the difference in the number of hops that the HTTP Get traverses to reach the server versus the number of hops that the ICN request travels to reach a closer source, s is the size of the request, and S is the size of the data. If we assume that M, that is, the number of parallel requests is very high, the gain could be considerably high.

19.5.3   Caching

ICN, by allowing for in-path caching as well as dedicated caching, allows recipients to access the data from closer sources. This could result in the reduction of the number of hops as well as the latency as mentioned earlier. For instance, in Figure 19.1, let us consider the scenario that multiple users are interested in the same content and their path to the source of the content is the same. In the case of IP, all the routers along the path will see the different connections and process it, but in fact the number of bits (i.e., content) transferred is the same. Whereas, in the case of ICN, the first hop router will realize that in fact all the connections are requesting for the same content and will either serve them from the cache or just forward one request upward. This saves computing resources and thereby energy on routers above it.

Furthermore, by allowing for large and concentrated caches on routers and services similar to CDN that are either in-path or closer to the source, one could also optimize energy by using means that are used in data-centers. For instance, servers, caching capability in routers can be brought up/down based on demand. By serving a large variety of content from such caches, the servers that are the actual source of content can be switched off. Efficient cooling and power mechanisms that are easily deployable in larger scales could be used, because these mechanisms are not straightforward for smaller content providers.

Caching policies could also be used to increase energy efficiency. For instance, policies that facilitate the switching ON/OFF caches on routers based on demand might be useful. Caching policies that are able to perform cooperative caching, that is, neighboring router caches serving complimentary data instead of the same data, could increase the likelihood of content hit in the vicinity, thereby reducing the number of hops.

It must be noted that caching introduces an additional burden of energy consumption on the routers, not only to store and save the data, but also to perform operations such as search for every request that passes it. Therefore, caching policies and algorithms must take this into account while devising the right solution.

Figure 19.4 illustrates an example scenario where multiple clients are seen to be requesting for the same data doc1.txt at different time periods in an IP as well as an ICN scenario. In the case of IP (Figure 19.4a), we can see that all the requests are being forwarded by the first hop IP router to the server mentioned in the IP header. It can also be seen that even in the presence of a cached version of that data in the IP router, the IP router is not able to serve that request because it is not aware of the requested data but just the IP address of the server. In the case of ICN (Figure 19.4b), we can observe that the first hop ICN router is adding the

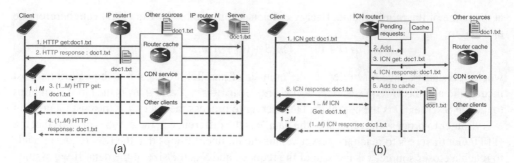

(a)   (b)

Figure 19.4 Message flow comparing a standard HTTP Get versus ICN Get. (a) IP message flow: Multiple requests for the same data `doc1.txt` is being forwarded to the server. The server therefore responds with a copy of the data for every single request received. (b) ICN message flow: Multiple requests for the same data `doc1.txt` is being served by the first hop ICN router because it has the data in its cache after the first interaction

data to its cache after the completion of the first interaction. It is, therefore, able to serve the requests arriving later from its own cache. This is again feasible because the ICN router does name-based forwarding and is aware that the multiple requests from different clients are for the data that it has in its cache. Energy efficiency represented by Eqs. (19.1) and (19.2) can, therefore, be updated to:

$$\eta \propto (n_1 rD + (M-1)nrS + (M-1)nrs - c) \tag{19.3}$$

where η is energy saved, c refers to the energy expended in the router's cache as well as to perform cache lookup. M refers to the number of parallel requests, n_1 is the number of hops to the DNS server, D is the cost at the DNS server to process the query, r is the cost incurred at every router to process the request or the data, n is the difference in the number of hops that the HTTP Get traverses to reach the server versus the number of hops the ICN request traveled to reach the first hop ICN router (i.e., 1), s is the size of the request, and S is the size of the data.

19.5.4 Seamless Support of Network Operations for Energy Efficiency

Currently, switching off nodes and/or network elements implies that a lot of signaling needs to be performed to ensure that the current state is transferred to another serving node or could result in the termination of the ongoing communication. For instance, in the case of Follow the Sun or Follow-me cloud service where data-centers that are closer to active parts of the world are switched on while data-centers away from active parts are switched off could result in the ongoing connection being terminated. Similarly, base stations, mobile gateways, service gateways, and other management servers in the case of a 3GPP network could be switched off when load is low. An ICN-based network allows for such network management operations by seamlessly adapting to the state of the network and forwarding the request to the suitable node.

Figure 19.5 illustrates an example scenario where a client is involved in a long-term transaction (e.g., a video stream or the transfer of a huge data file) in IP as well as ICN scenario. In the case of IP (Figure 19.5a), we can see that the appearance of a new data source at a closer location does not affect the ongoing communication. In the case of ICN (Figure 19.5b), we can observe that the requests are seamlessly sent to the new data source after time X. This allows

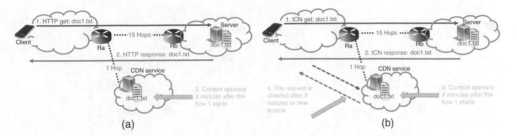

Figure 19.5   Message flow comparing a continuous HTTP Get versus ICN Get in the presence of a new data source (at Time *X*). (a) IP message flow: The requests are sent to the original server even after a new source for the data appears. (b) ICN message flow: The requests are forwarded to the CDN source after time *X* because it is closer in a seamless fashion

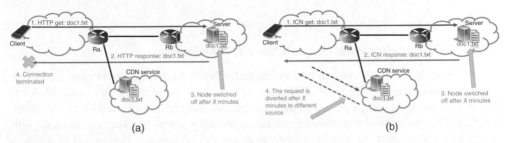

Figure 19.6   Message flow comparing a continuous HTTP Get versus ICN Get in the absence of the initial data source (at time *X*). (a) IP message flow: The connection is terminated once the source switches off after *X* minutes. (b) ICN message flow: The requests are forwarded to the CDN source after time *X* seamlessly

for greater flexibility in adapting to network conditions and making efficient use of network resources.

Figure 19.6 illustrates an example scenario where a client is involved in a long-term transaction (e.g., a video stream or the transfer of a huge data file) in IP as well as ICN scenario and the initial data source is switched off at time *X*. In the case of IP (Figure 19.6a), we can see that the disappearance of the initial data source at time *X* results in the termination of the ongoing communication. In the case of ICN (Figure 19.6b), we can observe that the requests are seamlessly sent to another data source after time *X* when the initial source is no longer available. This again illustrates the flexibility in adapting to network conditions and making efficient use of network resources.

Figure 19.7 illustrates an example scenario where a client is involved in a long-term transaction (e.g., a video stream or the transfer of a huge data file) in IP as well as ICN scenario and the client moves to a different base station/operator/location at time *X*. In the case of IP (Figure 19.7a), we can see that the connection to the initial data source is still maintained even if there is a significant increase in the number of hops. In the case of ICN (Figure 19.7b), we can observe that once the client moves at time *X*, the requests are seamlessly sent to the closer data source. This again illustrates the flexibility in adapting to network conditions and making efficient use of network resources to reduce the number of hops as well as latency for the end user.

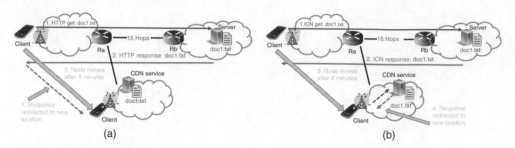

(a)   (b)

Figure 19.7 Message flow comparing a HTTP Get versus ICN Get when the node moves from one base-station/operator/location to another (at time *X*). (a) IP message flow: Though the client has moved, the connection to the initial data source is still maintained, resulting in inefficient usage. (b) ICN message flow: The requests are forwarded to the CDN source that is closer to the new location seamlessly

19.5.5   Coexistence with IP and Other Technologies

ICN is being designed not only to be incrementally deployed, but also to coexist [8, 9] with existing technologies to reap their advantages. ICN routers due to their increased complexity might consume more energy per query they handle compared to IP an router. This is due to the fact that ICN routers have to perform name-based lookup whereas IP routers do longest prefix match of an IP address. Moreover, a large amount of effort has been expended to optimize the IP routers to use the hardware very efficiently. ICN on the other hand is still in its nascent stage and would require a lot of efforts to standardize as well as optimize. During this period, as suggested in Ref. [8, 9], the ICN router could coexist with IP and use IP-based forwarding wherever possible. Note that though ICN might require more energy at a router level compared to that of IP, as we argue in this chapter, the ICN architecture on the whole might be more energy efficient.

19.6   Summary

This chapter introduces ICN, a future Internet architecture currently being developed. It details upon how the design of ICN is influenced by current workarounds that exist in the Internet infrastructure to provide content-centric features. The content-name-based routing allows for the possibility to make radical changes in how energy efficiency is achieved. The key reasons why ICN is a good candidate for energy-efficient protocol is that (i) ICN can react dynamically to the presence of data in a closer location, thereby reducing hops; (ii) ICN dynamically and seamlessly reacts to changes in path; and (iii) ICN's name-aware routing increases the flexibility of the network.

But one must also remember that the name-based routing might introduce additional work load on the routers, resulting in the need for more energy consumption. As technology advances, and the ICN protocol becomes standardized, routers will be designed to handle ICN more efficiently.

ICN and energy efficiency in ICN are at a very nascent stage. GreenICN [7], an EU project started recently, attempts to solve many of these issues and to obtain a better understanding of the energy saving potential of ICN.

References

[1] C. Perkins, "IP Mobility Support for IPv4," IETF Request for Comments (RFC 3220), 2002.
[2] V. Jacobson, D. K. Smetters, J. D. Thornton, M. F. Plass, N. H. Briggs, and R. L. Braynard, "Networking named content," in *CoNEXT*, Rome, Italy, 2009.
[3] L. Zhang, D. Estrin, J. Burke, V. Jacobson, and J. Thornton, "Named Data Networking (NDN) Project," PARC, Tech, https://www.parc.com/publication/2709/named-data-networking-ndn-project.html, 2010.
[4] T. Koponen, M. Chawla, B.-G. Chun, A. Ermolinskiy, K. H. Kim, S. Shenker, and I. Stoica, "A data-oriented (and beyond) network architecture," in *SIGCOMM*, Kyoto, Japan, 2007.
[5] J. Chen, M. Arumaithurai, L. Jiao, X. Fu, and K. K. Ramakrishnan, "COPSS: an efficient content oriented Pub/Sub system," in *ANCS*, Brooklyn, NY, USA, 2011.
[6] J. Seedorf and E. Burger, "Application-Layer Traffic Optimization (ALTO) Problem Statement," RFC 5693, October 2009.
[7] GreenICN, "Architecture. and Applications of Green Information Centric Networking (GreenICN)," project website: www.greenicn.org. [Accessed 10 January 2015].
[8] J. Chen, M. Arumaithurai, X. Fu, and K. K. Ramakrishnan, "Coexist: integrating content oriented publish/subscribe systems with IP," in *ANCS*, Austin, Texas, USA, 2012.
[9] J. Chen, M. Arumaithurai, X. Fu, and K. K. Ramakrishnan, "Coexist: a hybrid approach for content oriented publish/subscribe systems," in *ICN*, Helsinki, Finland, 2012.
[10] C. Esteve, F. Verdi, and M. Magalhaes, "Towards a new generation of information-oriented internetworking architectures," in *ReArch*, Madrid, Spain, 2008.
[11] PURSUIT. Publish-Subscribe Internet Technology, project website: www.fp7-pursuit.eu. [Accessed 10 January 2015].
[12] D. Lagutin, K. Visala, and S. Tarkoma, Valencia FIA Book 2010 Publish/Subscribe for Internet: PSIRP Perspective, IOS Press, 2010.
[13] P. Jokela, A. Zahemszky, C. E. Rothenberg, S. Arianfar, and P. Nikander. LIPSIN: line speed publish/subscribe internetworking. In *Proceedings of ACM SIGCOMM'09*, Barcelona, Spain, August 2009.
[14] V. Koptchev and V. Dimitrov, "Traffic and congestion control in a publish/subscribe network," in *Proceedings of CompSysTech'10*, Sofia, Bulgaria, June 2010.
[15] X. Vasilakos, V. A. Siris, G. C. Polyzos, and M. Pomonis, "Proactive selective neighbor caching for enhancing mobility support in information-centric networks," in *Proceedings of ACM SIGCOMM ICN*, Helsinki, Finland, August 2012.
[16] 4WARD, "The FP7 4WARD Project," http://www.4ward-project.eu/. [Accessed 10 January 2015].
[17] B. Ahlgren, M. D'Ambrosio, C. Dannewitz, M. Marchisio, I. Marsh, B. Ohlman, K. Pentikousis, R. Rembarz, O. Strandberg, and V. Vercellone, "Design considerations for a network of information," in *ReArch*, Madrid, Spain, 2008.
[18] "SAIL project website," http://www.sail-project.eu/.
[19] IRTF, "Internet Research Task Force (IRTF)," http://irtf.org. [Accessed 10 January 2015].
[20] IETF, "The Internet Engineering Task Force (IETF)," http://ietf.org. [Accessed 10 January 2015].
[21] IRTF, "Information Centric Networking Research Group (ICNRG)," http://irtf.org/icnrg.
[22] IETF, "Decoupled Application Data Enroute (DECADE)," https://datatracker.ietf.org/wg/decade/charter/. [Accessed 10 January 2015].
[23] IETF, Light-Weight Implementation Guidance (lwig), "http://datatracker.ietf.org/wg/lwig/charter/." [Accessed 10 January 2015].
[24] Z. Zhu, S. Wang, X. Yang, V. Jacobson, and L. Zhang. "ACT: audio conference tool over named data networks," in *Proceedings of ACM SIGCOMM ICN*, Toronto, Canada, August 2011.
[25] J. Chen, M. Arumaithurai, X. Fu, and K. K. Ramakrishnan, "G-COPSS: a content centric communication infrastructure for gaming," in *ICDCS*, Macau, China, 2012.

[26] Y. Zhu, M. Chen, and A. Nakao, "CONIC: content-oriented network with indexed caching," in *INFOCOM IEEE Conference on Computer Communications Workshops, 2010*, pp. 1–6, San Diego, CA, USA, 15–19 March 2010.

[27] L. Dong, H. Liu, Y. Zhang, S. Paul, and D. Raychaudhuri, "On the cache-and-forward network architecture," in *IEEE International Conference on Communications, ICC '09*, Dresden, Germany, June 2009.

[28] K. Katsaros, G. Xylomenos, and G. C. Polyzos, MultiCache: an overlay architecture for information-centric networking, Comput. Networks., vol. 55, no. 4, pp. 936–947, 2011.

20

Energy Efficiency Standards for Wireline Communications

Konstantinos Samdanis[1], Manuel Paul[2], Thomas Kessler[3] and Rolf Winter[1]

[1]NEC Europe Ltd, Heidelberg, Germany
[2]Deutsche Telekom AG, Berlin, Germany
[3]Deutsche Telekom AG, Darmstadt, Germany

20.1 Introduction

Given the exponential growth of network traffic and increasing network infrastructure for achieving higher speeds and capacity, energy consumption of networks has become a significant concern, from a business as well as environmental perspective. Besides the need for sustainable and interoperable solutions, regulatory initiatives are influencing standardization efforts enforcing energy conservation for network equipment and telecommunication systems. The majority of standardization bodies have nowadays adopted an energy efficiency or green agenda to address energy-saving mechanisms applicable to a wide set of network equipment. Among the most noticeable efforts by Standards Development Organizations (SDO) and consortia for energy efficiency in wireline networks are the ones from International Telecommunication Union Telecommunication Standardization Sector (ITU-T), Institute of Electrical and Electronics Engineers (IEEE), Internet Engineering Task Force (IETF), European Telecommunications Standards Institute (ETSI), Alliance for Telecommunication Industry Solutions (ATIS), and Broadband Forum (BBF).

The Telecommunication Standardization Sector of the International Telecommunication Union (ITU-T, a United Nations Agency) develops international standards, referred to as ITU-T recommendations. In the field of energy efficiency and green communications, ITU-T dedicated the environmental and climate change Study Group 5 (SG5), which investigates

Green Communications: Principles, Concepts and Practice, First Edition.
Edited by Konstantinos Samdanis, Peter Rost, Andreas Maeder, Michela Meo and Christos Verikoukis.
© 2015 John Wiley & Sons, Ltd. Published 2015 by John Wiley & Sons, Ltd.

energy efficiency and the environmental impact of Information Communication Technology (ICT) as well as low-cost sustainable communications, methodologies for assessment and power feeding. ITU-T transport, access and home Study Group 15 (SG15) concentrates on energy-saving mechanisms for metallic and optical access networks covering digital subscriber line (DSL) and passive optical network (PON) technologies. The future networks Study Group 13 (SG13) has introduced energy-saving frameworks for next-generation networks including cloud and data center environments.

IEEE is a professional association for advancing technological innovation, developing also standards. In the field of energy saving for wireline communications, IEEE has developed energy-efficient Ethernet (EEE), which is one of the most significant standards considering the broad industry adoption of Ethernet. EEE is analyzed in Chapter 14, while this chapter considers power over Ethernet (PoE), including IEEE 802.3af and IEEE 802.3at.

IETF and Internet Research Task Force (IRTF) are a part of an open international community of network designers, operators, vendors, and researchers concentrating on the evolution of the Internet architecture and its smooth operation. The IRTF, typically sharing meeting venue with the IETF, focuses on longer term research issues related to the Internet, while the IETF focuses on the shorter term issues of engineering and standards making. Regarding energy efficiency, main activities are carried out within the IETF Energy Management (EMAN) Working Group, the IETF Routing Area Working Group related to network transport and control, and the IRTF. ETSI, a European standards body, which addresses EU regulations regarding energy saving has introduced the Green Agenda that covers several aspects of wireline communication including energy efficiency for broadband and transport equipment as well as measurements and metrics. ATIS, a North American technology and solutions development organization, has launched the Green Initiative that focuses on power consumption measurements and reporting for Ethernet switches and routing equipment as well as metrics. The BBF is an industry organization, driving broadband wireline solutions, empowering converged multi-service packet networks addressing interoperability, architecture and management. Such converged multi-service network architectures can help to reduce the amount of network equipment providing energy saving in the network planning and deployment phase.

Current standardization efforts mainly concentrate on the device, equipment and network level, that is, the way equipment is organized or structured in order to accommodate energy efficiency and performance requirements, on device and equipment energy conservation modes (power saving states), as well as on network-based mechanisms as they need to interoperate across the network [1]. Energy-aware networking is enabled by mechanisms and protocols that support operating the network at a minimum level of aggregate power consumption while satisfying network coverage, robustness and performance, that is, service level agreements (SLAs). As the actual traffic load in a network varies, network elements can be adaptively operated in a mode with lower power consumption during off-peak periods, without causing service degradation or content restriction. Such a fundamental requirement influences the design parameters of certain standards, for example, the transition period that a device needs to take in order to change from a power saving state to a fully operational state, while for particular equipment or network-based standards complementary monitoring and control is recommended to ensure profitable energy saving periods and SLA assurance.

This chapter aims to provide an overview of the main energy saving standardization efforts for wireline communications. Section 20.2 concentrates on energy-efficient equipment considering power states based on the ITU-T framework and the EC code-of-conduct, followed by the IETF/IRFT efforts focusing on energy-aware control planes, routing and traffic engineering in Section 20.3. Section 20.4 analyzes the energy efficiency in network planning based on the converged multi-service broadband architecture of the BBF, while Section 20.5 addresses the ITU-T energy management framework, IETF EMAN and IEEE PoE. Section 20.6 contains energy measurements and metric standards according to ETSI, ATIS and ITU-T, as well as evaluation and testing procedures based on the ECR Initiative. Finally, Section 20.7 provides the concluding remarks.

20.2 Energy-Efficient Network Equipment

A comprehensive analysis of the power consumption for network equipment should consider the complete equipment life cycle. ITU-T [1] identifies the following life-cycle phases in relation with energy efficiency: (i) the production phase, which concentrates on preparing raw materials and individual components, (ii) manufacturing that includes construction and shipping, (iii) usage or equipment operation, and (iv) disposal/recycling. However, ICT and telecommunication standardization efforts concentrate on reducing the energy consumption related to the "always-on" operation via network architecture, equipment capabilities and dynamic operations that reflect the traffic load.

20.2.1 Power Modes/Power Saving States

The performance of networking equipment has evolved considerably over the recent years, similar to the growing dynamics of information technology (IT) systems; still, the efficiency (ratio of performance to consumed energy) has not kept pace, leading to a continued rise in absolute energy consumption. Two main counter-measures are considered: improvements in system design (e.g., improving energy saving considering the thermal design of a node [1]) – addressing the total power consumption of equipment and devices, as well as energy proportionality – addressing the equipment operation. A key enabler for energy-efficient networking is the capability of devices and equipment to support a load-adaptive and energy proportional operation, that is, support adjusting the offered capacities in order to match the actual traffic demand.

Design, engineering and operation have typically focused on service performance, scalability, and availability without taking energy expenses into account; thus, networks have commonly been operated at peak, that is, with full power, even during off-peak hours, where network resources are underutilized. Hence, the power consumption of network equipment remained almost constant at peak level, independent of the actual network traffic demand. As traffic demands vary, elements can be adaptively operated based-on energy conservation modes or states, lowering power consumption during off-peak periods without perceptible impacts on the user service.

ITU-T identifies two levels to address load proportionality in Ref. [1] centered on the device and equipment. On the device level, energy efficiency can be realized by optimizing the operation of large-scale integration (LSI) micro-fabrication, by introducing multi-core CPU. Energy consumption is proportional to clock frequency while dynamic control technologies such as clock gating and sleep mode control can be applied separately to each CPU and via power-aware on-demand virtual or cache memory.

In order to support load-proportional efficiency, allowing adapting energy consumption to variable-load conditions, power states are a prerequisite for the equipment level. In Ref. [1], a sleep mode applied on the equipment or network interface is introduced as a design concept considering the deployment requirements and limitations including the need for delivering control packets even at off-peak times, for example, for routing updates. For link and interface-specific mechanisms, adaptive link rate enables granular control of the bit rate based on the traffic load and dynamic voltage scaling, controlling the voltage of the equipment CPU, hard disk, and network interface cards (NIC) to reflect the expected processing load.

Besides the conceptual analysis of ITU-T, IETF EMAN introduces in Ref. [2] two generic energy saving states: (i) the sleep state or otherwise dosing state where the equipment is not functional but immediately available and (ii) the off state where the equipment requires a significant amount of time to return to the conventional operational state. In principle, equipment may adopt a number of different energy states with diverse properties as documented in Ref. [3]. In the simplest case, the equipment could support two extreme energy states, that is, powered-off state and fully operational state, while certain equipment may also support an additional sleep or dosing state as the main energy saving state. It should be noted that each different technology may support different energy states, which are specified explicitly for the associated standards.

Energy efficiency in network equipment can also be realized via techniques beyond power saving states, as documented in Ref. [1], via packet filtering that blocks inessential data traffic, traffic shaping that controls the output rate or using traffic engineering and multi-layer routing through optical instead of electronic tecnologies. Improvements of the energy efficiency of network equipment can also be achieved by re-engineering on the device or component level, for example, by advancements in chip design or via the use of power-adjustable components, leading to better energy utilization. While internal system design is typically beyond the scope of standardization, regulatory initiatives, such as the EU CoC, aim to promote energy conservation for ICT equipment, facilitating activities in SDOs with influence on the system development and deployment.

20.2.2   EC Code-of-Conduct (CoC)

The EC introduced certain regulations for ICT with the objective of reducing CO_2 emissions. The Commission already recognized early in 1999 that the standby mode of ICT end-use devices is not adequate for minimizing CO_2 emissions. Hence, a Commission Communication proposed actions to promote the efficient use of energy and recommended policies for reducing power consumption of consumer electronic and network equipment in order to reduce CO_2 emissions. As a result, the Commission introduced the instrument of the so-called CoC [4] on voluntary basis. The CoC contains the policy of maximizing energy efficiency of ICT equipment. Service providers, network operators, equipment and component manufacturers may voluntarily commit to the CoC by signing as individual companies. For the time being, five different ICT CoC documents are in force:

- Efficiency of External Power Supplies
- Energy Efficiency of Digital TV Service Systems
- Uninterruptible Power Systems
- Energy Consumption of Broadband Equipment
- Data Centers Energy Efficiency

The EC did not opt for a regulatory instrument, because the CoC is more flexible and can be progressed quicker than regulation. However, if the voluntary agreement is not effective, the CoC may still be transformed to a regulation. The EC Joint Research Center (JRC) is responsible for the development and review of the CoC documents. The targets of the CoC should be realistic and challenging at the same time. The performance of the systems should not be reduced. The CoC will be regularly reviewed and updated in cooperation with all relevant stakeholders including individual companies and fora. The standardization fora BBF, Home Gateway Initiative (HGI) and ETSI TCs EE and ATTM cooperate via liaisons. The industry should be stimulated to optimize the systems and equipment. Procurement specifications should comply with the CoC.

The CoC on Energy Consumption of Broadband Equipment and on Energy Efficiency of Digital TV Service Systems are related to the relevant network equipment, while the CoC on Energy Consumption of Broadband Equipment covers both customer premises equipment (CPE) and network equipment. Covered CPEs are DSL modems, cable modems, optical CPEs, Ethernet routers and wireless user equipment. Types of considered network equipment are digital subscriber line access multiplexers (DSLAMs), multi-service access node (MSAN), optical line terminals (OLTs), base stations, access points, and CMTS. For each type of equipment, power consumption limits depending on the operation state and on the timescale are specified. For CPEs, the operation states "on," "idle" and "off" are defined. Defined network states are full-load, medium-load, low-load, and standby. In the off state, the CoC on efficiency of external power supplies must be met. The power consumption limits differ depending on the year the equipment is brought to market or purchased or procured or tendered. The definition of the relevant date is still under discussion. Test methods are also defined.

The CoC is regularly reviewed in order to update power consumption targets for future time periods. The actual power consumption of the equipment available in the market is yearly measured in order to monitor the effectiveness of the CoC in achieving the goals. After these activities, a new version of the CoC may be released, which supersedes the old version. Furthermore, targets for future technologies, for example, VDSL vectoring and G.fast, are developed, which will be incorporated into the CoC. A reduction of power consumption of broadband equipment in Europe from 50 to 25 TWh per year is estimated to be achieved by obeying the CoC on Energy Consumption of Broadband Equipment [5].

The CoC on Energy Efficiency of Digital TV Service Systems covers set-top boxes, digital TV sets, computers with digital TV tuners or TV add-in cards, digital receivers with recording function and so on. Values of "Annual Energy Allowance" and "Total Energy Consumption" are given depending on the effective date. Test procedures for different operation modes and types of equipment are specified. A reduction of power consumption of digital TV service systems in Europe from 23 to 15 TWh per year is estimated to be achieved by obeying the CoC on Energy Efficiency of Digital TV Service Systems [5].

20.3   Network-Based Energy Conservation

Network-based energy conservation involves mechanism and protocols that stretch beyond the equipment level combining switches/routers, links, and interfaces as well as data transport and routing protocols. The objective is to achieve low-power consumption operation in a coordinated manner considering a set of equipment, for example, for how long a switch/router and an interface can transmit data or sleep, or which paths and network resources should be used to enable network-wide energy saving. Well-established standardized network-based energy saving protocols concentrate on the network access, such as ITU-T gigabit passive optical network (GPON) [6, 7] and IEEE EEE [8], which are described in detail in chapters 15 and 14, respectively. However, energy efficiency has also become important for wide area networks, including IP core and Internet backbone networks that typically rely on IP packet routing and dynamic control plane protocols. In Ref. [1], ITU-T identifies among the key enablers for achieving network-based energy conservation, energy-aware routing and traffic engineering, transmission scheduling and the use of lightweight protocols.

Energy awareness in network routing and transport is enabled by concepts, mechanisms, and protocols that support operating the network at a minimum level of aggregate power consumption while satisfying the requirement levels for network coverage, robustness, and performance. Overall, energy-efficient network operation depends on the network elements' and the network control's ability in adapting capacities to current demand, and in managing quality of service (QoS), to meet service levels and resilience levels. This relates to the IP networking protocols in scope of the IETF, covering network control, routing, metrics/profiles, and traffic engineering, in order to optimize network resources and operation in terms of energy efficiency. Specific topics currently being addressed include considerations and requirements for energy-aware control plane design, protocol extensions to exchange power ratio metrics, and power-aware networking allowing path selection and traffic steering based on energy profiles.

While research-driven proposals for routing and control adaptations have existed since around 2010, broader activities for energy efficiency in routing and transport have been seen since the IETF84 meeting, including efforts to form a new working group addressing energy-aware networking. The community has elaborated the area of network routing and transport in greater detail over recent years, but it needs to be noted that development and standardization efforts for energy-aware networking within the IETF are still to be considered as being in early stages at the time of this writing. Currently, discussions have been held within the IETF Routing Area Working Group [9], while no new working group has been formed specific to this topic yet.

Regarding the adoption of transmission scheduling, the objective is to minimize buffering on network nodes and provide the means for controlling the amount and timing of packet transmission in order to minimize per packet waiting time at each node. Operating with fewer buffer resources can save energy according to Ref. [1], but no standard has yet been developed to address such an issue, though there is potential for software defined network (SDN) protocols like OpenFlow [10], which can program the use of network resources. Lightweight protocols seek to save energy via simplifying operations or processing data traffic faster at lower layers. Currently IETF is considering the design of new lightweight communication protocols for low power and constraint network environments, for example, battery-powered devices, analyzing also routing, transport, and application layer as well as cross-layer optimization opportunities in Refs. [11].

20.3.1   Energy-Aware Control Planes

Energy-aware control planes are work in progress at the time of this writing in the IETF routing area focusing on network control, closely related to energy conservation and network performance. In large-scale carrier IP and Internet backbone networks, designed for highest availability and resiliency, it is imperative to consider how energy saving procedures, resulting in network resource adaptations, could affect network operation and applications running on the network. The Internet draft [12] provides considerations, use cases, and requirements for energy-aware control planes including operational impacts. Based on high-level business and network application requirements, the document encourages efforts for energy-efficient networking, considering how to balance efficiency and performance in practice, elaborating on effects and trade-offs potentially arising from energy reduction, and giving guidance for energy-aware control plane protocol design. Beyond basic insights on what designers of energy-aware control plane protocols ought to take into consideration, it analyses potential impacts on network QoS metrics.

Considering the main network performance drivers: bandwidth, delay, and jitter, the Internet draft [12] analyzes how these are generally affected by network control operations, related to network stretch and network convergence. Stretch in a network is to be understood as a path extension, resulting in additional hops on the packets' route compared to the shortest path, while network convergence is an effect seen during the distribution of network database updates and re-synchronization of the topology view across all network devices. Currently, Ref. [12] lists the following four ways for energy conservation in a network:

- Removing redundant links from the network topology
- Removing redundant network equipment from the network topology
- Reducing the amount of time equipment or links are operational
- Reducing the link speed or processing rate of equipment

Analyzing use cases and a sample network scenario, [12] elaborates effects potentially resulting from energy conservation actions including bandwidth reduction, increased network stretch, network convergence/recovery, and jitter, revealing to protocol designers the affected operational aspects and limitations. In particular, considering the impact on network performance, that is, bandwidth, jitter, and delay, it proposes to exploit the capability of setting parameters for a minimum expected level of performance, and to enforce it, when selecting elements to be powered down or be removed from the network for achieving energy conservation. As for network stretch, introduced when traffic is steered along a so-called loop-free alternate path for reasons of energy efficiency, it is suggested that developers should include an analysis as part of the protocol design and consider making the maximum allowed additional stretch configurable. In addition, it is suggested to provide the opportunity to maintain a minimum level of redundancy when the network is modified for energy conservation.

Concerning delays induced by the local transition from low-power conservation states to full power states of network equipment or links, Ref. [12] points out that the resulting jitter, subject to accumulation across a longer network path, needs to be considered, as well as options to coordinate the packet transmission considering sleep states or cycle network operations. The aforementioned ways of reducing energy are not dependent on whether the network control is distributed, as it is typically the case in today's Internet backbone networks, or logically centralized, that is, using a path computation engine (PCE) or a network controller. Furthermore, it is assumed that controlling local energy saving mechanisms could be left out of scope unless coordination on the level of the network-wide IP control plane is provided. It should be noted that inter-domain applications are currently not in scope.

Recently, the IETF Internet draft [13] addresses the impact of energy-aware network operation on network performance and, in turn, on the service quality perceived by the user, focusing on mobile, heterogeneous, and hybrid packet access networks. In particular, it elaborates the concept of load-proportional operation and load-adaptive network reconfiguration considering also requirements and basic approaches for network and service management, for managing service quality aspects and for counteracting potential impacts of energy-aware network operation in mobile systems.

20.3.2 Power-Aware Routing and Traffic Engineering

Power-aware routing and traffic engineering is motivated by the Internet draft [14], which provides the problem statement detailing how power awareness can be improved on a network-wide level, that is, beyond a node and link level techniques, and discusses the main technical development and operational practices. In particular, it highlights the components, for example, hardware and software, designs and operational issues, to be considered when developing energy-aware protocols and suggests categorizing potential solutions according to three dimensions: link sleep versus rate adaptation, configured versus adaptive, distributed versus centralized. The proposed problem statement was taken as the basis for a discussion about the problem space and potential solutions, to be in scope of the IETF work on power-aware networking, at the IETF86 meeting [15]. Solutions covered by the discussion concentrated on extensions to Link State Databases (LSDB), enabling layer 3 awareness by a so-called routing adjacency for sleeping links, and allowing component links to enter a sleeping state, while maintaining connectivity of an entire composite link.

As part of the power-aware networking discussion at the IETF86 meeting, proposals for a metric-based approach for reducing power consumption in the Internet routing were analyzed in the Routing Area Working Group [16]. A metric-based hierarchical approach to reduce power consumption in core and edge networks was introduced, covering both the Intra-Autonomous System (Inter-AS) case as well as a collaborative approach between Autonomous Systems (Inter-AS). The main objective concentrated on providing a comprehensive and globally applicable solution beyond powering off resources locally, supporting distributed network environments, and providing operational feasibility and benefits for a fast and widespread industry adoption. While solutions to monitor the network load and to adaptively power down unused network resources can be seen as effective to reduce energy consumption locally, more advanced approaches were recommended to improve energy efficiency in global, large-scale routing systems.

For unicast routing, a metric based on consumed power to available bandwidth was proposed, to determine a low-power path between sources and destinations, while for multicast routing, the proposed metric is based on consumed power to available multicast replication capacity, in order to allow identifying both low-power multicast paths as well as multicast replication points. Beyond energy-efficiency routing metrics, the proposal also covered related modifications to routing topology databases, that is, OSPF/ISIS Link State Database and OSPF/ISIS Traffic Engineering database, routing algorithms, and Traffic Engineering protocols, for example, RSVP-TE, in order to enable energy awareness in intra-domain and inter-domain routing. Furthermore, by introducing the notion of "TCAM power ratio" to tackle the issue of the disproportionately high power consumption of Ternary Content Addressable Memory (TCAM) components – a specialized type of high-speed memory used in network routers – energy efficiency can be achieved by enabling selective use of TCAM, allowing unused TCAM components to be powered down.

Beyond activities for energy-aware routing and transport, proposals have been made to generally investigate IETF protocols with regard to energy efficiency within the IRTF and consider forming an Energy Efficiency Research Group [17]. Particular areas of interest were energy efficiency in cloud networks by optimizing virtual machine (VM) allocations and traffic aggregation/steering in order to power-down network devices, while other topics included higher-layer/application-layer awareness, end users experience and common metrics for energy in network applications. Although aiming to reduce energy consumption from network inactivity is seen as straightforward, two potential major work areas may include: (i) exploring how existing protocols could be made more energy efficient and (ii) ensuring that new protocols support energy saving modes. While characteristics of "traditional" distributed IP control planes are being discussed with regard to energy efficiency, alternative broker-based approaches, for example, based on PCE or based on logically centralized SDN controllers, may more systematically steer traffic based on energy profiles and may be able to resolve routing database update challenges.

20.4   Energy-Aware Network Planning

The initial step toward energy-efficiency networking is network planning, which typically concentrates on performance and reliability issues without considering energy saving until lately. Network planning consists of network design, that is, dimensioning of physical network resources, routing policies and other predetermined network operations. Despite the fact that the actual placement of network resources is simply subject to network operator's service needs, convergent technologies that allow equipment consolidation in the network design phase and the arrangement of interoperable overlay or virtualized networks have been accountable for standardization. Consolidation can be achieved by converging a large number of different access equipment and their associated interfaces, gaining higher energy and space efficiency, which is also typically understood as "removing stovepipes." Network virtualization and overlay technologies allow networking to inherently become more energy efficient due to sharing of network equipment for different services, re-enforcing consolidation by reducing further network physical "boxes" and interfaces.

Consolidation in network design is achieved by standardizing network architecture and transport, an effort that is performed by BBF focusing on broadband infrastructures and services. In particular, BBF [18, 19] has specified the architecture for DSL and GPON-based access with Ethernet aggregation creating broadband loop carriers and multi-service access platforms. Such platforms combine legacy digital loop carriers (DLCs), optical add-drop multiplexers (ADMs), DSLAMs, OLTs, aggregation, and transport elements into single equipment, enhancing energy efficiency [20]. Early standards enabling multi-service overlay networking concentrated on point-to-point protocol (PPP), frame relay and later asynchronous transfer mode (ATM). Nowadays, multi-protocol label switching (MPLS) has been widely adopted to provide overlay networks unifying the priori PPP, frame relay and ATM technologies with the evolving IP and Ethernet, enabling the support of different network services [21]. Such efforts are carried out at the BBF supporting converged packet networks. Specifically BBF has progressed the multi-service broadband network architecture [22], considering the latest infrastructures, topologies and deployment scenarios, while specifying nodal requirements.

BBF has also specified a set of recommendations and best practices for the Mobile Backhaul in Ref. [48], where network planning is particularly challenging due to the diversity of

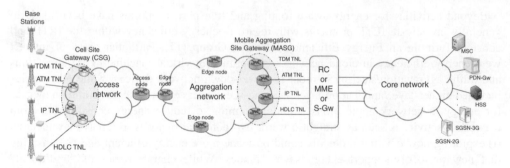

Figure 20.1   An overview of the BBF mobile Backhaul architecture based on Ref. [6]

radio access technologies (RAT) and the progressive development of the network infrastructure using different transport technologies for the access and aggregation networks. Early Mobile Backhaul deployments supporting the 2nd Generation (2G) of mobile systems employed time division multiplexing (TDM) or high-level data link control (HDLC), while later systems adopted ATM and frame relay transport to assure delay and loss for evolving multimedia applications. Nowadays, Mobile Backhaul infrastructures need to support such legacy RATs and transport services, while integrating them with the Long Term Evolution (LTE) and IP/Ethernet transport, which can handle the increasing data traffic growth and application services.

The BBF Mobile Backhaul architecture [23] unifies via the use of converged transport over MPLS diverse transport solutions, enabling a single backhaul access and aggregation network, which combines 2G, with Universal Mobile Telecommunication System (UMTS), High Speed Downlink Packet Access (HSDPA) and LTE, as illustrated in Figure 20.1, allowing a substantial reduction of the number of network elements and communication links, while increasing the network utilization via statistical multiplexing [20]. Consolidating multiple networks does not only save energy but can also ease network sharing/wholesale models, simplify network management, and reduce operational complexity.

Routing and other network policies that address energy efficiency should also be considered in the network planning phase. Beyond energy-aware routing and traffic engineering, which are "on-line" solutions, energy saving may be achieved by defining a policy regarding network equipment that lay within backup paths, which are used for resiliency as introduced in [24] for generalized-MPLS (GMPLS) label switched paths (LSPs) considering examples for 1 + 1 and 1:N protection. In addition, policies regarding content delivery, caching, and content optimization may help avoiding over-provisioning network resources, while reducing the costs of data transfer and, in turn, energy consumption. Such policies may be part of the network planning phase, though the actual mechanisms are typically application-based. IETF application-layer traffic optimization (ALTO) within the Transport Working Group specifies content optimization solutions addressing the problems and key use cases in Ref. [25].

20.5   Energy Saving Management

The energy-efficient operation of network systems requires network management processes, responsible for collecting information regarding the network status, analyzing such information and taking decisions on how to configure and control the network for achieving optimal energy savings. Such network status information and control decisions may correspond into different technology levels according to [1] including device, equipment and network. Network management processes may be distributed, that is, located in every equipment and/or be centralized, that is, performed on a controller or management server, requiring in either case coordination among the devices, equipment and network.

Network status information created on equipment basis is typically forwarded and stored in a database, which offers optimization functions related to performance and service quality profiles as well as energy consumption measures. Traditionally, network management has not included energy monitoring or control processes, but has typically concentrated on fault, configuration, accounting, performance and security management. To address such a need for energy management in a way that provides interoperability for different types of equipment, IETF formed the EMAN working group. For relatively low power edge equipment IEEE 802.3 working group developed PoE, a technology that enables remote monitoring and power control.

20.5.1 ITU-T Energy Control Framework

Managing the energy efficiency in network systems is a process that provisions equipment enabling dynamic control that reflects evolving service demands. Although energy saving management is subject to the operator's needs, certain guidelines from standards bodies provide an insight view of network operations and management principles. ITU-T in [1] documents a network management framework for supporting energy saving on devices, equipment and network systems as illustrated in Figure 20.2, which is composed by the following functions:

- *Energy management function* provides the optimal energy-efficient network state based on information retrieved from the network status information base and issues energy saving decisions toward devices, equipment and network systems. It consists of the Data Collecting sub-function, which interfaces with the status information base, the Optimization sub-function that contains the energy saving algorithms and the Operating sub-function that actuates the energy control decisions and performs potential alternations on the measurement parameters and methods.
- *Energy control and measurement function* is responsible for operating devices, equipment and network systems based on the request of the energy management function, while it also enables feeding measurements related to energy optimization corresponding to the device, equipment and network level toward the network status information base.
- *Status information base* is a database that maintains data traffic, energy consumption, and service quality, information related to the status of different devices, equipment and network systems, and provides the energy management function with the appropriate information in order to perform the corresponding energy optimization.

Such energy management framework can be performed distributed, that is, on local equipment including all described functions, and/or alternatively some of the functions can be performed remotely on a centralized controller or management server as described in Ref. [1]. Distributed arrangements allow self-optimization for network equipment focusing on the device and equipment level, while centralized models execute on a remote controller or server the energy management function, which impacts the network level. Hybrid approaches may

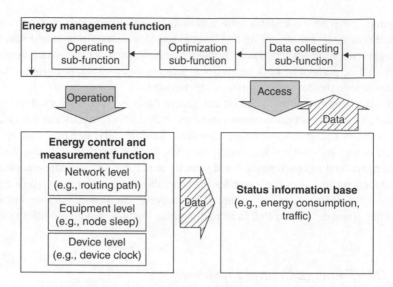

Figure 20.2   ITU-T energy control framework [Y.3021]

combine both distributed and centralized schemes in parallel, optimizing energy consumption in both local and global scale considering the device, equipment, and network level.

20.5.2   IETF Energy Management (EMAN)

There is a long history of managing IT equipment using the simple network management protocol (SNMP) [26] together with management information base (MIB) modules that provide data about and control over a set of manageable device properties. Energy management of IP networks and networked equipment at first seems just like any other network management task. On the monitoring side, instead of reading out things like packet counters, energy-related data is being monitored such as the charging state of a battery. On the control side, instead of setting elements like network protocol parameters, energy-related parameters are being controlled such as a device's power state. If the above were true, the only thing that needs to be done to support energy management using SNMP was to define MIB modules that express energy-related device properties and information.

On a closer look, however, energy management can actually be quite different from "traditional" network management. It is not so much that existing protocols are ill-suited, but the particularities of energy management require special attention. Typical network management tasks require only direct communication with the device to be managed. That might not be enough or possible when managing energy-related properties of a device, for example, many devices do not measure their power consumption simply because they lack the proper instrumentation. That does not necessarily mean that the power consumption of the device is unknown. A power distribution unit (PDU) to which the device is attached might be instrumented to deliver this data. The same applies to control. Switching off a port on a PoE switch will result in shutting down the device powered by that port. In other words, there are intrinsic relationships amongst devices along a power distribution tree that are vital to be understood when managing a network.

For all of this, it is important to note that the power distribution topology and the network topology do not always correlate, for example, when a device is attached to a PDU, that does not imply that the PDU can directly communicate with the device (i.e., there's a link between the two). It could correlate, of course, for example, in the PoE case, but this assumption does not always and generally hold. This makes energy management more complex than traditional network management tasks. Another aspect that makes energy management more complex is that quite a number of non-IP communication networks exist that are already used for energy management, for example, building management systems often use specialized building automation networks to read out energy-related information. It seems wrong to restrict new energy management standards to only IP-enabled devices from the onset as more and more devices will require being energy managed, but not all of them can support an IP stack. Therefore, being able to connect these devices to IP energy management systems is necessary, for example, through a gateway device, while once again a capability to report on other devices is necessary.

The aforementioned issues drove the IETF EMAN working group's [27] effort to define energy management standards using the SNMP. It turned out during the work that existing standards needed to be altered in order to accommodate some of the new requirements. The first standards document – Request For Comment (RFC) – the working group produced was therefore an update to the Entity MIB [28]. The changes to the Entity MIB allow resource-constraint devices to use the Entity MIB (a number of which were assumed to exist in the energy management context), while an Internet Assigned Numbers Authority (IANA) registry was created, so new general hardware types can be registered without having to change the Entity MIB in the future. The latter change was triggered by the EMAN work, because it became evident that the existing types were not sufficient, for example, a power supply type existed, but a battery did not quite fit that definition since at specific times it is a receiver of power rather than a supplier. The new mechanism of an IANA registry now decouples the registration process of new hardware types from the standards document, a process that may also be useful for further standards work.

The working group's priority when chartered was to list the requirements for new standards [2] and to describe a framework [3], where all particularities of network management were captured. A number of documents were also initiated concentrating on defining the actual energy-related managed objects [29], describing the relationships amongst managed devices [30], while batteries were segregated out into a separate document [31] since they are special from an energy management perspective, for example, they age, have a limited capacity, are temperature sensitive, and so on. Since energy management was a new topic to the IETF, it took a considerable amount time to model relationships, to capture a wide range of metrics to make the standard applicable beyond IT equipment and to understand requirements coming from other groups such as the IEEE/ISTO Printer Working Group. What emerged after a number of iterations was a concept quite familiar to IETF participants–power interfaces. A power interface can be an inlet (where a device is supplied with electric power) or an outlet (where a device supplies electric power to other devices); a concept quite similar to network interfaces, which shaped the problem in much easier terms for the EMAN working group. IETF is also working toward a document that contains an applicability statement [32], which describes how the EMAN work can be applied in various scenarios ranging from data centers to industrial automation network.

At the time of this writing, the first implementations of the Internet drafts have been reported, while there is a significant interest in the technology with progress made after a large number of iterations with significant changes. When the set of EMAN documents are finalized, it will be easy to perform energy management with existing IP network management systems as another important aspect of IT management.

20.5.3   IEEE Power over Ethernet (PoE)

IEEE PoE transfers electrical power along with data on a single standard Ethernet cable, empowering remote network equipment, referred to as powered devices (PD), without the need for a conventional alternating current (AC) power supply. Effectively, PoE reduces cabling, eliminating the need for AC outlets, while simplifying installation and maintenance, by using an Ethernet switch, termed as power source equipment (PSE), to provide power for attached equipment. Such a centralized power supply scheme establishes energy efficiency introducing 0.6–2.1 W of power conservation per interface [33], while it also provides the means for "power backup", ensuring a continuous full operation for PDs despite power interruptions.

The original version of PoE, IEEE 802.3af-2003 [34] provides 15.4 W per port; however, only 12.95 W is available at the device as some power is dissipated in the cable. IEEE 802.3af enables PSEs to automatically discover attached PDs and determine their power class. Currently, IEEE 802.3af can be used with 10BASE-T and 100BASE-TX and supports four power classes that correspond to Type 1 PD power levels, as illustrated in Table 20.1. Power originated from a PoE switch is supplied by end-span, that is, directly from the powered port, or by mid-span via another PoE supply. When a PoE connection is initiated, the PD may communicate its power class indicating to the PSE the amount of needed power via a 1-Event Physical Layer Classification. The updated IEEE 802.3at-2009 [35] known as PoE Plus offers up to 25.5 W and it can also be used with 1000BASE-T. The classification scheme of IEEE 802.3at, shown in Table 20.1, is the same as the one defined for IEEE P802.3af, to ensure backwards compatibility, while an additional class is specified for PDs that require more than 12.95 W, referred to as Type 2. PDs can be classified by the IEEE 802.3at PSE by a 2-Event Physical Layer Classification, data link layer classification, for example, Link Layer Discovery Protocol (LLDP) or a combination of both.

PDs may stretch up to 100 m from a PSE, which can empower a wide variety of equipment including Voice over IP (VoIP) phones, wireless access points, Ethernet hubs, security pan-tilt zoom cameras, print servers, and so on. PoE enables remote power management of such PDs,

Table 20.1   PoE PD classification and power supply

Class	PD classification	Power available for PD (W)
0	Default / type 1	0.44–12.95
1	Type 1	0.44–3.84
2	Type 1	3.84–6.49
3	Type 1	6.49–12.95
4	Type 2	12.95–25.5

enabling remote power control of specific ports. Specifically, PoE may enable scheduled, that is, time-based or event-based power-on/off control of particular ports reducing in this way energy consumption related with the attached PDs. PoE may also be combined with the EEE for further energy savings providing a higher reduction on energy per interface.

20.6 Energy-Efficiency Metrics, Measurements, and Testing

Energy-efficiency metrics, measurements and testing procedures are key enablers for providing energy saving in network equipment and telecommunication systems. Metrics can provide a quantified indication of energy efficiency, which once standardized can enable comparisons of different equipment, network of equipment or equipment component s of the same type. Equally, testing procedures and measurements should be performed under identical conditions, which are subject to standardization, to ensure equivalent procedures for assessing the energy efficiency of equipment and networks. Among the different standardization bodies, ETSI, ATIS, ITU-T and the ECR Initiative have produced comprehensive concepts and principles and have specified standards that can be used for measuring, reporting and assessing energy efficiency.

ETSI has introduced energy-efficiency metrics and measurements for broadband equipment, identifying reference models and key performance indicators (KPI) as a part of the Green Agenda. The main Technical Committees (TCs) handling energy efficiency are the environmental engineering (EE) TC, which deals with definitions of energy efficiency, measurement methods and indicators and the access, terminals, transmission and multiplexing (ATTM) TC that focuses on energy-efficiency metrics, KPIs and recommendations. Early efforts on energy efficiency such as [36] have concentrated on defining a power consumption model per line, considering the bit rate and line length, and on specifying measurements and test conditions focusing on digital subscriber line access multiplexer (DSLAM) equipment. An analysis indicating the energy-efficiency factor (EEF), defined as the energy usage to data rates, and KPI figures for access network broadband equipment including metallic loop solutions, for example, asymmetric digital subscriber line (ADSL), and optical fiber access solutions including fiber to the cabinet (FTTC), fiber to the building (FTTB) and fiber to the home (FTTH) is documented in Ref. [37], providing also a summary of power requirement metrics related to each aforementioned technology. Global KPIs in relation to energy efficiency are also specified in Ref. [38] considering broadband infrastructure scalability as well as measurement points and procedures.

ETSI ES 203 215 [39] specifies more advanced measurement methods and test conditions for broadband equipment considering DSLAM, MSAN and GPON OLT, while it also provides, as informative data, power consumption limits corresponding to each priori analyzed technology. Energy-efficiency measurements for IP routers and Ethernet switches are defined in ETSI ES 203 136 [40], which also specifies test suits and the equipment energy efficiency ratio (EEER) that indicates energy efficiency per throughput. ETSI ES 203 184 [41] defines a methodology and test conditions based on EEER considering transport equipment, that is, connected to the network by copper or fiber. It focuses on the physical layer and on equipment running at data link layer, which were not included in Ref. [40], considering switches, multi-service transport platforms, (e.g., combinations of SDH and Ethernet), DWDM multiplexers/demultiplexers, optical amplifiers, transponders, and so on. The ETSI European Standard EN 301.575 [42] provides energy consumption measurement methods for CPE and test conditions concentrating on broadband equipment, LAN and WAN, that is, Ethernet, considering different power modes, including disconnected mode, off mode, standby, idle state and low-power state.

ATIS has also defined energy measurements as a part of its Green Initiative, introducing within the Network Power and Protection (NIPP) Committee the Telecommunications Energy Efficiency Ratio (TEER) standard to measure and report the energy efficiency of telecommunication equipment. ATIS documents a base standard in Ref. [43], which specifies a methodology for deriving TEER, while establishing uniform means for measuring energy consumption and reporting applied to different types of equipment, for example, core, transport, access. Supplementary standards subject to particular types of equipment specify details of measurement configurations and reporting for formulating TEER. In particular, Ref. [44] defines TEER for routers and switches considering fixed and modular equipment, including also the case where power saving is applied on component subsets, while it specifies test procedures, relating energy expenditure with equipment load. In Ref. [45], a set of guidelines for specifying TEER for transport equipment is documented, including testing procedures considering equipment configuration, data rates for typical transport interfaces (TDM/PDH, optical transport, Ethernet packet data, storage area networking and DWDM), and measurements methods.

ITU-T specifies in Ref. [46] the principles and concepts of energy-efficiency metrics and summarizes testing procedures and measurement methodologies for assessing the energy efficiency of network equipment and small networking equipment based on ETSI and ATIS documentation, enabling comparison among equipment that belong within the same class, for example, equipment of the same technology. The network equipment considered includes DSLAM, MSAN, GPON, gigabit Ethernet PON (GEPON), routers, switches, small network devices, WDM/TDM/OTN transport. A framework for approximating energy efficiency as the ratio of power consumption to transmission bandwidth is documented by ECR Initiative in Ref. [47], covering various operation conditions and practical considerations, including peak, variable-load, and idle energy efficiency. Such a framework defines a measurement methodology, which is applicable to many types of packet-oriented networks and equipment, including, but not limited to, core/edge routers, L2/L3 switches, and so on.

20.7   Conclusions

Rising energy costs, environmental policies, as well as equipment scaling and operation issues are driving the need for energy-efficient networking. This chapter overviews the main wireline standardization efforts considering network equipment, network-based mechanisms, planning and management operations, as well as energy metrics, measurements, and evaluation. It summarizes the EC CoC and provides insight views of energy proportionality on network equipment and telecommunication systems. The support of power saving states for network equipment is analyzed, considering functional capabilities, which may vary across different types of equipment. While power states and control mechanisms are typically defined as part of particular technology standards, SDN APIs may provide a generic means to allow more flexibility and control. Network-oriented efforts require coordination to address automated, interoperable ways of adapting network configuration and, in turn, energy consumption considering network performance and service levels. IETF energy-aware routing and control are in early stages where several issues, such as traffic steering, energy profiles, operational overhead, are still open. Logically centralized control and path computation based on SDN paradigms may provide enhancements in terms of flexibility and scalability. For network planning, equipment consolidation and overlay/converged multi-service networking, for example, as specified by BBF, is seen as key enabler for achieving energy efficiency. Considering network management, ITU-T introduces a framework that considers energy saving at different levels of devices, equipments, and networks, while IETF EMAN provides means for energy monitoring, reporting, and control. For small edge equipment, PoE provides installation efficiency and remote power control. In general, energy-saving mechanisms and OAM tools need to be coordinated, aligning the transition into a power saving state with running continuity checks and/or protection mechanisms. Finally, metrics and measurement methodologies introduced by ETSI, ATIS, ITU-T, and the ECR Initiative provide the foundation for comparing and evaluating energy efficiency among network equipment of the same type. Following recent trends toward cloud-centric networking, holistic strategies for energy saving will become more important. While IT/server consolidation is one of the key energy saving approaches in today's data centers, virtual resource optimization and migration in combination with energy-aware, network-wide, interoperable control solutions can become an important topic in data center-centric network architectures.

References

[1] ITU-T Y.3021, Framework of Energy Saving for Future Networks, 2012.
[2] J. Quittek, M. Chandramouli, R. Winter, T. Dietz, B. Claise, Requirements for energy Management, IETF RFC 6988, 2013.
[3] J. Parello, B. Claise, B. Schoening, J. Quittek, Energy management framework, IETF RFC 7326, Sep., 2014.
[4] European Commission, Code of conduct on energy consumption of broadband equipment, Version 5.0, 2013.
[5] http://re.jrc.ec.europa.eu/energyefficiency/html/standby_initiative.htm.
[6] ITU-T, Series G Sup. 45, GPON Power Conservation, 2009.
[7] ITU-T Series G.987.3, Chapter 16, ONU Power Management, 2010.
[8] IEEE Std 802.3az: Energy Efficient Ethernet-2010.
[9] IETF, "Routing Area Working Group", http://tools.ietf.org/wg/rtgwg.
[10] Open Flow Foundation, SDN Architecture, Issue 1, Jun. 2014.
[11] Z. Cao, C. Gomez, M. Kovatsch, H. Tian, X. He, Energy efficient implementation of IETF constrained protocol suite, IETF Internet Draft, version 2, 2015.
[12] A. Retana, R. White, M. Paul, A framework and requirements for energy aware control planes, IETF Internet Draft, version 3, 2014.
[13] D. von Hugo, N. Bayer, C. Lange, Energy aware control approach for QoS in heterogeneous packet access networks, IETF Internet Draft, version 3, 2014.
[14] B. Zhang, J. Shi, J. Dong, M. Zhang, M. Boucadair, Power-aware networks (PANET): problem statement, IETF Internet Draft, version 3, 2013.
[15] IETF, Routing Area Working Group IETF-86 Proceedings, 2013.
[16] S. Raman, B. Venkat, G. Raina, "Using routing protocols for reducing power consumption – a metric based approach," IETF-86 Proceedings, 2013.
[17] IETF Energy-Efficient Internet Research Group (EEIRG), 2011.
[18] DSL Forum TR-101, Migration to Ethernet-based DSL aggregation, 2006.
[19] BBF TR-156, Using GPON Access in the Context of TR-101, 2008.
[20] BBF MR-204, Energy efficiency, dematerialization and the role of the Broadband Forum, 2009.
[21] MFA Forum 16.0.0, Multi-service interworking – IP over MPLS, 2007.
[22] BBF TR-145, Multi-service broadband network functional modules and architecture, 2012.
[23] BBF TR-221, Technical Specification for MPLS in Mobile Backhaul Networks, 2011.
[24] S. Okamoto, Requirements of GMPLS extensions for energy efficiency traffic engineering, IETF Internet Draft, version 2, 2013.
[25] J. Seedorf, E. Burger, Application-layer traffic optimization (ALTO) problem statement, IETF RFC 5693, 2009.

[26] J. Case, M. Fedor, M. Schoffstall, J. Davin, A Simple Network Management Protocol (SNMP), IETF RFC 1157, 1990.

[27] IETF Energy Management Working Group (EMAN WG), https://datatracker.ietf.org/wg/eman/charter/.

[28] A. Bierman, D. Romascanu, J. Quittek, M. Chandramouli, Entity MIB (Version 4), IETF RFC 6933, 2013.

[29] M. Chandramouli, B. Claise, B. Schoening, J. Quittek, T. Dietz, Monitoring and Control MIB for Power and Energy, IETF RFC 7460, 2015.

[30] J. Parello, B. Claise, M. Chandramouli, Energy object MIB, IETF RFC 7461, 2015.

[31] J. Quittek, R. Winter, T. Dietz, Definition of managed objects for battery monitoring, IETF Internet Draft, version 20, 2015.

[32] E. Tychon, B. Schoening, M. Chandramouli, B. Nordman, Energy management (EMAN) applicability statement, IETF Internet Draft, version 10, 2015.

[33] Ethernet Alliance, Energy efficient power over Ethernet, 2011.

[34] IEEE Std 802.3af: Power over Ethernet-2003.

[35] IEEE Std 802.3at: Power over Ethernet Plus-2009.

[36] ETSI TS 102 553, Environmental engineering (EE); measurements methods and limits for energy consumption in broadband telecommunication network equipment, v.1.1.1, 2008.

[37] ETSI TR 105 174–4, Access, terminals, transmission and multiplexing (ATTM); broadband deployment – energy efficiency and key performance indicators; part 4: access networks, v1.1.1, 2009.

[38] ETSI ES 205 200–1, Access, terminals, transmission and multiplexing (ATTM); energy management; global KPIs; operational infrastructures; part 1: general requirements, v1.1.0, 2013.

[39] ETSI ES 203 215, Environmental engineering (EE); measurement methods and limits for power consumption in broadband telecommunication network equipment, v1.2.1, 2011.

[40] ETSI ES 203 136, Environmental engineering (EE); measurement methods for energy efficiency of routers and switch equipment, v1.1.1, 2013.

[41] ETSI ES 203 184, Environmental engineering (EE); measurement methods for power consumption in transport telecommunication network equipment, v1.1.1, 2013.

[42] ETSI EN 301 575, Environmental engineering (EE); measurement method for energy consumption of customer premises equipment (CPE), v.1.1.1, 2012.

[43] ATIS-0600015.2009, Energy efficiency for telecommunication equipment: methodology for measurement and reporting – general requirements, 2009.

[44] ATIS-0600015.03.2009, Energy efficiency for telecommunication equipment: methodology for measurement and reporting for router and Ethernet switch products, 2009.

[45] ATIS-0600015.02.2009, Energy efficiency for telecommunication equipment: methodology for measurement and reporting – transport requirements, 2009.

[46] ITU-T L.1310, Energy efficiency metrics and measurement methods for telecommunication equipment, 2011.

[47] A. Alimian, B. Nordman, D. Kharitonov, Network and telecom equipment energy and performance assessment – test procedures and measurement methodologies, ECR Initiative, Draft 3.0.1, 2010.

[48] BBF TR-293, Energy Efficient Mobile Backhaul, Sep. 2014.

21

Conclusions

Yinan Qi, Muhammad Ali Imran and Rahim Tafazolli
Institute for Communication Systems (ICS), University of Surrey, Guildford, Surrey, UK

21.1 Summary

This book is divided into two main parts: In the first part, the wireless network aspects of green communication are addressed, including the fundamental, that is, categorization of the concepts, trade-off between energy efficiency and other efficiency metrics, the scales of achieved energy efficiency and the relationship between the embodied energy and the energy saving potentials of the current networks. It also covers not only base station considerations, network planning and management, but other emerging paradigms, such as vehicular networks and Internet of Things (IoT). In the second part, we focus on the energy efficient design and management approaches for wired networks, where Ethernet and optical networks are also discussed. The application, transport and network layer solutions are presented and the standardization efforts are introduced. Generally speaking, almost every individual aspect of green communications from wired and wireless perspectives is touched.

21.1.1 Green Communications in Wireless Networks

Three fundamental approaches for energy saving are identified: introduction of new energy-efficient network elements, improvement of dynamics of the network and employment of sleep mode. However, most of the strategies stemming from these basic approaches only consider the operating energy consumption but the manufacture energy is ignored. In this regards, an extended energy consumption model is proposed, where energy consumed by all the production processes of a network element, also referred as embodied energy, is taken into consideration [1].

Once a comprehensive energy model is defined, the metrics to evaluate the energy efficiency by the research communities are presented. The energy efficiency metrics can be categorized

Green Communications: Principles, Concepts and Practice, First Edition.
Edited by Konstantinos Samdanis, Peter Rost, Andreas Maeder, Michela Meo and Christos Verikoukis.
© 2015 John Wiley & Sons, Ltd. Published 2015 by John Wiley & Sons, Ltd.

into three main classes: facility-level metrics, equipment-level metrics and network-level metrics [2–4]. Then the trade-off between energy and other costs is discussed from a variety of perspectives because energy efficiency can only be achieved by spending more on the other "costs," which can be spectrum efficiency, deployment efficiency and so on.

For wireless networks, today's existing various established and emerging system capabilities (features) towards green communications can be generally classified into three different elemental areas:

- **Radio component level**. A network node, such as a base station, comprises lots of sub-systems including radio unit, antennas, main power supply, baseband processing unit and cooling equipments. Solutions for reducing energy consumption of radio components exploit the potential of adapting to the load situations. In medium or low traffic load, some of the components can be deactivated, such as radio unit and power amplifier and some components can apply power scaling enablers to scale their power consumption according to the traffic load [23]. Other potential energy efficiency improving methods include reconfigurable antennas, where the antenna parameters are adaptive specifically according to spatial changes in traffic situation and changes in traffic load, and low loss antennas, where new foam substrate is used to improve the energy efficiency of the printed antenna structure [24]. Furthermore, antenna muting in smart MIMO systems, where the number of active antennas is adaptive by switching on antennas only when motivated by traffic load, is able to reduce energy consumption. Discontinuous transmission in time domain in radio part of a transmitter is also an enabler for energy efficiency improvement when there is not data to transmit [25].
- **Network elements level**. For each network element, the ability of sensing its environment gives the possibility of controlling the transmitting power, obtaining better management of interference. Recent research in cognitive radio exploits this possibility and addresses the potential of delivering green network nodes. A cognitive radio node exploits and uses the free spectrum available and power efficient modulation whenever possible. By doing this, power efficiency can be improved with little or no compromise on data rate [26]-[27]. In addition, the ability of dynamically adapting its resource usage according to the traffic level allows to reduce the energy consumption.
- **System architecture level**. Recent research on energy-aware system topology models and approaches for cellular networks focuses on two aspects: management issues such as switching on and off network elements based on the variation of traffic and radio planning issues such as deployment strategies and heterogeneous networks. Green Communications implements distributed and centralized algorithms, and network virtualization. Distributed algorithms allow for network management in different nodes so that each individual node is capable of taking a decision alone to reduce its energy consumption without degrading network performance. Centralized algorithms enables the operations and maintenance (O&M) to supervise and control network elements. Virtualization offers the possibility to share the same physical hardware to multiple instances of network services [28]. Heterogeneous networks (HetNets) are also proposed as a solution to improve overall energy efficiency, especially for cellular networks, where a mixture of Macro cells and small but agile cells is deployed [29]. In addition, the network management feature, i.e. node-activation/deactivation, has been proposed in order to improve energy efficiency.

Recent advances in the area of self-organized network (SON) reveals its potential to provide higher performance but also more efficient O&M [30]. SON offers main features such as self-configuration, self-optimization and self-healing and is part of the move to the next-generation of radio technology, aiming to leapfrog to a higher level of automated operation in mobile networks. It adapts itself to the fast changing traffic pattern in cellular networks and quickly and autonomously optimizes itself to sustain both network quality and a satisfying user experience.

These aforementioned green communication solutions/enablers can be regarded as a toolbox, meaning that it is unnecessary to apply all of them simultaneously because they are targeting various application scenarios. Actually, some of the solutions are not compatible with each other, thus cannot be employed at the same time. An integration of a carefully chosen subset of available solutions should be tailored based on applicable scenarios to achieve maximum reduction of energy expenditure.

21.1.2 Green Communication in Wired Networks

Network design and planning should consider the physical placement of devices, network topology, Quality of Service (QoS) and resilience. The design of converged networks that combine different transport technologies via network virtualization by introducing fewer network devices and network sharing is shown to be energy efficient in Ref. [5] and references therein. In addition to optimized network design, the network management with adaptability for new applications and paradigms in the network can further improve energy efficiency. One of the most promising solutions is software-defined networks, which allow for administration of network services more easily through abstraction of lower level functionality into virtual services and features the separation of the network elements' control plane to a central external entity [6]. An energy-efficient SDN-based network architecture with a newly defined Green Abstraction Layer (GAL) is presented for the power management of an individual device, where three main control plane processes are used including LCPs, NCPs & monitoring and OAM [7].

Besides the approaches in network level, the network devices can also be adaptive in modularity and rate. For the most common computer networking technology – LAN, the energy efficient enhancement of Ethernet is introduced, that is Energy Efficient Ethernet, also known as green Ethernet [8, 9]. The concept was presented in 2005 and substantially developed into IEEE 802.3az in 2007 [10]. In energy-efficient Ethernet, some network devices are put into sleep mode during the period of low data rate and they are only activated when the data is due to be sent.

In addition to computer networks, the study of energy-efficient strategies for optical networks is also important, as they are the backbone of core networks and an evolving solution providing high speed access. Optical networks are evolving into complex interconnected circuit-switched networks due to the continued growth in high-bandwidth applications. New cluster-based architecture, in which the nodes of the optical backbone network are divided into disjoint sets, is being discussed as a potential solution to improve the energy efficiency [11]. Each set consists of more than one node to form a single cluster. These clusters can be set to adopt a sleep mode initiated by the optical control plane (OCP) to save energy in low traffic scenarios.

An overview of energy-aware routing algorithms traffic engineering methods in backbone networks is also provided. Two potential approaches are identified: *switch off* and *energy proportional*. In the first approach, some devices are put into sleep mode to save energy. For instance, the routing algorithms and traffic engineering methods that perform aggregation of the current network load into as few paths as possible during off-peak periods are proposed to power off underutilized router/switches and links [12]. A solution named Table Lookup Bypass (TLB), based on the deactivation of forwarding functionalities inside a router line card, is elaborated in Ref. [13]. On the contrary, the second approach adapts to the speed and capacity of the devices to the actual load over relatively short timescales [14].

One of the most essential parts in today's backbone infrastructure is the modern data centres whose energy consumption can be decoupled into data server/storage power, network infrastructure power and cooling facilities power. These three main targets can be individually or jointly optimized to save a data centre's energy consumption. The state-of-the-art solutions to improve the energy efficiency of a data centre are focused on the computing and storage power mainly from three perspectives: dynamic link rate adaptation, link and switch sleep mode and improvement of the network topology [15–19]. Most importantly, these various approaches can be combined to be jointly optimized subject to some constraints on the network performance to achieve further energy efficiency improvement.

The energy consumption can be divided into two parts for core networks and edge devices, that is user equipments in homes and offices, respectively. In core networks, a continuous increase in traffic by a factor of 10 every 5 years has made TCP/IP-based core network suffer from severe scalability problems with respect to power. The current transport protocols can also be modified to become more energy efficient. For the TCP protocol, efforts have been made to make TCP aware of burst error, which is a major cause of packet loss, and act in a more energy-efficient manner [20]. The retransmission protocol can be further optimized to limit unwanted retransmission to save the wasted energy. Besides, the headers of protocols can be compressed to minimize the energy consumed for transmission of headers [21]. For edge devices, several energy saving protocols are introduced including *on-demand wakeup*, *proxying*, *content-aware power management* and *power-aware protocols*, each addressing the energy saving problem for edge devices from different perspectives.

From the perspective of Internet, the content delivery network (CDN) plays an important role in saving energy by resorting to location and mobility content update. However, CDN might not be efficient if the intended content for the user is not stored in the server nearby. Since users are more interested in obtaining the contents and do not care the establishment and maintenance of the connections, content-centric future Internet architecture is instigated, where the routing is influenced by the content request and the request is sent to the closest server to the user [22].

21.2 Green Communication Effects on Current Networks

Global warming has become a more and more serious challenge across the whole world in the past century mostly caused by human activities that increase concentrations of greenhouse gases in the atmosphere. Some scientific findings reveal that the CO_2 emission of the Information and Communications Technology (ICT) industry has contributed to a considerable percentage to the world energy consumption budget and global warming [31, 32]. In order to meet the exponentially growing demand, nearly 120,000 new base stations are deployed every year, each consuming an average power of 1 KW and using a total of 8,800 KW-hours each year. A typical medium-sized network constituting 12,000–15,000 cell sites, each equipped with two technologies (2G and 3G) and around three antennas per technology, accounts to an energy use of 736,000 MW-hours [33]. In particular, ICT across a wide range of applications currently accounts for 5.7% of the world's electricity consumption and 1.8% of CO_2 emissions [SBI Bulletin] SBI Bulletin: energy efficiency technologies in information and communication industry 2005-2015. Apparently, this revolutionary growth will be accompanied by a huge increase of the energy consumed by the telecommunication industry.

Other than the CO_2 emissions, there are other metrics to evaluate the energy expenditure of dissimilar components, individual network nodes and the whole network. In the node level, the Energy Consumption Index (ECI) can be used to measure the energy efficiency of a network node [25]. In the network level, consumed power per unit area is a common measurement metric.

The current research on green communication has successfully created innovations to dramatically reduce the energy consumption as well as the CO_2 emissions by bringing the current networks closer to its practical energy efficiency limit. The new developments in hardware have defined new innovative architectural concepts and are proved to be able to reduce the energy consumption of a network node by more than 40%. From the network perspective, the network optimization and management in the individual node level and in the system level, respectively, bring additional 30% energy savings all together in terms of consumed energy per information bit [25]. These analysis and values reveal the effects that the state-of-the-art green communication technologies bring to the current implemented networks. In the following section, we will exploit its future developing path.

21.3   Future Developments

Wired/wireless communication proliferates into nearly every aspect of life, leading to exploding traffic demands due to meteoric growth of static/mobile broadband services and as such is an essential driver for economic growth and improvement of welfare and well-being. For operators, equipment suppliers and service providers, it is becoming more and more challenging to achieve revenue growth in the same pace of the traffic growth. In this regard, the design of future communication systems not only needs to address the capacity crunch but it also has to deal with ecological concerns to reduce the cost per bit.

21.3.1   Future Network Requirements

As a consequence of the evolving global trends for advancing the efficiency of wired/wireless communication networks beyond the established state-of-the-art ones, the following objectives are broadly identified in order to enhance the sustainability of the future networks:

1. *Improved spectral and energy efficiency.* Since the current growth in traffic demand and the number of connected user terminals is inevitably limited by spectrum and energy consumption, a primary goal to be addressed by future communication networks is to provide an economic communication network with increased spectral and energy efficiency due to flexible infrastructure.

2. *Improvement of quality of experiences (QoE).* High bandwidth-demanding applications such as HD-video and 3D-video are becoming more and more popular. The actual throughput experienced by every individual user is essential and needs to be improved. In addition, users have increasing expectation in terms of quality of experience including transfer rate, responsiveness, accessibility, integrity, usability and so on. However, the increased QoE should be subject to the cost constraint and not bring huge energy expenditure.
3. *Reduction of cost of ownership and operation.* Huge investments are expected to deploy new networks in order to grantee satisfactory coverage and QoS for all customers. For example, in the radio access network side, it is expected that nearly 120,000 new base stations are deployed every year, let alone the associated core network enhancements. Since most of the telecommunication markets all over the world are shared by several main operators, it is compelling to share the network infrastructures for reduced total cost of ownership. Joint infrastructure sharing agreements between operators also reduces the operation cost in terms of energy consumption.
4. *Fusion of new and legacy networks.* Current deployed legacy networks employ different paradigms, spectrum bands and technologies from future networks. However, it would be a great waste to completely abandon these legacy networks. Network operators should ensure backwards compatibility, while exploiting multiple technologies and spectrum bands to establish an integrated interface across both legacy and future networks to support customers with different requirements.

21.3.2   Towards Holistic Energy Efficient Networking

There might be more potentials and objectives unveiled in addition to the aforementioned ones. For instance, improving the signalling efficiency is becoming increasingly important for wireless networks due to the dramatic increase of small data traffic. These objectives provide a direction where the current industry trends and research should be aligned with. It is an enormous challenge taking into account every aspect but the following topics/areas are considered as the most promising ones:

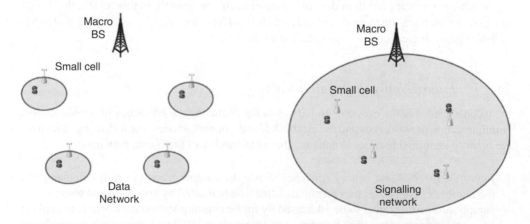

Figure 21.1   Splitting data and signalling

- *Architecture evolution.* Most of the aforementioned green technologies are built upon current enhancements networks and standardizations, which could be bottlenecks to constrain the architecture and functionalities. Revolutionary architecture enhancements without legacy network dependence needs to be studied to provide 10-fold or even more capacity improvement along with increased efficiency. Heterogeneous networks (HetNets), where large cells and small cells are mixed, have been studied to handle the traffic growth with densified spectrum reuse [34, 35]. However, the disadvantages of HetNets are obvious: Firstly, the constant association changes of a moving user generate mobility problems; secondly, inter-cell interferences are increased; thirdly, high densification of cell deployment causes additional energy consumption requiring a tighter alignment, i.e. a joint coordination, with the backhaul. As shown in Figure 21.1, innovative adaptable splitting between signalling and user traffic via their association to different cells is proposed to differentiate functionalities and remove the capacity and energy bottlenecks [25, 36]. The essence of this concept is to allow the terminals to exchange signalling traffic with large cells and data traffic with small cells, respectively. It provides a new degree of freedom to optimize and decouple the connectivity from the capacity. The mobility management can be easily handled by the large cells. Since the signalling is taken care of by the large cells, small cells are dedicated for data transmission and no longer need periodic signalling exchange, new network management schemes can be applied to put the small cells into "deep sleeping" status to save energy.
- *Extreme agile networks.* Agile networks with intelligence capabilities offer great adaptation and reconfiguration opportunities and make possible an unprecedented improvement of the energy efficiency through an extremely agile management of network elements and radio resources [37]. The agile efficient management is built upon context awareness and cognition of the network and comes at the cost of a higher complexity with respect to traditional systems. The context on which the network takes decision is not stable, but is rapidly changing. The agile management is based on a context-aware approach that makes use of a set of information elements creating a context framework and it includes the selection and activation of data access points, the dynamic and coordinated management of sleep modes, traffic balancing, and mobility and traffic variations management, taking into account different degrees of signalling and data access separation. In order to perform agile management, the following three steps need to be employed: First, identify, define and classify the set of information elements that can be used for creating the context in the network, identify possible information sources and define mechanisms for extracting information and making it available. Second, define context elaboration algorithms that allow obtaining high-level information from raw data exploiting historical traces, data caching, environment models and so on. Third, design algorithms and high-level signalling protocols for the activation of data sessions and the management of terminal mobility, and design optimization algorithms for energy-efficient, dynamic and automatic configuration of the network.
- *Orchestration.* An orchestration framework for a holistic and inclusive situation-aware management across cell sizes, wireless and backhaul technologies, spectrum opportunities and operators has been proposed to constitute the "brain" and "memory" of a network. This framework is designed to optimize the infrastructure behaviour across all current and foreseen scenarios for global handling of data/signalling traffic, mobility, interferences and so on. This framework considers both the network management transparent to the terminals and necessary interfaces between the terminals and the network(s). The framework itself should

consist of orchestration enablers including context-gathering, multi-technology and virtualization for both infrastructure and spectrum. The Orchestration intelligence exploited by multiple operators at a time is the core part and consists of situation awareness, knowledge, memory, reasoning and optimization.

- *Service-tailored network programmability.* Enable a network programmability allocating a resource slice that is tailored to optimize the energy consumption for a particular service by virtualizing and flexibly placing core network and radio functions at selected locations. This aims to facilitate a joint energy optimization of the radio, backhaul and core networks, while at the same time providing resource efficient and energy aware cloud, Over-The-Top, social applications and other emerging services. The main technologies for supporting this approach are the Network Function Virtualization (NFV) and Software Defined Networks (SDN) that can enable additionally the support for network sharing and multi-tenancy. SDN can also bring the advantage of considering the energy efficiency for using the underlying transport and backbone networks and forward requests towards the appropriate data center optimizing energy consumption.

- *Network federation.* Federating networks means to share resources among multiple independent networks in order to optimize the use of those resources, improve the quality of network-based services and/or reduce costs [38–41]. The sharing concept introduces management functionalities for transparently mapping logical networks (each associated with one operator) to (a) physical network(s). The management functionality will offer optimal cost reduction in combination with ample opportunities for competition through service level differentiation between operators, while guaranteeing fairness and security. For the users, this leads to transparent access to the shared physical networks. The sharing concept will enable fundamentally new business models for infrastructure and resource sharing (and federations thereof), including new business roles and actors in networks and services. The network segments can be jointly owned by several operators and each network segment covers a subset of the area can be owned by one operator, and is open to other operators.

- *Reconfigurable hardware.* The low power, load adaptive and reconfigurable hardware are essential to implement any of aforementioned solutions. The new generation hardware should go beyond state-of-the-art and possess the capability of adapting themselves to different activity levels and optimizing their performance with finer granularity, which facilitate seamless transitions between different network operation states and maximize the energy efficiency improvements.

There will be more directions for green communication with the continuously evolving markets. These no-stopping researches will guide us to the inspiring new generation of products and networks and eventually change the world as well as our lives significantly.

References

[1] I. Nawaza and G. N. Tiwarib, "Embodied energy analysis of photovoltaic (PV) system based on macro- and micro-level," Energy Policy, vol. 34, no. 17, pp. 3144–3152, 2006.

[2] A. P. Bianzino, A. K. Raju, and D. Rossi, "Apple-to-Apple: A framework analysis for energy-efficiency in networks," Proc. of SIGMETRICS, 2nd GreenMetrics workshop, 2010.

[3] T. Chen, H. Kim, and Y. Yang, "Energy efficiency metrics for green wireless communications," 2010 International Conference on Wireless Communications and Signal Processing (WCSP), pp. 1–6, 2010.

[4] C. Belady, et al., Green Grid Data Center Power Efficiency Metrics: PUE and DCIE, The Green Grid, 2008.

[5] N. M. M. K. Chowdhury and R. Boutaba, "Network virtualization: state of the art and research challenges," IEEE Wireless Commun. Mag., vol. 47, pp. 20–26, 2009.

[6] D. Mcdyson, "Software defined networking opportunities for transport," IEEE Wireless Commun. Mag., vol.51, pp. 28–31, 2013.

[7] Advanced Configuration & Power Interface (ACPI), URL: http://www.acpi.info.

[8] K. Christensen, et al., "IEEE 802.3az: The Road to Energy Efficient Ethernet," IEEE Wireless Commun. Mag., vol. 48, pp. 50–56, 2010.

[9] P. Reviriego, K. Christensen, J. Rabanillo, and J. A. Maestro, "An initial evaluation of energy efficient Ethernet," IEEE Commun. Lett., vol. 15, no. 5, pp. 578–580, 2011.

[10] "IEEE Std 802.3az: Energy Efficient Ethernet-2010".

[11] X. Niu, F. Yuan, S. Huang, B. Guo, and W. Gu, "Dynamic Clustering Scheme Based the Coordination of Management and Control in Multi-layer and Multi-region Intelligent Optical Network," in Proc. of Communications and Photonics Conference and Exhibition, Shanghai, China, pp. 1–7, 2011.

[12] C. Alippi, and L. Sportiello, "Energy-aware wireless-wired communications in sensor networks," in Proc. of IEEE Sensors, Christchurch, pp. 83–88, 2009.

[13] Coiro, A.; Polverini, M.; Cianfrani, A.; and Listanti, M., "Energy saving improvements in IP networks through table lookup bypass in router line cards," 2013 International Conference on Computing, Networking and Communications (ICNC), pp. 560–566, 2013.

[14] R. Bolla, R. Bruschi, A. Cianfrani, and M. Listanti, "Enabling backbone networks to sleep,"IEEE Netw., vol. 25, no. 2, pp. 26–31, 2011.

[15] D. Abts, M. R. Marty, P. M. Wells, P. Klausler, and H. Liu, Energy Proportional Datacenter Networks, Proceedings of the 37th Annual International Symposium on Computer Architecture (ISCA'10), Saint-Malo, France, pp. 338–347, 2010.

[16] M. Al-Fares, A. Loukissas, A. Vahdat, A Scalable, Commodity Data Center Network Architecture, ACM SIG-COMM Comput. Commun. Rev., vol. 38, no. 4, pp. 63–74, 2008.

[17] P. Reviriego, J. A. Maestro, J. A. Hernández, D. Larrabeiti, "Burst transmission for energy-efficient Ethernet," IEEE Internet Comput., vol. 14, no. 4, pp. 50–57, 2010.

[18] C. Gunaratne, K. Christensen, B. Nordman, and S. Suen, "Reducing the energy consumption of Ethernet with Adaptive Link Rate (ALR)," IEEE Trans. Comput., vol. 57, no. 4, pp. 448–461, 2008.

[19] B. Heller, S. Seetharaman, P. Mahadevan, Y. Yiakoumis, P. Sharma, S. Banerjee, N. McKeown, ElasticTree: Saving Energy in Data Center Networks, Proceedings of the 7th ACM/USENIX Symposium on Networked Systems Design and Implementation (NSDI'10), San Jose, California, USA, 2010.

[20] L. Donckers, P. J. M. Havinga, G. J. M. Smit, and L. T. Smit, "Energy efficient TCP," in Proc. of the 2nd Asian International Mobile Computing Conference, Langkawi, Malaysia, 2002.

[21] CISCO white paper, "Configuring TCP Header Compression," [online]. Available: http://www.cisco.com/en/US/docs/ios/qos/configuration/guide/config_tcp_hdr_comp.pdf.

[22] A. Detti, "CONET: A Content Centric Inter-Networking Architecture," in Proc. of SIGCOMM.

[23] EARTH project, "D4.3 – Final report on green radio technologies," [online]. Available: https://www.ict-earth.eu/publications/publications.html.

[24] EARTH project, "D4.2 – Green radio technologies," [online]. Available: https://www.ict-earth.eu/publications/publications.html.

[25] EARTH project, "D3.3 – Final report on green network technologies" [online]. Available: https://www.ict-earth.eu/publications/deliverables/deliverables.html.

[26] F. Bouali, O. Sallent, J. Perez-Romero, and R. Agusti, "Exploiting knowledge management for supporting spectrum selection in Cognitive Radio Networks", in Proc. Cognitive Radio Oriented Wireless Networks and Communications (CROWNCOM), 2012, Stockholm, Sweden, 2012.

[27] F. Bouali, O. Sallent, J. Perez-Romero, and R. Agusti, "A framework based on a fittingness factor to enable efficient exploitation of spectrum opportunities in Cognitive Radio networks", in Proc. 14th International Symposium on Wireless Personal Multimedia Communications (WPMC), Brest, France, 2011.

[28] X. Wang, P. Krishnamurthy, and D. Tipper, "Wireless Network Virtualization," in Proc. of International Conference on Computing, Networking and Communications, pp. 818–822, San Diego, USA, 2013.

[29] L. Falconetti, P. Frenger, H. Kallin, and T. Rimhagen, "Energy Efficiency in Heterogeneous Networks," in Proc. of IEEE Online Conference on Green Communications, pp. 98–103, 2012.

[30] S. Mumtaz, V. Monteiro, J. Rodriguez, and C. Politic, "Energy Efficient Load Balancing in Self Organized Shared Network," in Proc. of International Conference on Telecommunications and Multimedia, Chania, pp. 37–42, 2012.

[31] CISCO VNI Mobile Forecast 2013.

[32] http://www.greentouch.org/, accessed 5 May 2015.

[33] Ericsson press release, 2008.

[34] Shu-Ping Y.; Talwar, S.; Geng W.; Himayat, N.; Johnsson, K.; "Capacity and coverage enhancement in hetero-geneous networks", IEEE Wireless Commun. Mag., vol. 18, no. 3, pp. 32–38, 2011

[35] T. Nakamura, S. Nagata, A. Benjebbour, Y. Kishiyama, Tang Hai, X. Shen, N. Yang, and N. Li, "Trends in small cell enhancements in LTE advanced", IEEE Commun. Mag., vol. 51, no. 2, pp. 98–105, 2013.

[36] H. Ishii, Y. Kishiyama; and H. Takahashi, "A Novel Architecture for LTE-B, C-Plane/U-Plane Split and Phantom Cell Concept", in Proc. IEEE GLOBECOM 2012, Anaheim, California, 2012.

[37] OneFIT project, website: www.ict-onefit.eu, Accessed: January 2013.

[38] 3GPP TR23.851 "Network sharing; architecture and functional description", http://www.3gpp.org/ftp/Specs /html-info/23251.htm, Accessed: January 2013.

[39] Study on "RAN sharing enhancements", [online] http://www.3gpp.org/ftp/Specs/html-info/22852.htm, Accessed: January 2013.

[40] F. Berkers, et al., "To share or not to share?," in Proc. 14th International Conference on Intelligence in Next Generation Networks (ICIN), 2010.

[41] Meddour, D. E., et al., "On the role of infrastructure sharing for mobile network operators in emerging markets", Comput. Netw., vol. 55, no. 7, pp. 1576–1591, 2011.

Index

Green Communications: Principles, Concepts and Practice, First Edition.
Edited by Konstantinos Samdanis, Peter Rost, Andreas Maeder, Michela Meo and Christos Verikoukis.
© 2015 John Wiley & Sons, Ltd. Published 2015 by John Wiley & Sons, Ltd.